EXTREMES IN NATURE

Water Science and Technology Library

VOLUME 56

The titles published in this series are listed at the end of this volume.

EXTREMES IN NATURE

An Approach Using Copulas

by

GIANFAUSTO SALVADORI
Department of Mathematics, Università del Salento, Italy

CARLO DE MICHELE*,

NATHABANDU T. KOTTEGODA*,

and

RENZO ROSSO
**Department of Hydraulic, Environmental and Surveying Engineering, Politecnico di Milano, Italy*

Springer

A C.I.P. Catalogue record for this book is available from the Library of Congress.

ISBN 978-1-4020-4414-4 (HB)
ISBN 978-1-4020-4415-1 (e-book)

Published by Springer,
P.O. Box 17, 3300 AA Dordrecht, The Netherlands.

www.springer.com

Cover image: Calculations drawn on the blackboard. A part of the original results outlined in Illustration 3.9.

Printed on acid-free paper

CONTENTS

PREFACE

The most powerful earthquake in 40 years occurred on 26th December 2004 off the west coast of Sumatra, Indonesia. The tsunami it generated turned into one of the worst known natural disasters when walls of water crashed across the Indian Ocean, causing waves to reach Somalia in Africa. The death toll, mainly in Indonesia and Sri Lanka, exceeded 200,000. Nine months later, hurricane Katrina devastated the southern coast of USA along the Gulf coast. Winds reached 281 kilometers per hour and the storm surge of over nine meters was the highest recorded in the United States. It brought destruction to New Orleans when portions of the 563 kilometers of levees surrounding the city were suddenly breached. Nearly 1700 people died and damages are currently estimated at $100 billion, the costliest natural disaster in the United States. Within days hurricane Rita, another maximum category hurricane, struck the same coastal region damaging Texas and other states, followed soon afterwards by hurricane Wilma. Then on October 8th 2005 an earthquake in Kashmir, part of northern Pakistan and India, killed 75,000 inhabitants when innumerable buildings collapsed. Simultaneously, hurricane Stan led to costly landslides and more than 2000 deaths in Central America. To highlight the major catastrophes of nature during the previous decade, Cyclone Gorky and its storm surge caused 139,000 deaths in coastal Bangladesh during 1991. Insurance claims amounted to $17 billion consequent to Hurricane Andrew on the Gulf Coast of USA in 1992, and the Kobe earthquake in Japan during 1995 led to economic losses of $100 billion. These events were followed by the earthquake in northern Turkey in 1999 and the floods in southern Mosambique the following year.

Extremes in nature, such as those described, occur almost worldwide and cause incalculable human losses, in addition to billions of dollars in damages each year. For example, 622,000 people died in natural disasters in a ten year period from 1992. It is reported that more than half of the fifty most significant of such disasters recorded in 2001, a typical year, were flood or storm events. The effects of other weather-related hazards such as windstorms, hurricanes, typhoons and landslides are also known to have been severe; whereas earthquakes, nature's most destructive force, caused the highest economic losses. Droughts and unusually low flows, at the other end of the spectrum of unusual events, resulted in undesirable hardships and environmental conditions such as high levels of pollution concentrations.

Disaster loss from acts of nature is increasing fast. There has been a rising trend in the number of events and economic and insurance losses from the 1960's to the 1990's. For example, the past 15 years has seen at least a 36 percent increase in the number of hurricanes of category 4 and 5 in each of the oceans of the world, compared to those in a similar period previously; the respective numbers for the West Pacific are 116 and 85. Furthermore, on 23rd October 2005 the Gulf of Mexico had the seventh hurricane in 18 months. Regarding the cost of natural disasters to the United States, current estimates range from $6 billion to $10 billion per year. Such events are beyond our control.

Where the next disaster will strike, its nature and time of occurrence is unpredictable. Amongst possibilities, an instantaneous collapse of the Cumbre Vieja volcano in the Canary Islands, for instance, can cause a tsunami bringing death and destruction in its wake from the West Indies to the Atlantic coastline of the USA. A giant earthquake off Japan may cause similar devastation to Asia.

We consider the study of the statistics of extreme events to be a first step in the mitigation of these national disasters. For example, the parameters of structural designs need to cater for rare flood flows. As stated, flood and storm events comprise a major area of concern. This may involve a spillway, bridge, culvert or flood defense structure. Besides floods, the damage to a structure with loss of life and injury may be caused by an earthquake, windstorm or other type of natural hazard. Knowledge of its mechanism and of the probability distribution is essential for making risk assessments.

Alternatively, the purpose of the study can be for the calculation of insurance risks. For instance, in October 2005, the United Nations' World Food Program invited bids from ten of the world's largest insurance companies to insure Ethiopia against drought. They will use an index based on rainfall depths in some of the most vulnerable areas of the country. Droughts have also been experienced recently in Brazil, Kenya, Queensland in Australia, South Africa, Vietnam and Wyoming in the United States.

In these situations, historical records containing observations from the past are usually the only source of information. However, there are exceptions such as tsunamis, for which water depths and velocities need to be estimated from geological records. Let us consider the flood problem which, as stated, concerns an average close to 50 percent of annual natural disasters globally. Using limited samples of data and imperfect knowledge of the processes involved one is expected to develop and apply statistical techniques in order to estimate the degree of risk involved.

The immediate problem pertains to the way in which the probabilities are estimated and what levels of accuracy are associated with such probabilities. Firstly, what types of distributions and peaks do we consider? When do we limit the analysis to annual maximums and when is account taken of additional information such as the peaks during the year? What joint bivariate and multivariate distributions of the extremes are of physical interest and feasible for analysis (of wave heights and storm occurrences, for instance)? Currently a regional approach is adopted in the estimation of extreme quantiles, although at best this is somewhat subjective. How does one deal with observations that do not conform with other events? These are some of the questions to which we attempt to provide answers. In addition to floods and rainstorms there is a wide range of geological and climatic hazards to deal with.

Chapter 1 pertains to univariate extreme value distributions. We investigate initially the basic concepts from which extremes in nature derive. There are in-depth discussions on the concepts of extreme value theory, commencing with order statistics. Using asymptotic arguments and by specifying rules this leads to types of models that enable extrapolation, under certain conditions. Here we adopt a suitable theoretical level but without oversimplification. The assumptions on which the theory is based are emphasized throughout. We consider the distribution of the largest value and that of the smallest value. Probability distributions treated include the Gumbel, Fréchet, Weibull, and general extreme value types. We also demonstrate the use of contagious extreme value distributions.

Under extreme value theory we give examples that concern the analysis of various natural events apart from the perennial flood problem. As already discussed these are hazardous to human life and tend to destruct national economic and social infrastructures. We begin with geological hazards. Initial topics are earthquakes, volcanic eruptions and tsunamis. The related subjects of landslides and avalanches follow. These have links with the weather. We then focus on climatic and associated hazards such as windstorms, extreme sea levels and high waves, droughts, floods and wildfires. We provide substantial discussion of the nature of each hazard, give details of past events and discuss the problems involved and some social issues.

The analysis of an annual maximum series ignores vital information on high flow occurrences. An alternative method devised to overcome such problems and make better use of information on extreme values is the Peaks-Over-Threshold method, a part of Chapter 1. In the United States and in some countries this is known as the Partial Duration Series. The classical approach is based on a Poisson distributed times of exceedance over a high threshold and generalized Pareto distributed exceedances; a particular case is the exponential distribution.

In addition, recent advances concerning the Generalized Extreme Value and the Generalized Pareto distributions are given. Their relationships and scaling properties, which provide a link with stochastic fractals and multifractals, are investigated. Also the construction of suitable derived distributions is investigated and fully discussed. A special section deals with the concepts of hazard, return period, and risk. Rigorous general definitions, based on marked point processes and Measure Theory, are introduced, and the advantages offered by such an approach are illustrated via simple examples.

Whereas Chapter 1 concerns the univariate case, the important subject of multivariate extreme value analysis is treated in Chapter 2. The motivation is that quite often the joint occurrences of high values are of practical importance. We show the laws governing multivariate extremes and how they are characterized. In addition, several concepts of multivariate dependence are illustrated. A special section deals with the extension of the concept of return period to a multivariate setting, and shows how dimensionality may create a full variety of events having different return periods.

The two opening Chapters 1 and 2 constitute an important prelude to the main part of the book, in which we illustrate the approach based on copulas ranging from the bivariate to the multi-dimensional cases. By using copulas we describe exactly, and model the dependence structure between random variables independently of the marginal laws involved. The mathematical theory is presented and applications of copulas are emphasized. This approach simplifies the analysis of phenomena involved and makes it possible to introduce new functions and parameters for the characterization of the extreme behavior of a system.

Contrary to traditional approaches (for which extensions to the multivariate case often are not clear), the copula-approach is easy to generalize to a multi-dimensional framework. Moreover, a few multivariate distributions seen in the literature are direct extensions of well known univariate cases. Their possible limitations and constraints are that the marginal distributions belong to the same probability family, and that the parameters of the marginals may also rule the dependence between the variables considered. Copulas do not have these drawbacks. In addition, they have the advantage that complex marginal distributions, such as finite mixtures that are receiving increased attention, can be easily incorporated. Furthermore, all multivariate distributions can be described in a straightforward way in terms of suitable copulas.

Readers will become aware of the uniqueness and scope of copulas and the diverse real-life situations concerning natural hazards in which they can play an important role. Whereas the subject has long been in the domain of the mathematicians, many possibilities for their practical use are opening, as demonstrated in this book. Previously, applications have been confined to the financial sector. This is the first text in which copulas are utilized in geophysics.

Chapter 3 pertains to bivariate analysis via copulas. Natural events are often characterized by the joint behavior of several random variables. Copulas provide a very useful mathematical tool to model such multivariate types. Here we deal with 2-copulas. Archimedean types form a particular group that suffice for modeling quite a few phenomena. Also, univariate and bivariate frequency analyses are contrasted. The conventional analysis of return periods in univariate cases may lead to errors in the estimation of risks associated with events. We extend the concept of return period to the conditional case. Furthermore, in dealing with different classes of events according to their severity, we define primary and secondary return periods that provide a tool for investigating outliers or potentially destructive events. At the same time we model the possible tail dependence of the extreme events in multivariate frequency analysis, a fundamental step in estimating the risk adequately.

Chapter 4 pertains to multivariate analysis in d-dimensional space, $d > 2$, via copulas. We commence with definitions and properties of the multivariate types.

Archimedean and multidimensional copulas are analyzed, and special constructions, based on conditional mixtures, are presented. One example of the applications is the intensity and duration of rainfall (through a statistical characterization of the storm structure), another is the investigation of sea storms in terms of wave height–duration–interarrival times–direction.

Chapter 5 deals with extreme value analysis making use of the concepts previously discussed. This provides a different, but powerful, way to deal with extreme events in a bivariate or multivariate context. Extreme value copulas are introduced. The bivariate and general cases are included with an analysis of tail dependence. Several construction methods are illustrated, and a few examples show the features of some families of extreme value copulas. An application to regionalization procedures is also given.

In order to facilitate applications we provide three appendices. First, Appendix A concerns simulation of copulas in the two-dimensional and general cases. Second, Appendix B deals with investigations of dependence; this includes several notions of dependence, and measures of association and concordance (such as Kendall's τ_K and Spearmans's ρ_S) are illustrated. Third, families of copulas are described in detail in Appendix C: this includes numerous well-known types. Most importantly, we give construction methods, transformation and composition of copulas, which show how to generate the majority of the copulas that can be found in the literature.

ACKNOWLEDGMENTS. The Authors wish to express their gratitude to Prof. C. Sempi (Department of Mathematics, University of Salento, Italy) for his support and encouragement throughout the preparation of the manuscript.

Many thanks to Dr. F. Durante (formerly at the Department of Mathematics, University of Salento, Italy, and now at the Department of Knowledge-Based Mathematical Systems, Johannes Kepler University, Linz, Austria) for writing Appendices B and C.

The following offered their expertise in copulas in different ways: R.B. Nelsen, B. Schweizer, A. Sklar, C. Genest, A.-C. Favre, P. Embrechts, J.J. Quesada Molina, B. Lafuerza Guillén, M. Úbeda Flores, J.A. Rodríguez Lallena, R. Mesiar and E.P. Klement.

In the geophysical sciences and engineering the following shared their knowledge with us: J. Corominas (UPC, Barcelona, Spain), P. Gulkan (METU, Ankara, Turkey), G. Solari (University of Genova, Italy), G. Vannucchi (University of Firenze, Italy), M.C. Rulli and D. Bocchiola (Polytechnic of Milano, Italy).

Computer support was provided by M. Pini (University of Pavia, Italy).

To Petra D. van Steenbergen, Ria Balk, and the Springer Editorial Board, we appreciate your patience.

This book would not have been completed without the support of our families: Antonella (GS); Federica, Federico, and Ginevra (CDM); Mali, Shani, Siraj, and Natasha (NTK); Donatella and Riccardo (RR).

G. Salvadori
Department of Mathematics
University of Salento, Lecce (Italy)

C. De Michele
Department of Hydraulic, Environmental, Roads and Surveying Engineering
Polytechnic of Milano, Milano (Italy)

N.T. Kottegoda
Department of Hydraulic, Environmental, Roads and Surveying Engineering
Polytechnic of Milano, Milano (Italy)

R. Rosso
Department of Hydraulic, Environmental, Roads and Surveying Engineering
Polytechnic of Milano, Milano (Italy)

CHAPTER 1

UNIVARIATE EXTREME VALUE THEORY

The classical univariate theory of extreme values was developed by Fréchet [100], and Fisher and Tippett [95]. Gnedenko [118] and Gumbel [124] showed that the largest or smallest value from a set of independently distributed random variables tends to an asymptotic distribution that only depends upon that of the basic variable. After standardization using suitable norming and centering constants, the limit distribution is shown to belong to one of three types, as pointed out by Gnedenko [118] and deHaan [59].

In this chapter we study the distributions of the largest and the smallest values of a given sample. Two different approaches are presented. In Subsection 1.2.1 the limit distributions of maxima and minima are calculated by using the "block" method. The *Generalized Extreme Value* distribution is derived, which includes all the three types of limit distributions (i.e., the *Gumbel*, the *Fréchet*, and the *Weibull*). In Subsection 1.2.2 the extremes are studied by considering their exceedances over a given threshold, corresponding to the "Peaks-Over-Threshold" method. The *Generalized Pareto* distribution is derived. The scaling features of the *Generalized Extreme Value* and the *Generalized Pareto* distributions are then investigated in Subsection 1.2.3, providing a useful tool for characterizing in a synthetic way the probabilistic structure of the extremes. In addition, the *Contagious Extreme Value* distributions are studied in Subsection 1.2.4. General definitions of *Hazard*, *Return period*, and *Risk*, based on marked point processes and Measure Theory, are given in Section 1.3. Finally, examples of *Natural Hazards* illustrate the theory in Section 1.4. We commence by studying the *Order Statistics* associated with a given sample.

1.1. ORDER STATISTICS

Order statistics are one of the fundamental tools in non-parametric statistics, inference, and analysis of extremes. After a formal definition, we now provide an introduction — see [58, 36] for further details. First of all, let us define order statistics.

DEFINITION 1.1 (Order Statistics (OS)). *Let $X_1, X_2, \ldots, X_n$ denote a random sample of size n extracted from a distribution F. Arranging the sample in ascending order of magnitude generates a new family of observations, written as $X_{(1)} \le X_{(2)} \le \cdots \le X_{(n)}$, called order statistics associated with the original sample. In particular, the r.v. $X_{(i)}$, $i = 1, \ldots, n$, denotes the i-th order statistic.*

NOTE 1.1. In general, even if the r.v.'s X_i's are i.i.d., the corresponding OS $X_{(i)}$'s are *not independent*: in fact, if $X_{(1)} > x$, then $X_{(i)} > x$, $i = 2, \ldots, n$. As we shall see in Subsection 1.1.3, the OS are *not identically distributed*.

ILLUSTRATION 1.1 (Water levels). ►

The water level of a stream, river or sea is a variable of great interest in the design of many engineering works: irrigation schemes, flood protection systems (embankments, polders), river regulated works (dams), and maritime dykes. Let X_i, $i = 1, \ldots, n$, represent a sample of water levels. The first k OS, $X_{(1)}, \ldots, X_{(k)}$, are the smallest observations, and represent the behavior of the system during the drought or "calm" periods. Instead, the last k OS, $X_{(n-k+1)}, \ldots, X_{(n)}$, i.e. the largest values, represent the behavior of the system during the flood or "storm" periods. ◄

1.1.1 Distribution of the Smallest Value

The first order statistic, $X_{(1)}$, represents the minimum of the sample:

$$X_{(1)} = \min\{X_1, \ldots, X_n\}. \tag{1.1}$$

Let $F_{(1)}$ be the c.d.f. of $X_{(1)}$.

THEOREM 1.1 (Distribution of the smallest value). *Let $X_1, \ldots, X_n$ be a generic random sample. Then $F_{(1)}$ is given by*

$$F_{(1)}(x) = \mathbb{P}\left\{X_{(1)} \le x\right\} = 1 - \mathbb{P}\{X_1 > x, \ldots, X_n > x\}. \tag{1.2}$$

Since the two events $\left\{X_{(1)} > x\right\}$ and $\{X_1 > x\} \cap \cdots \cap \{X_n > x\}$ are equivalent, the proof of Theorem 1.1 is straightforward. The statement of the theorem can be made more precise if additional assumptions on the sample are introduced.

COROLLARY 1.1. *Let $X_1, \ldots, X_n$ be a random sample of independent r.v.'s with distributions F_i's. Then $F_{(1)}$ is given by*

$$F_{(1)}(x) = 1 - \prod_{i=1}^{n} \mathbb{P}\{X_i > x\} = 1 - \prod_{i=1}^{n} (1 - F_i(x)). \tag{1.3}$$

COROLLARY 1.2. *Let $X_1, \ldots, X_n$ be a random sample of i.i.d. r.v.'s with common distribution F. Then $F_{(1)}$ is given by*

$$F_{(1)}(x) = 1 - \prod_{i=1}^{n} \mathbb{P}\{X_i > x\} = 1 - (1 - F(x))^n. \tag{1.4}$$

COROLLARY 1.3. *Let $X_1, \ldots, X_n$ be a random sample of absolutely continuous i.i.d. r.v.'s with common density f. Then the p.d.f. $f_{(1)}$ of $X_{(1)}$ is given by*

$$f_{(1)}(x) = n\,(1 - F(x))^{n-1}\,f(x). \tag{1.5}$$

ILLUSTRATION 1.2 (Smallest OS of Geometric r.v.'s). ►

Let $X_1, \ldots, X_n$ be a sample of discrete r.v.'s having a Geometric distribution. The p.m.f. of the i-th r.v. X_i is

$$\mathbb{P}\{X_i = k\} = p_i\, q_i^{k-1}, \quad k = 1, 2, \ldots \tag{1.6}$$

where $p_i \in (0, 1)$ is the parameter of the distribution, and $q_i = 1 - p_i$. The c.d.f. of X_i is

$$F_i(x) = \mathbb{P}\{X_i \le x\} = 1 - q_i^{\lfloor x \rfloor}, \quad x > 0, \tag{1.7}$$

and zero elsewhere. Using Corollary 1.1, the distribution of $X_{(1)}$, for $x > 0$, is

$$F_{(1)}(x) = 1 - \prod_{i=1}^{n}(1 - F_i(x)) = 1 - \prod_{i=1}^{n}\left(1 - 1 + q_i^{\lfloor x \rfloor}\right) = 1 - q^{\lfloor x \rfloor}, \tag{1.8}$$

where $q = q_1 \cdots q_n$. ◄

Illustration 1.2 shows that thc *minimum* of independent Geometric r.v.'s is again a r.v. with a Geometric distribution. In addition, since $q < q_i$ for all i's, then

$$\mathbb{P}\left\{X_{(1)} > k\right\} = q^k < q_i^k = \mathbb{P}\{X_i > k\}, \quad k = 1, 2, \ldots \tag{1.9}$$

Thus, the probability that $X_{(1)}$ is greater than k is less than the probability that the generic r.v. X_i is greater than k. This fact is important in reliability and risk analysis.

ILLUSTRATION 1.3 (Smallest OS of Exponential r.v.'s). ►

Let $X_1, \ldots, X_n$ be a sample of continuous r.v.'s having an Exponential distribution. The c.d.f. of the i-th r.v. X_i is

$$F_i(x) = \mathbb{P}\{X_i \le x\} = 1 - e^{-x/b_i}, \quad x > 0, \tag{1.10}$$

and zero elsewhere. Here $b_i > 0$ is the distribution parameter. Using Corollary 1.1, the distribution of $X_{(1)}$, for $x > 0$, is

$$F_{(1)}(x) = 1 - \prod_{i=1}^{n}(1 - F_i(x)) = 1 - \prod_{i=1}^{n}\left(1 - 1 + e^{-x/b_i}\right) = 1 - e^{-x/b}, \tag{1.11}$$

where $1/b = 1/b_1 + \cdots + 1/b_n$. ◀

Illustration 1.3 shows that the *minimum* of independent Exponential r.v.'s is again a r.v. with an Exponential probability law. In addition, since $1/b > 1/b_i$ for all i's, then

$$\mathbb{P}\left\{X_{(1)} > x\right\} = e^{-x/b} < e^{-x/b_i} = \mathbb{P}\left\{X_i > x\right\}, \quad x > 0. \tag{1.12}$$

Thus, the probability that $X_{(1)}$ is greater than x is less than the probability that the generic r.v. X_i is greater than x. Again, this fact is important in reliability and risk analysis.

1.1.2 Distribution of the Largest Value

The last order statistic, $X_{(n)}$, represents the maximum of the sample:

$$X_{(n)} = \max\left\{X_1, \ldots, X_n\right\}. \tag{1.13}$$

Let $F_{(n)}$ be the c.d.f. of $X_{(n)}$.

THEOREM 1.2 (Distribution of the largest value). *Let $X_1, \ldots, X_n$ be a generic random sample. Then $F_{(n)}$ is given by*

$$F_{(n)}(x) = \mathbb{P}\left\{X_{(n)} \le x\right\} = \mathbb{P}\left\{X_1 \le x, \ldots, X_n \le x\right\}. \tag{1.14}$$

Since the two events $\left\{X_{(n)} \le x\right\}$ and $\left\{X_1 \le x\right\} \cap \cdots \cap \left\{X_n \le x\right\}$ are equivalent, the proof of Theorem 1.2 is straightforward. The statement of the theorem can be made more precise if additional assumptions on the sample are introduced.

COROLLARY 1.4. *Let $X_1, \ldots, X_n$ be a random sample of independent r.v.'s with distributions F_i's. Then $F_{(n)}$ is given by*

$$F_{(n)}(x) = \prod_{i=1}^{n}\mathbb{P}\left\{X_i \le x\right\} = \prod_{i=1}^{n}F_i(x). \tag{1.15}$$

COROLLARY 1.5. *Let $X_1, \ldots, X_n$ be a random sample of i.i.d. r.v.'s with common distribution F. Then $F_{(n)}$ is given by*

$$F_{(n)}(x) = \prod_{i=1}^{n}\mathbb{P}\left\{X_i \le x\right\} = \left(F(x)\right)^n. \tag{1.16}$$

COROLLARY 1.6. *Let $X_1, \ldots, X_n$ be a random sample of absolutely continuous i.i.d. r.v.'s with common density f. Then the p.d.f. $f_{(n)}$ of $X_{(n)}$ is given by*

$$f_{(n)}(x) = n\,(F(x))^{n-1}\,f(x). \tag{1.17}$$

ILLUSTRATION 1.4 (Largest OS of Exponential r.v.'s). ►

Let $X_1, \ldots, X_n$ be a sample of i.i.d. unit Exponential r.v.'s. Using Corollary 1.5, the c.d.f. of $X_{(n)}$ is, for $x > 0$,

$$F_{(n)}(x) = (1 - e^{-x})^n, \tag{1.18}$$

and the corresponding p.d.f. is

$$f_{(n)}(x) = n(1 - e^{-x})^{n-1} e^{-x}. \tag{1.19}$$

Note that, as $n \to \infty$, the c.d.f. of $X_{(n)} - \ln n$ tends to the limiting form $\exp[-\exp(-x)]$, for $x \in \mathbb{R}$ — see Section 1.2. ◄

1.1.3 General Distributions of Order Statistics

In this section we study the distribution of the generic order statistic.

THEOREM 1.3 (Distribution of the i-th OS). *Let $X_1, \ldots, X_n$ be a random sample of size n extracted from F. Then the i-th OS $X_{(i)}$ has c.d.f. $F_{(i)}$ given by*

$$F_{(i)}(x) = \mathbb{P}\left\{X_{(i)} \le x\right\} = \sum_{j=i}^{n} \binom{n}{j} (F(x))^j\,(1 - F(x))^{n-j}. \tag{1.20}$$

The proof of Theorem 1.3 is based on the observation that the event $\left\{X_j \le x \text{ for at least } i \text{ out of } n \text{ r.v.'s } X_j\text{'s}\right\}$ is equivalent to $\left\{X_{(i)} \le x\right\}$.

Putting $p = F(x)$ in Eq. (1.20), this can be rewritten as

$$F_{(i)}(x) = \sum_{j=i}^{n} \binom{n}{j} p^j\,(1-p)^{n-j}. \tag{1.21}$$

Then, using the functions *Beta* and *Incomplete Beta*, it is possible to write $F_{(i)}$ as

$$F_{(i)}(x) = \frac{\mathrm{B}_{F(x)}(i, n-i+1)}{\mathrm{B}(i, n-i+1)}. \tag{1.22}$$

If the sample of Theorem 1.3 consists of absolutely continuous r.v.'s, then the following result holds.

COROLLARY 1.7. *Let $X_1, \ldots, X_n$ be absolutely continuous with common p.d.f. f. Then $X_{(i)}$ has p.d.f. $f_{(i)}$ given by*

$$f_{(i)}(x) = \frac{1}{\mathrm{B}(i, n-i+1)} (F(x))^{i-1} (1-F(x))^{n-i} f(x). \tag{1.23}$$

NOTE 1.2. Differentiating Eq. (1.22) with respect to x yields

$$f_{(i)}(x) = \frac{1}{\mathrm{B}(i, n-i+1)} \frac{\mathrm{d}}{\mathrm{d}x} \int_{0+}^{F(x)} t^{i-1} (1-t)^{n-i+1-1} \, \mathrm{d}t, \tag{1.24}$$

which, in turn, leads to Eq. (1.23).

Alternatively, the result of Corollary 1.7 can be heuristically derived as follows. The event $\{x < X_{(i)} \leq x + \mathrm{d}x\}$ occurs when $i-1$ out of n r.v.'s X_i's are less than x, one r.v. is in the range $(x, x+\mathrm{d}x]$, and the last $n-i$ r.v.'s are greater than $x+\mathrm{d}x$. As the random sample $X_1, \ldots, X_n$ is composed of i.i.d. r.v.'s, the number of ways to construct such an event is

$$\frac{n!}{(i-1)!\,1!\,(n-i)!} = \binom{n}{i-1}\binom{n-i+1}{1}\binom{n-i}{n-i}, \tag{1.25}$$

each having probability

$$(F(x))^{i-1} (F(x+\mathrm{d}x) - F(x)) (1 - F(x+\mathrm{d}x))^{n-i}. \tag{1.26}$$

Multiplying Eq. (1.25) and Eq. (1.26), dividing the result by $\mathrm{d}x$, and taking the limit $\mathrm{d}x \to 0$, yields the density given by Eq. (1.23). Clearly, Corollary 1.3 and Corollary 1.6 are particular cases of Corollary 1.7.

An important property of order statistics concerns their link with quantiles. Let us consider an absolutely continuous r.v. X, with strictly increasing c.d.f. F. In this case the quantile x_p associated with the probability level $p \in (0, 1)$ satisfies the relationship $F(x_p) = p$. The event $\{X_{(i)} \leq x_p\}$ can then be written as

$$\{X_{(i)} \leq x_p\} = \{X_{(i)} \leq x_p, X_{(j)} \geq x_p\} \cup \{X_{(i)} \leq x_p, X_{(j)} < x_p\}, \tag{1.27}$$

where $i < j$. Since the two events on the right side of Eq. (1.27) are disjoint, and $X_{(j)} < x_p$ implies $X_{(i)} \leq x_p$, then

$$\mathbb{P}\{X_{(i)} \leq x_p\} = \mathbb{P}\{X_{(i)} \leq x_p \leq X_{(j)}\} + \mathbb{P}\{X_{(j)} < x_p\}, \tag{1.28}$$

or

$$\mathbb{P}\{X_{(i)} \leq x_p \leq X_{(j)}\} = \mathbb{P}\{X_{(i)} \leq x_p\} - \mathbb{P}\{X_{(j)} < x_p\}. \tag{1.29}$$

Now, using Theorem 1.3, the following important result follows.

PROPOSITION 1.1. *The random interval* $\left(X_{(i)}, X_{(j)}\right)$, $i < j$, *includes the quantile* x_p, $p \in (0, 1)$, *with probability*

$$\begin{aligned}\mathbb{P}\left\{X_{(i)} \le x_p \le X_{(j)}\right\} &= F_{(i)}\left(x_p\right) - F_{(j)}\left(x_p\right) \\ &= \sum_{k=i}^{j} \binom{n}{k} p^k \left(1-p\right)^{n-k},\end{aligned} \tag{1.30}$$

where n *is the sample size.*

NOTE 1.3. The probability calculated in Proposition 1.1 is independent of the distribution F of X. Thus the random interval $\left(X_{(i)}, X_{(j)}\right)$ is a "weak estimate" of x_p independent of F, i.e. a "non-parametric" estimate of x_p.

Similarly, following the above rationale, it is possible to calculate the *bivariate* or *multivariate* distribution of order statistics, as well as the *conditional* one.

THEOREM 1.4 (Bivariate density of OS). *Let* $X_1, \ldots, X_n$ *be a random sample of absolutely continuous i.i.d. r.v.'s with common density* f. *Then the joint p.d.f.* $f_{(i,j)}$ *of* $(X_{(i)}, X_{(j)})$, $1 \le i < j \le n$, *is given by*

$$\begin{aligned}f_{(i,j)}(x, y) = &\frac{n!}{(i-1)!(j-i-1)!(n-j)!} \left(F(x)\right)^{i-1} f(x) \\ &\cdot \left(F(y)-F(x)\right)^{j-i-1} \left(1-F(y)\right)^{n-j} f(y),\end{aligned} \tag{1.31}$$

for $x < y$, *and zero elsewhere.*

Using Theorem 1.4, it is then possible to derive the *conditional* density of $X_{(i)}$ given the event $\left\{X_{(j)} = y\right\}$, with $i < j$. Note that this is fundamental for performing computer simulations.

THEOREM 1.5 (Conditional density of OS). *Let* $X_1, \ldots, X_n$ *be a random sample of absolutely continuous i.i.d. r.v.'s with common density* f. *Then the conditional p.d.f.* $f_{(i|j)}$ *of* $X_{(i)}$ *given the event* $\left\{X_{(j)} = y\right\}$, $1 \le i < j \le n$, *is given by*

$$\begin{aligned}f_{(i|j)}(x \mid y) &= \frac{f_{(i,j)}(x, y)}{f_{(j)}(y)} \\ &= \frac{(j-1)!}{(i-1)!(j-i-1)!} \left(F(x)\right)^{i-1} f(x) \cdot \\ &\qquad \left(F(y)-F(x)\right)^{j-i-1} \left(F(y)\right)^{1-j},\end{aligned} \tag{1.32}$$

where $x < y$, *and zero elsewhere.*

NOTE 1.4. An alternative interpretation of Eq. (1.32) shows that it equals the density of $X_{(i)}$ in a sample of size $j-1$ extracted from F and censored to the right at y.

The calculation of the *joint* density of all the OS $X_{(1)}, \ldots, X_{(n)}$ stems from the observation that any of the $n!$ permutations of the n r.v.'s X_i, $i = 1, \ldots, n$, has the same probability of occurrence.

THEOREM 1.6 (Density of all the OS). *Let $X_1, \ldots, X_n$ be a random sample of absolutely continuous i.i.d. r.v.'s with common density f. Then the joint p.d.f. $f_{(1,\ldots,n)}$ of all the OS is given by*

$$f_{(1,\ldots,n)}(x_1, \ldots, x_n) = n! \prod_{i=1}^{n} f(x_i), \tag{1.33}$$

for $x_1 < \cdots < x_n$, and zero elsewhere.

NOTE 1.5. Using the result of Theorem 1.6, it is possible to calculate the joint density of any subset of OS simply by integrating $f_{(1,\ldots,n)}$ with respect to the variables to be eliminated.

From Theorem 1.4 it is easy to calculate the distribution of several functions of OS.

PROPOSITION 1.2 (Law of the difference of OS). *Let $X_1, \ldots, X_n$ be a random sample of absolutely continuous i.i.d. r.v.'s with common density f. Then the probability law of the difference $V = X_{(j)} - X_{(i)}$, $1 \le i < j \le n$, is given by*

$$\begin{aligned} f_V(v) = {} & \frac{n!}{(i-1)!(j-i-1)!(n-j)!} \cdot \\ & \int_{\mathbb{R}} (F(u))^{i-1} f(u) \left(F(u+v) - F(u)\right)^{j-i-1} \cdot \\ & \qquad (1 - F(u+v))^{n-j} f(u+v)\, du, \end{aligned} \tag{1.34}$$

for $v > 0$, and zero elsewhere.

The statement of Proposition 1.2 can be demonstrated by using the method of the *auxiliary variable*. Let us consider the invertible transformation

$$\begin{cases} U = X_{(i)} \\ V = X_{(j)} - X_{(i)} \end{cases} \iff \begin{cases} X_{(i)} = U \\ X_{(j)} = U + V \end{cases}, \tag{1.35}$$

having Jacobian $J = \left|\begin{smallmatrix} 1 & 0 \\ 1 & 1 \end{smallmatrix}\right| = 1$. The joint density g of (U, V) is then given by $g(u, v) = J \cdot f_{(i,j)}(u, u+v)$, i.e.

$$\begin{aligned} g(u, v) = {} & \frac{n!}{(i-1)!(j-i-1)!(n-j)!} \cdot \\ & (F(u))^{i-1} f(u) \left(F(u+v) - F(u)\right)^{j-i-1} \cdot \\ & \qquad (1 - F(u+v))^{n-j} f(u+v), \end{aligned} \tag{1.36}$$

for $v > 0$. Then, integrating Eq. (1.36) with respect to u yields Eq. (1.34).

ILLUSTRATION 1.5 (Distribution of the sample range). ►

In applications it is often of interest to estimate the *sample range* R given by

$$R = X_{(n)} - X_{(1)}. \tag{1.37}$$

This provides a sample estimation of the variability of the observed phenomenon. For example, the so-called Hurst's effect [143, 167] stems from sample range analysis of long-term geophysical data. Using Proposition 1.2 it is easy to calculate the distribution of R. Putting $i = 1$ and $j = n$ in Eq. (1.34) yields

$$f_R(r) = n\,(n-1) \int_{\mathbb{R}} f(u)\,(F(u+r) - F(u))^{n-2}\, f(u+r)\, \mathrm{d}u \tag{1.38a}$$

and

$$F_R(r) = \int_{-\infty}^{r} f_R(t)\, \mathrm{d}t = n \int_{\mathbb{R}} f(u)\,(F(u+r) - F(u))^{n-1}\, \mathrm{d}u, \tag{1.38b}$$

where $r > 0$. ◄

1.1.4 Plotting Positions

In practical applications it is often necessary to compare the "theoretical" (or expected) probability distribution with the "empirical" (or observed) frequency of the sample values. More precisely, the theoretical values $F(x_{(i)})$'s (i.e., the model evaluated at all the observed OS), are compared with a surrogate of the empirical distribution, denoted by $\widehat{F}_i$'s, called *plotting position*.

ILLUSTRATION 1.6 (Weibull plotting position). ►

Let $X_1, \ldots, X_n$ be a random sample of continuous r.v.'s, and let $X_{(1)}, \ldots, X_{(n)}$ be the corresponding OS. The ordered sequence of ordinates

$$\widehat{F}_i = \frac{i}{n+1}, \tag{1.39}$$

with $i = 1, \ldots, n$, defines the so-called *Weibull* plotting position. Note that $0 < \widehat{F}_i < 1$. The interesting point is that, using Corollary 1.7, it is possible to show that $\mathbb{E}\left(F\left(X_{(i)}\right)\right)$ is given by

$$\begin{aligned}
\mathbb{E}\left(F\left(X_{(i)}\right)\right) &= \int_{\mathbb{R}} F(x)\, f_{(i)}(x)\, \mathrm{d}x \\
&= \frac{1}{\mathrm{B}(i, n-i+1)} \int_{\mathbb{R}} (F(x))^{i}\, (1-F(x))^{n-i}\, f(x)\, \mathrm{d}x \\
&= \frac{1}{\mathrm{B}(i, n-i+1)} \int_{0}^{1} t^{i}\, (1-t)^{n-i}\, \mathrm{d}t \\
&= \frac{\mathrm{B}(i+1, n-i+1)}{\mathrm{B}(i, n-i+1)} \\
&= \frac{\Gamma(i+1)\Gamma(n-i+1)}{\Gamma(n+2)} \frac{\Gamma(n+1)}{\Gamma(i)\Gamma(n-i+1)} \\
&= \frac{i!\,(n-i)!}{(n+1)!} \frac{n!}{(i-1)!\,(n-i)!} \\
&= \frac{i}{n+1}.
\end{aligned}$$

Thus, *on average*, $F\left(X_{(i)}\right)$ is equal to $\frac{i}{n+1}$. This explains why the *Weibull* plotting position are so appealing in practice: for this reason they are usually known as *standard* plotting position. ◀

In general, the plotting positions provide a "non-parametric" estimate of the (unknown) c.d.f. of the sample considered, for they are independent of F. In the literature many plotting positions are available [56, 44, 229]. Some of these are given in Table 1.1.

Note how all the expressions shown in Table 1.1 are particular cases of the general formula

$$\widehat{F}_i = \frac{i-a}{n+1-b}, \tag{1.40}$$

where a, b are suitable constants.

Table 1.1. Formulas of some plotting positions

Plotting position	$\widehat{\mathrm{F}}_{\mathrm{i}}$	Plotting position	$\widehat{\mathrm{F}}_{\mathrm{i}}$
Adamowski	$\frac{i-0.26}{n+0.5}$	**Gringorten**	$\frac{i-0.44}{n+0.12}$
Blom	$\frac{i-3/8}{n+1/4}$	**Hazen**	$\frac{i-1/2}{n}$
California	$\frac{i-1}{n}$	**Hosking (APL)**	$\frac{i-0.35}{n}$
Chegodayev	$\frac{i-0.3}{n+0.4}$	**Tukey**	$\frac{i-1/3}{n+1/3}$
Cunanne	$\frac{i-1/5}{n+2/5}$	**Weibull**	$\frac{i}{n+1}$

1.2. EXTREME VALUE THEORY

In this section we illustrate the main results of *classical* Extreme Values theory. According to Section 1.1, one could argue that the problems concerning extreme values (namely, maximum and minimum) can be solved if (*a*) the distribution F of X is known, and (*b*) the sample size n is given. However, these conditions seldom occur in practice. For instance, consider the following cases:

1. the distribution F of X is not known;
2. the sample size n is not known;
3. the sample size n diverges.

Case (1) is quite common in practical applications: only rarely is the "parent" distribution F of X known. In order to bypass the problem, generally the distribution of X is fixed *a priori*. However, the question "How to proceed when F is not known?" is of some interest in its own right, both from a theoretical and a practical point of view.

Case (2) is somewhat paradoxical: n should be perfectly known once the data have been collected. However, if the instruments only sample above (or below) a given threshold — i.e., the data are *censored* — a certain amount of information is lost. Thus, "How do the missed data influence the distribution of extreme values?" and "How do we proceed in this situation?" are again questions of some interest.

Case (3) arises by considering the following limits:

$$\lim_{n\to\infty} F^n(x) = \begin{cases} 1, & \text{if } F(x) = 1 \\ 0, & \text{if } F(x) < 1 \end{cases}, \tag{1.41a}$$

$$\lim_{n\to\infty} 1 - (1 - F(x))^n = \begin{cases} 0, & \text{if } F(x) = 0 \\ 1, & \text{if } F(x) > 0 \end{cases}. \tag{1.41b}$$

Apparently, the *limit* distributions of the maximum and the minimum are degenerate. Thus, the following questions arise: (*a*) Does the limit distribution of the maximum (or minimum) exist? and if the answer is positive (*b*) What is the limit distribution? (*c*) May different "parent" distributions have the same limit? (*d*) Is it possible to calculate analytically the limit distribution associated with a given "parent" distribution?

In this section we provide some answers to the above questions. Further details can be found in [36, 87, 13, 168, 83, 169, 45], and references therein. In order to simplify the discussion, we consider continuous distributions, and samples of i.i.d. r.v.'s.

NOTE 1.6. Below we focus the attention on the analysis of maxima only. In fact, the results obtained for the maxima can be easily extended to the analysis of minima, via the following transformation:

$$Y_i = -X_i, \quad i = 1, \ldots, n, \tag{1.42}$$

Then, if $X_{(1)} = \min\{X_1, \dots, X_n\}$ and $Y_{(n)} = \max\{Y_1, \dots, Y_n\}$, it is evident that

$$X_{(1)} = -Y_{(n)}. \tag{1.43}$$

Thus, the analysis of the maxima suffices.

DEFINITION 1.2 (Lower and upper limit of a distribution). *The lower limit α_F of the distribution F is defined as*

$$\alpha_F = \inf\{x : F(x) > 0\}. \tag{1.44a}$$

The upper limit ω_F of the distribution F is defined as

$$\omega_F = \sup\{x : F(x) < 1\}. \tag{1.44b}$$

Clearly, α_F and ω_F represent, respectively, the minimum and maximum attainable values of the r.v. X associated with F. Obviously, $\alpha_F = -\infty$ in case of a lower-unbounded r.v., and $\omega_F = +\infty$ for an upper-bounded one.

1.2.1 "Block" Model

The subject of this section concerns a procedure widely used to analyze the extrema (maxima and minima) of a given distribution. In particular, the term *block model* refers to the way in which the data are processed, as explained in Note 1.11 which follows shortly.

Eq. (1.41) shows how the search for the limit distributions requires some caution, so that it does not result in degenerate forms. A standard way to operate is to search for the existence of two sequences of constants, $\{a_n\}$ and $\{b_n > 0\}$, such that the function

$$G(x) = \lim_{n\to\infty} F^n(a_n + b_n x), \qquad x \in \mathbb{R}, \tag{1.45}$$

is a non-degenerate distribution. This procedure (similar to the one used in deriving the *Central Limit Theorem*), aims to identify a non-degenerate limit distribution after a suitable *renormalization* of the variables involved.

DEFINITION 1.3 (Maximum domain of attraction (MDA)). *The distribution F is said to belong to the maximum domain of attraction of the distribution G if there exist sequences of constants, $\{a_n\}$ and $\{b_n > 0\}$, such that Eq. (1.45) is satisfied.*

An analogous definition can be given for minima. The calculation of the limit distribution of the maximum (or minimum) can be summerized as follows: (*a*) check the validity conditions of Eq. (1.45); (*b*) provide rules to construct the sequences $\{a_n\}$, $\{b_n\}$; (*c*) find the analytical form of the limit distribution G in Eq. (1.45).

THEOREM 1.7 (Asymptotic laws of maxima). *Let $X_1, \ldots, X_n$ be a sample of i.i.d. r.v.'s, and let $M_n = \max\{X_1, \ldots, X_n\}$. If norming sequences $\{a_n\}$ and $\{b_n > 0\}$ exist such that*

$$\lim_{n\to\infty} \mathbb{P}\left\{\frac{M_n - a_n}{b_n} \leq z\right\} = G(z), \quad z \in \mathbb{R}, \tag{1.46}$$

where G is a non-degenerate distribution, then G belongs to one of the following three types of limit (or asymptotic) distributions of maxima:

1. ***Type I (Gumbel)***

$$G(z) = \exp\left\{-\exp\left(-\frac{z-a}{b}\right)\right\}, \quad -\infty < z < \infty; \tag{1.47}$$

2. ***Type II (Fréchet)***

$$G(z) = \begin{cases} 0, & z \leq a \\ \exp\left\{-\left(\frac{z-a}{b}\right)^{-\gamma}\right\}, & z > a \end{cases}; \tag{1.48}$$

3. ***Type III (Weibull)***

$$G(z) = \begin{cases} \exp\left\{-\left[\left(-\frac{z-a}{b}\right)^{\gamma}\right]\right\}, & z < a \\ 1, & z \geq a \end{cases}. \tag{1.49}$$

Here, $a \in \mathbb{R}$ is a position parameter, $b > 0$ a scale parameter, and $\gamma > 0$ a shape parameter.

The above result is known as the *Fisher-Tippet Theorem*. Note that the convention on the parameters a, b, γ in Theorem 1.7 is only one among those available in literature. Theorem 1.7 is an analogue of the *Central Limit Theorem* within the Theory of Extreme Values: in fact, it states that the limit distribution of M_n (once rescaled via suitable affine transformations) can take only one out of three specific types. However, the following distribution provides a counter-example:

$$F(x) = 1 - \frac{1}{\ln x}, \quad x > e. \tag{1.50}$$

In fact, F has a heavier tail than Pareto-like distributions, and there is no extreme value limit based on linearity.

NOTE 1.7. An analysis of the asymptotic behavior of G in a suitable neighbourhood of ω_G shows how the three distributions given inTheorem 1.7 are quite different in terms of their *right tail* structure. Without loss of generality, let us consider the *canonical form* of G (i.e., with $a = 0$ and $b = 1$).

1. For the Gumbel distribution $\omega_G = +\infty$, and its p.d.f. g is given by

$$g(z) \propto \exp\{-\exp(-z)\} \exp(-z), \tag{1.51}$$

with an "exponential" fall-off for $z \gg 1$. G is called "light tailed" distribution.
2. For the Fréchet distribution $\omega_G = +\infty$, and its p.d.f. g is given by

$$g(z) \propto \exp\{-z^{-\gamma}\} z^{-\gamma-1}, \tag{1.52}$$

with an "algebraic" fall-off for $z \gg 1$. G is called "heavy tailed" (or "fat-tailed") distribution. In this case, the statistical moments of order $m \geq \gamma > 0$ of G do not exist.
3. For the Weibull distribution $\omega_G < +\infty$, and its density is of course null for $z > \omega_G$.

Although the three limit laws for maxima are quite different, from a mathematical point of view they are closely linked. In fact, let Z be a positive r.v.: then

Z has a Fréchet distribution with parameter γ
if, and only if,
$\ln Z^\gamma$ has a Gumbel distribution
if, and only if,
$-Z^{-1}$ has a Weibull distribution with parameter γ.

For the sake of completeness, we now give the analogue of Theorem 1.7 for minima.

THEOREM 1.8 (Asymptotic laws of minima). *Let $X_1, \ldots, X_n$ be a sample of i.i.d. r.v.'s, and let $\widetilde{M}_n = \min\{X_1, \ldots, X_n\}$. If norming sequences $\{\tilde{a}_n\}$ and $\left\{\tilde{b}_n > 0\right\}$ exist such that*

$$\lim_{n\to\infty} \mathbb{P}\left\{\frac{\widetilde{M}_n - \tilde{a}_n}{\tilde{b}_n} \leq z\right\} = \widetilde{G}(z), \quad z \in \mathbb{R}, \tag{1.53}$$

where $\widetilde{G}$ is a non-degenerate distribution, then $\widetilde{G}$ belongs to one of the following three types of limit (or asymptotic) distributions of minima:

1. ***Type I (Converse Gumbel)***

$$\widetilde{G}(z) = 1 - \exp\left\{-\exp\left(\frac{z-\tilde{a}}{\tilde{b}}\right)\right\}, \quad -\infty < z < \infty; \tag{1.54}$$

2. ***Type II (Converse Fréchet)***

$$\widetilde{G}(z) = \begin{cases} 1 - \exp\left\{-\left(-\frac{z-\tilde{a}}{\tilde{b}}\right)^{-\tilde{\gamma}}\right\}, & z \leq \tilde{a} \\ 1, & z > \tilde{a} \end{cases}; \tag{1.55}$$

3. ***Type III (Converse Weibull)***

$$\widetilde{G}(z) = \begin{cases} 0, & z < \tilde{a} \\ 1 - \exp\left\{-\left(\frac{z-\tilde{a}}{\tilde{b}}\right)^{\tilde{\gamma}}\right\}, & z \geq \tilde{a} \end{cases}. \tag{1.56}$$

Here, $\tilde{a} \in \mathbb{R}$ is a position parameter, $\tilde{b} > 0$ a scale parameter, and $\tilde{\gamma} > 0$ a shape parameter.

Theorem 1.7 can be reformulated in a compact form by introducing the *Generalized Extreme Value* (GEV) distribution, also known, in practical applications, as the *von Mises* or *von Mises-Jenkinson* probability law.

THEOREM 1.9 (GEV distribution of maxima). *Under the conditions of Theorem 1.7, G is a member of the GEV family of maxima given by, for* $\left\{z \colon 1+\gamma' \frac{z-a'}{b'} > 0\right\}$,

$$G(z) = \exp\left\{-\left[1+\gamma' \frac{z-a'}{b'}\right]^{-1/\gamma'}\right\}, \tag{1.57}$$

where $a' \in \mathbb{R}$ is a position parameter, $b' > 0$ a scale parameter, and $\gamma' \in \mathbb{R}$ a shape parameter. The limit case $\gamma' = 0$ yields the Gumbel distribution given by Eq. (1.47).

NOTE 1.8. Note that the Weibull family is obtained for $\gamma' < 0$, whereas the Fréchet family applies for $\gamma' > 0$.

The c.d.f. and p.d.f. of GEV distribution of maxima (using the canonical form) are shown, respectively, in Figures 1.1–1.2, considering different values of the shape parameter. In addition, in Figure 1.3 we show the quantile function z_G in the plane $(-\ln(-\ln(G)), z)$: thus plotted is the *Gumbel* (or *reduced*) variate for maxima.

For the sake of completeness, we now give the version of Theorem 1.9 for minima.

THEOREM 1.10 (GEV distribution of minima). *Under the conditions of Theorem 1.8, $\widetilde{G}$ is a member of the GEV family of minima given by, for* $\left\{z \colon 1-\tilde{\gamma}' \frac{z-\tilde{a}'}{\tilde{b}'} > 0\right\}$,

$$\widetilde{G}(z) = 1 - \exp\left\{-\left[1-\tilde{\gamma}' \frac{z-\tilde{a}'}{\tilde{b}'}\right]^{-1/\tilde{\gamma}'}\right\}, \tag{1.58}$$

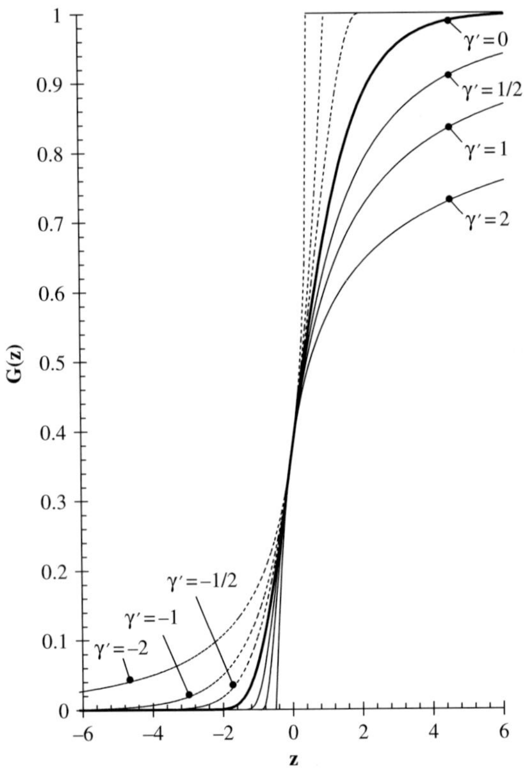

Figure 1.1. The c.d.f. of the GEV law of maxima, in the canonical form ($a' = 0$ and $b' = 1$). The shape parameter takes on the values $\gamma' = -2, -1, -1/2, 0, 1/2, 1, 2$

where $\tilde{a}' \in \mathbb{R}$ is a position parameter, $\tilde{b}' > 0$ a scale parameter, and $\tilde{\gamma}' \in \mathbb{R}$ a shape parameter. The limit case $\tilde{\gamma}' = 0$ yields the Gumbel distribution of minima given by Eq. (1.54).

NOTE 1.9. Note that the Weibull family for minima is obtained for $\tilde{\gamma}' < 0$, whereas the Fréchet family for $\tilde{\gamma}' > 0$.

The c.d.f. and p.d.f. of GEV distribution of minima (using the canonical form) are shown, respectively, in Figures 1.4–1.5, considering different values of the shape parameter. In addition, in Figure 1.6 we show the quantile function $z_{\tilde{G}}$ in the plane $(-\ln(-\ln(1-\tilde{G})), z)$: thus plotted is the *Gumbel* (or *reduced*) variate for minima.

In Table 1.2 we show the domain of attraction of maxima and minima for some "parent" distributions frequently used in applications.

The proofs of Theorems 1.7–1.10 are quite advanced [83], and are beyond the scope of this book. However, some informal justifications and heuristic details

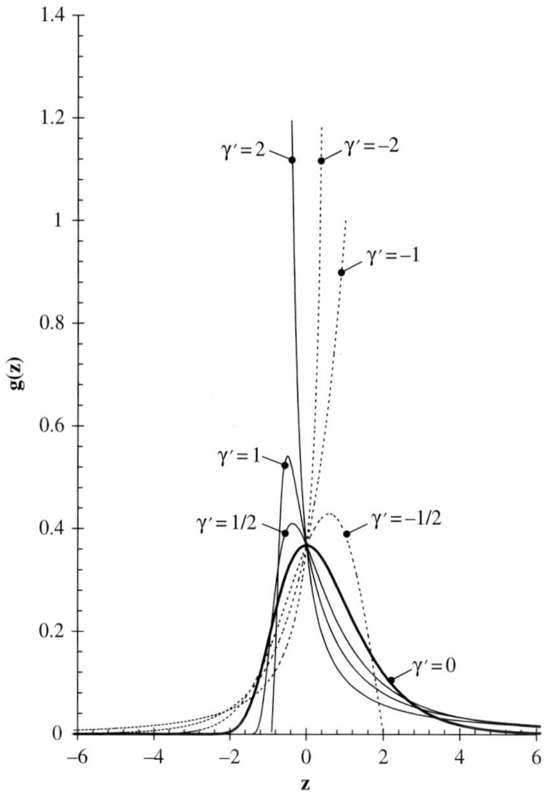

Figure 1.2. The p.d.f. of the GEV law of maxima, in the canonical form ($a' = 0$ and $b' = 1$). The shape parameter takes on the values $\gamma' = -2, -1, -1/2, 0, 1/2, 1, 2$

will follow. We need first to introduce the following important notion, which conceals a *postulate of stability* (see also Definition 5.2 and the ensuing discussion in Chapter 5).

DEFINITION 1.4 (Max-stable distribution). *If for all* $n \in \mathbb{N}$ *there exist constants* a_n *and* $b_n > 0$ *such that, for all* $z \in \mathbb{R}$,

$$G^n(a_n + b_n z) = G(z), \tag{1.59}$$

then the distribution G is called max-stable.

NOTE 1.10. Since G^n is the distribution of the maximum of n i.i.d. r.v.'s having a common probability law G, then the property of *max-stability* is satisfied by all those distributions characterized by a limit probability law for maxima equal to the "parent" distribution itself (except for possible affine transformations).

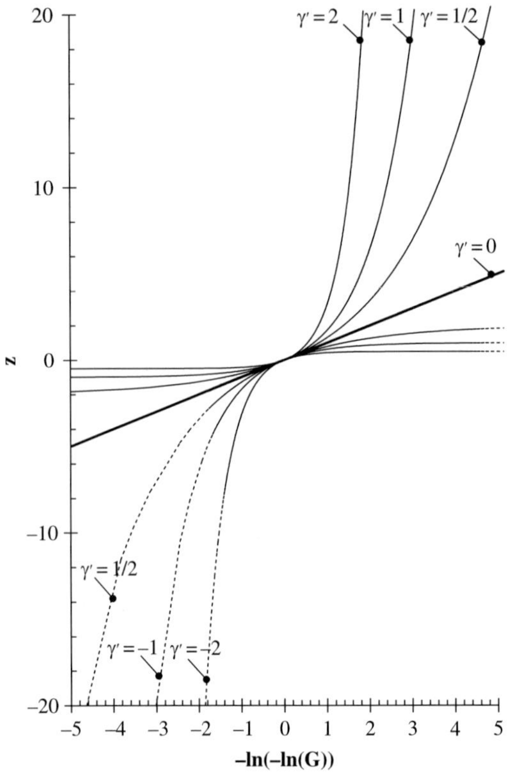

Figure 1.3. Quantiles of the GEV distribution of maxima, in the canonical form ($a' = 0$ and $b' = 1$). The shape parameter takes on the values $\gamma' = -2, -1, -1/2, 0, 1/2, 1, 2$

The link between the GEV probability law and *max-stable* distributions is given by the following theorem, that plays an important role in the proofs of Theorems 1.7–1.10.

THEOREM 1.11. *A distribution is max-stable if, and only if, it belongs to the GEV family.*

It is easy to show that the members of the GEV family feature the max-stability property. In fact, without loss of generality, let us consider the canonical form of the GEV probability law. Then Eq. (1.59) yields

$$\exp\left\{-n\,[1+\gamma'\,(a_n+b_n z)]^{-1/\gamma'}\right\} = \exp\left\{-[1+\gamma'\,z]^{-1/\gamma'}\right\}, \tag{1.60}$$

or

$$n^{-\gamma'}\,(1+\gamma'\,(a_n+b_n z)) = 1+\gamma'\,z. \tag{1.61}$$

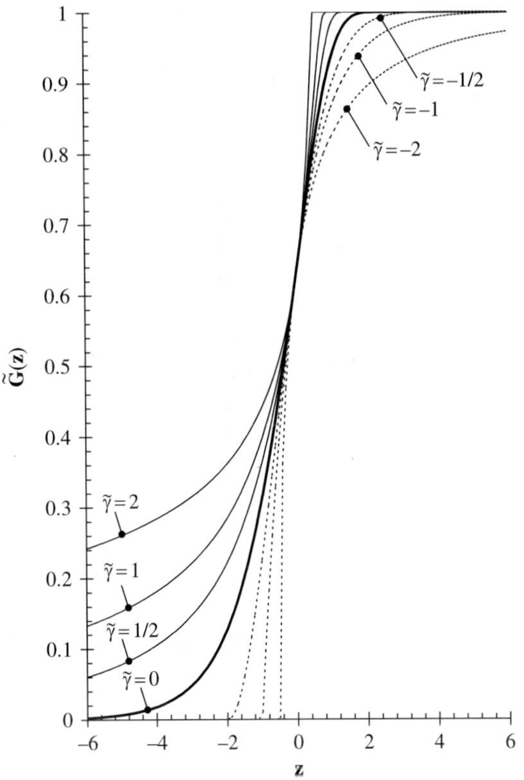

Figure 1.4. The c.d.f. of the GEV law of minima, in the canonical form ($\tilde{a}' = 0$ and $\tilde{b}' = 1$). The shape parameter takes the values $\tilde{\gamma}' = -2, -1, -1/2, 0, 1/2, 1, 2$

In turn, the constants a_n and b_n can be calculated as, respectively, $a_n = (1 - n^{-\gamma'})/(\gamma' n^{-\gamma'})$ and $b_n = 1/n^{-\gamma'}$. Conversely, let us consider the maximum M_{nk} of a random sample of size nk, with $n \gg 1$ and $k \in \mathbb{N}$. Clearly, M_{nk} can be viewed as (*a*) the maximum of nk r.v.'s, or (*b*) the maximum of k maxima from samples of size n. Let the (limit) distribution of $(M_n - a_n)/b_n$ be G. Then, for $n \gg 1$,

$$\mathbb{P}\left\{\frac{M_n - a_n}{b_n} \le z\right\} \approx G(z). \tag{1.62}$$

Since $nk \gg 1$,

$$\mathbb{P}\left\{\frac{M_{nk} - a_{nk}}{b_{nk}} \le z\right\} \approx G(z). \tag{1.63}$$

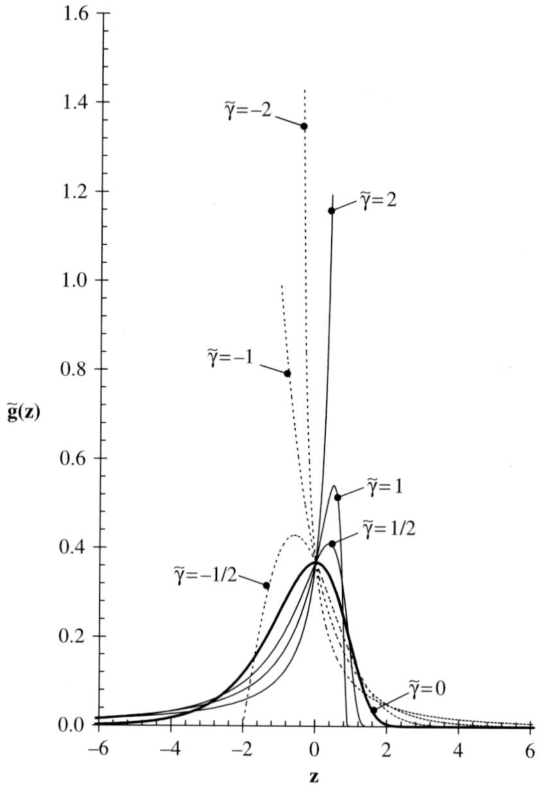

Figure 1.5. The p.d.f. of the GEV law of minima, in the canonical form ($\tilde{a}' = 0$ and $\tilde{b}' = 1$). The shape parameter takes the values $\tilde{\gamma}' = -2, -1, -1/2, 0, 1/2, 1, 2$

However, since M_{nk} is the maximum of k r.v.'s having the same distribution as M_n, it follows that

$$\mathbb{P}\left\{\frac{M_{nk}-a_{nk}}{b_{nk}} \le z\right\} \approx \left(\mathbb{P}\left\{\frac{M_n-a_n}{b_n} \le z\right\}\right)^k. \tag{1.64}$$

As a consequence,

$$\mathbb{P}\{M_{nk} \le z\} \approx G\left(\frac{z-a_{nk}}{b_{nk}}\right) \quad \text{and} \quad \mathbb{P}\{M_{nk} \le z\} \approx G^k\left(\frac{z-a_n}{b_n}\right), \tag{1.65}$$

and thus G satisfies the *postulate of stability*, i.e. G and G^k are equal (except for an affine transformation). Consequently, G is max-stable, and according to Theorem 1.11, G is a member of the GEV family.

NOTE 1.11. Generally, geophysical data are collected using a daily or sub-daily (e.g., one hour, one minute, …) temporal resolution. For instance, consider a sample of maximum (or minimum, or average) daily temperatures, or a sample of daily

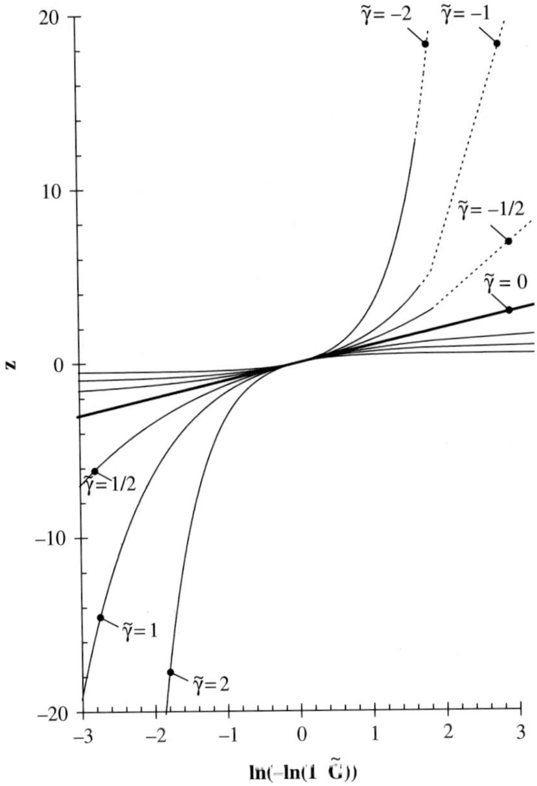

Figure 1.6. Quantiles of the GEV distribution of minima, in the canonical form ($\tilde{a}' = 0$ and $\tilde{b}' = 1$). The shape parameter takes the values $\tilde{\gamma}' = -2, -1, -1/2, 0, 1/2, 1, 2$

Table 1.2. Domain of attraction of maxima and minima of some "parent" distributions

Distribution	**MAX**	**MIN**	**Distribution**	**MAX**	**MIN**
Normal	Gumbel	Gumbel	**Rayleigh**	Gumbel	Weibull
Exponential	Gumbel	Weibull	**Lognormal**	Gumbel	Gumbel
Gamma	Gumbel	Weibull	**Uniform**	Weibull	Weibull
Cauchy	Fréchet	Fréchet	**Pareto**	Fréchet	Weibull
Gumbel (Max)	Gumbel	Gumbel	**Gumbel (Min)**	Gumbel	Gumbel
Fréchet (Max)	Fréchet	Gumbel	**Fréchet (Min)**	Gumbel	Fréchet
Weibull (Max)	Weibull	Gumbel	**Weibull (Min)**	Gumbel	Weibull

total precipitation. In practical applications, the interest is often focussed on the *annual maxima*. Consequently, the observations can be partitioned into k consecutive independent blocks, where k is the number of observed years, and each block contains $n = 365$ independent observations if the temporal scale is daily (or $n = 8760$ in case of hourly samples, or $n = 525600$ if the temporal scale is one minute, and so on). Then, the maximum observation is calculated for each block: in turn, a sample of maxima of size k can be collected. This sample is then used to estimate the parameters of the distribution of maxima.

The approach described above represents the *standard* procedure to collect and analyze, say, annual maxima (or minima). However, it has some drawbacks, for this strategy might discard important sample information. In fact, if the extremal behavior of a phenomenon persists for several days in a given year, only the observation corresponding to the most "intense" day will generate the annual maximum. As a consequence, all the remaining information concerning the extremal dynamics developed during the preceding and succeeding days will be discarded. In Subsection 1.2.2 we give an alternative approach to the analysis of this type of extremal behavior.

The notion of *slowly varying* function [92] is important in extreme value analysis.

DEFINITION 1.5 (Slowly varying function). *A positive, Lebesgue measurable, function L on $(0, \infty)$ is slowly varying at infinity if*

$$\lim_{x \to \infty} \frac{L(tx)}{L(x)} = 1 \tag{1.66}$$

for all $t > 0$.

The following theorem provides a characterization of the maximum domain of attraction. The Fréchet and Weibull cases are dealt with the help of slowly varying functions. Instead, the Gumbel case is more complex, for it would require the introduction of the *von Mises functions* (see, e.g., [83]); therefore, we only provide a sufficient condition [296].

THEOREM 1.12. *The distribution F belongs to the domain of attraction for maxima of the family*

1. ***Type I (Gumbel)*** *if*

$$\lim_{t \to \omega_F} \frac{\mathrm{d}}{\mathrm{d}t} \frac{1}{r(t)} = 0, \tag{1.67}$$

where $r(t) = F'(t)/(1 - F(t))$ is the hazard rate of F;

2. ***Type II (Fréchet)*** *if, and only if, $\omega_F = +\infty$ and*

$$1 - F(x) = x^{-\gamma} L(x), \quad \gamma > 0, \tag{1.68}$$

for some slowly varying function L;

3. ***Type III (Weibull)*** *if, and only if,* $\omega_F < +\infty$ *and*

$$1 - F(\omega_F - x^{-1}) = x^{-\gamma} L(x), \quad \gamma > 0, \tag{1.69}$$

for some slowly varying function L.

The following theorem provides necessary and sufficient conditions for the distribution F to belong to a given domain of attraction for maxima.

THEOREM 1.13 (Max-domain of attraction for maxima). *The distribution F belongs to the domain of attraction for maxima of the family*

1. ***Type I (Gumbel)*** *if, and only if,*

$$\lim_{n\to\infty} n\left[1 - F\left(x_{1-1/n} + x(x_{1-1/(ne)} - x_{1-1/n})\right)\right] = e^{-x}; \tag{1.70}$$

2. ***Type II (Fréchet)*** *if, and only if,* $\omega_F = +\infty$ *and*

$$\lim_{t\to\infty} \frac{1 - F(tx)}{1 - F(t)} = x^{-\gamma}, \quad x, \gamma > 0; \tag{1.71}$$

3. ***Type III (Weibull)*** *if, and only if,* $\omega_F < +\infty$ *and*

$$\lim_{t\to\infty} \frac{1 - F\left(\omega_F - 1/(tx)\right)}{1 - F\left(\omega_F - 1/t\right)} = x^{-\gamma}, \quad x, \gamma > 0. \tag{1.72}$$

Here x_q *is the quantile of order q of F.*

The conditions given in Theorem 1.13 were introduced in [118]. The following theorem provides a general criterion to calculate the norming constants a_n and b_n.

THEOREM 1.14. *The norming constants* a_n *and* b_n *in Eq. (1.45) can be calculated as follows.*

1. ***Type I (Gumbel):***

$$a_n = F^{-1}\left(1 - \frac{1}{n}\right), \quad b_n = F^{-1}\left(1 - \frac{1}{ne}\right) - a_n. \tag{1.73}$$

2. ***Type II (Fréchet):***

$$a_n = 0, \quad b_n = F^{-1}\left(1 - \frac{1}{n}\right). \tag{1.74}$$

3. ***Type III (Weibull):***

$$a_n = \omega_F, \quad b_n = \omega_F - F^{-1}\left(1 - \frac{1}{n}\right). \tag{1.75}$$

Here F^{-1} *is the quantile function associated with the distribution F.*

The following result shows how the choice of the sequences $\{a_n\}$ and $\{b_n\}$ is not unique.

PROPOSITION 1.3. *If* $\{a_n\}$ *and* $\{b_n\}$ *are two sequences satisfying Eq. (1.45), and* $\{a'_n\}$ *and* $\{b'_n\}$ *are sequences such that*

$$\lim_{n\to\infty} \frac{a_n - a'_n}{b_n} = 0 \quad and \quad \lim_{n\to\infty} \frac{b'_n}{b_n} = 1, \tag{1.76}$$

then also $\{a'_n\}$ *and* $\{b'_n\}$ *satisfy Eq. (1.45).*

ILLUSTRATION 1.7 (Norming constants). ►

Here the norming constants for Gumbel, Fréchet, and Weibull distributions are calculated.

1. **Type I (Gumbel).** Let us consider the Gumbel probability law as the "parent" distribution F. Then, if $z \in \mathbb{R}$ and $t \in (0, 1)$, $F(z) = \exp\{-\exp(-z)\}$ and $F^{-1}(t) = -\ln(-\ln(t))$. The quantiles $z_{1-1/n}$ and $z_{1-1/(ne)}$ of Eq. (1.70) are, respectively, $z_{1-1/n} = -\ln(-\ln(1-1/n))$ and $z_{1-1/(ne)} = -\ln(-\ln(1-1/(ne)))$. For $n \gg 1$, $z_{1-1/n} \approx \ln(n)$ and $z_{1-1/(ne)} \approx \ln(n)+1$. Consequently, the limit in Eq. (1.70) is e^{-z}. Thus, the Gumbel distribution belongs to its own domain of attraction. Then, using Eq. (1.73), it is possible to calculate the constants a_n and b_n as

$$a_n = -\ln\left(-\ln\left(1-\frac{1}{n}\right)\right),$$
$$b_n = -\ln\left(-\ln\left(1-\frac{1}{ne}\right)\right) + \ln\left(-\ln\left(1-\frac{1}{n}\right)\right).$$

Substituting a_n and b_n in $F^n(a_n + b_n z)$, and taking the limit, after some algebra we obtain $\lim_{n\to\infty} F^n(a_n+b_n z) = \exp\{-\exp(-z)\}$, in agreement with Theorem 1.7.
2. **Type II (Fréchet).** Let us consider the Fréchet probability law as the "parent" distribution F. Then, if $z > 0$ and $t \in (0, 1)$, $F(z) = \exp\{-z^{-\gamma}\}$ and $F^{-1}(t) = (-\ln(t))^{-1/\gamma}$. Here $\omega_F = +\infty$, and the limit in Eq. (1.71) is $x^{-\gamma}$. Thus, the Fréchet distribution belongs to its own domain of attraction. Then, using Eq. (1.74), it is possible to calculate the constants a_n and b_n as

$$a_n = 0,$$
$$b_n = \left(-\ln\left(1-\frac{1}{n}\right)\right)^{-1/\gamma}.$$

Substituting a_n and b_n in $F^n(a_n + b_n z)$, and taking the limit, after some algebra we obtain $\lim_{n\to\infty} F^n(a_n+b_n z) = \exp\{-z^{-\gamma}\}$, in agreement with Theorem 1.7.

3. **Type III (Weibull).** Let us consider the Weibull probability law as the "parent" distribution F. Then, if $z<0$ and $t\in(0,1)$, $F(z)=\exp\{-[(-z)^{\gamma}]\}$ and $F^{-1}(t)=-(-\ln(t))^{1/\gamma}$. Here $\omega_F<+\infty$, and the limit in Eq. (1.72) is $x^{-\gamma}$. Thus, the Weibull distribution belongs to its own domain of attraction. Then, using Eq. (1.75), it is possible to calculate the constants a_n and b_n as

$$a_n=0,$$
$$b_n=0+\left(-\ln\left(1-\frac{1}{n}\right)\right)^{1/\gamma}.$$

Substituting a_n and b_n in $F^n(a_n+b_nz)$, and taking the limit, after some algebra we obtain $\lim_{n\to\infty}F^n(a_n+b_nz)=\exp\{-[(-z)^{\gamma}]\}$, in agreement with Theorem 1.7.

The above results show the max-stability of the GEV probability law. ◀

Alternatively, the norming constants a_n and b_n for Gumbel, Fréchet, and Weibull distributions can be calculated as follows.

ILLUSTRATION 1.8 (Norming constants (cont.)). ▶
According to Theorem 1.11, and taking into account Proposition 1.3, the sequences $\{a_n\}$ and $\{b_n\}$ can be determined by comparing the functions $G(z)$ and $G^n(a_n+b_nz)$, where G is written in the canonical form.

1. For the **Gumbel** distribution, the equation $G^n(a_n+b_nz)=G(z)$ reads as $\exp\{-n\exp[-(a_n+b_nz)]\}=\exp\{-\exp(-z)\}$, and thus $ne^{-a_n}e^{-b_nz}=e^{-z}$. The two conditions $ne^{-a_n}=1$ and $e^{-b_nz}=e^{-z}$ then follow, yielding $a_n=\ln n$ and $b_n=1$.
2. For the **Fréchet** distribution, the equation $G^n(a_n+b_nz)=G(z)$ reads as $\exp\{-n(a_n+b_nz)^{-\gamma}\}=\exp\{-z^{-\gamma}\}$, or $n(a_n+b_nz)^{-\gamma}=z^{-\gamma}$. The two conditions $n^{-1/\gamma}a_n=0$ and $n^{-1/\gamma}b_nz=z$ then follow, yielding $a_n=0$ and $b_n=n^{1/\gamma}$.
3. For the **Weibull** distribution, the equation $G^n(a_n+b_nz)=G(z)$ reads as $\exp\{-n(-a_n-b_nz)^{\gamma}\}=\exp\{-(-z)^{\gamma}\}$, or $n(-a_n-b_nz)^{\gamma}=(-z)^{\gamma}$. The two conditions $n^{1/\gamma}a_n=0$ and $n^{1/\gamma}b_nz=z$ then follow, yielding $a_n=0$ and $b_n=n^{-1/\gamma}$.

As explained in Proposition 1.3, the norming sequences a_n and b_n are not unique. By varying the sequences, the rate of convergence of $G^n(a_n+b_nz)$ to $G(z)$ also changes. However, the shape parameter γ remains the same. Note that, taking $n\to\infty$, the norming constants just calculated asymptotically match those derived in Illustration 1.7, and both satisfy Eq. (1.76). ◀

The following theorem provides a useful rule to calculate the domain of attraction for maxima.

THEOREM 1.15. *Let F be a continuous distribution. Then F belongs to the max-domain of attraction of the distribution G if, and only if,*

$$\lim_{\epsilon \to 0} \frac{F^{-1}(1-\epsilon) - F^{-1}(1-2\epsilon)}{F^{-1}(1-2\epsilon) - F^{-1}(1-4\epsilon)} = 2^c, \tag{1.77}$$

where $c \in \mathbb{R}$. In particular, G belongs to the family

1. ***Type I (Gumbel)*** *if $c = 0$;*
2. ***Type II (Fréchet)*** *if $c > 0$;*
3. ***Type III (Weibull)*** *if $c < 0$.*

As a summary of the results given up to this point, let us state the following partial conclusions (true under the assumptions mentioned earlier).

- Only three families of distributions (namely, the *Gumbel*, the *Fréchet*, and the *Weibull*) model the law of maxima (minima) of i.i.d. sequences.
- There exist rules to verify whether a given distribution follows in the domain of attraction of a suitable limit law.
- There exist rules to calculate the norming constants.
- If $\omega_F = +\infty$, then F cannot lie in the Weibull max-domain of attraction.
- If $\omega_F < +\infty$, then F cannot lie in the Fréchet max-domain of attraction.

However, the fact that a "parent" distribution has a bounded left (or right) tail, does not imply that it lies in the Weibull domain of attraction: as a counter-example, consider the Lognormal law in Table 1.2. Some applications to distributions widely used in practice now follow.

ILLUSTRATION 1.9 (GEV law of maxima for Exponential variates). ▶
Let us consider the standard Exponential distribution with c.d.f.

$$F(x) = 1 - e^{-x}$$

for $x > 0$, and zero elsewhere. The quantile function is $F^{-1}(t) = -\ln(1-t)$, with $t \in (0, 1)$. Substituting for F^{-1} in Eq. (1.77) yields

$$\lim_{\epsilon \to 0} \frac{-\ln(1-(1-\epsilon)) + \ln(1-(1-2\epsilon))}{-\ln(1-(1-2\epsilon)) + \ln(1-(1-4\epsilon))} =$$

$$\lim_{\epsilon \to 0} \frac{-\ln(\epsilon) + \ln 2 + \ln(\epsilon)}{-\ln 2 - \ln(\epsilon) + \ln 4 + \ln(\epsilon)} = 1 = 2^0,$$

and thus $c = 0$. Then, the Exponential distribution belongs to the domain of attraction of the Gumbel family. On the other end, it is easy to show that the limit in Eq. (1.70) is e^{-x} as given. Figure 1.7 shows the analysis of maxima of a sample of size $n \times k = 3000$, obtained using $k = 30$ independent simulations of size $n = 100$ extracted from the standard Exponential distribution.

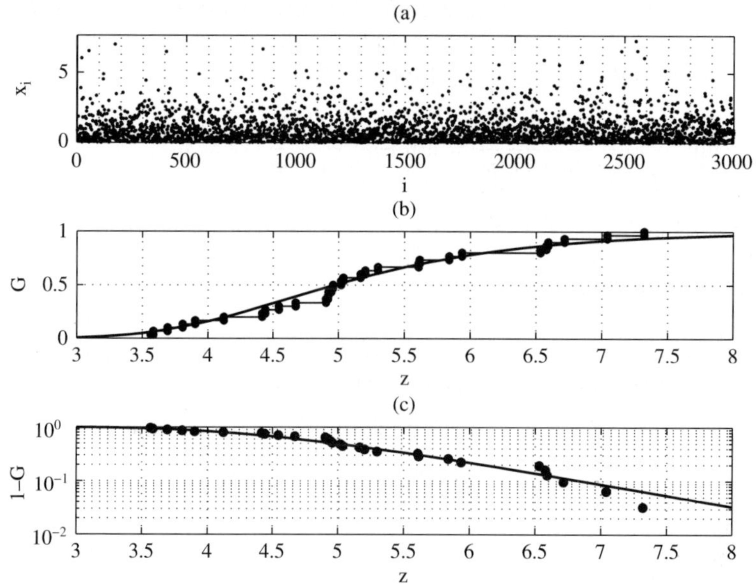

Figure 1.7. Analysis of maxima using the "block" method. (*a*) Simulated sample of size 3000 extracted from the standard Exponential distribution. (*b*) Comparison between the empirical c.d.f. of the maxima (*marked line*) and the Gumbel probability law (*line*). (*c*) Comparison between empirical (*markers*) and theoretical (*line*) *survival* functions of the maxima

The parameters a_n and b_n are given by Eq. (1.73): $a_n = \ln n$ and $b_n = 1$, where $n = 100$. Note that in Figure 1.7*c* the vertical axis is logarithmic, and the right tail exhibits an asymptotic linear behavior. This means that the distribution has an "exponential" fall-off, i.e. it is "light tailed", as shown in Note 1.7. ◀

ILLUSTRATION 1.10 (GEV law of maxima for Cauchy variates). ▶
Let us consider a standard Cauchy distribution with c.d.f.

$$F(x) = 1/2 + \arctan(x)/\pi$$

for $x \in \mathbb{R}$. The quantile function is $F^{-1}(t) = \tan((t-1/2)\pi)$, with $t \in (0,1)$. Substituting F^{-1} in Eq. (1.77) yields

$$\lim_{\epsilon \to 0} \frac{\tan((1-\epsilon-1/2)\pi) - \tan((1-2\epsilon-1/2)\pi)}{\tan((1-2\epsilon-1/2)\pi) - \tan((1-4\epsilon-1/2)\pi)} = \lim_{\epsilon \to 0} \frac{\tan((1/2-\epsilon)\pi) - \tan((1/2-2\epsilon)\pi)}{\tan((1/2-2\epsilon)\pi) - \tan((1/2-4\epsilon)\pi)} = 2^1,$$

and thus $c = 1 > 0$. Then, the Cauchy distribution belongs to the domain of attraction of the Fréchet family. In addition, it is possible to calculate the parameter γ of Eq. (1.71). Since $\arctan(y) \approx (\pi/2 - 1/y)$ for $y \gg 1$, then Eq. (1.71) yields

$$\lim_{t\to\infty} \frac{1/2 - \arctan(tx)/\pi}{1/2 - \arctan(t)/\pi} = \lim_{t\to\infty} \frac{\pi - 2\arctan(tx)}{\pi - 2\arctan(t)} = x^{-1},$$

and thus $\gamma = 1$. Figure 1.8 shows the analysis of maxima of a sample of size $n \times k = 3000$, obtained using $k = 30$ independent simulations of size $n = 100$ extracted from the standard Cauchy distribution.

The parameters a_n and b_n are given by Eq. (1.74): $a_n = 0$ and $b_n = \tan(\pi(1/2 - 1/n))$, where $n = 100$. Note that in Figure 1.8*c* the axes are logarithmic, and the right tail exhibits an asymptotic linear behavior. This means that the distribution has an "algebraic" fall-off, i.e. it is "heavy tailed", as shown in Note 1.7.

In Figure 1.8*a* it is evident how the sample is characterized by a few "outliers". This happens because the Cauchy distribution is "heavy tailed", and the same property is featured by the Fréchet family. This makes the Fréchet probability law appealing for the analysis of extremal or catastrophic phenomena like many of those observed in geophysics. ◄

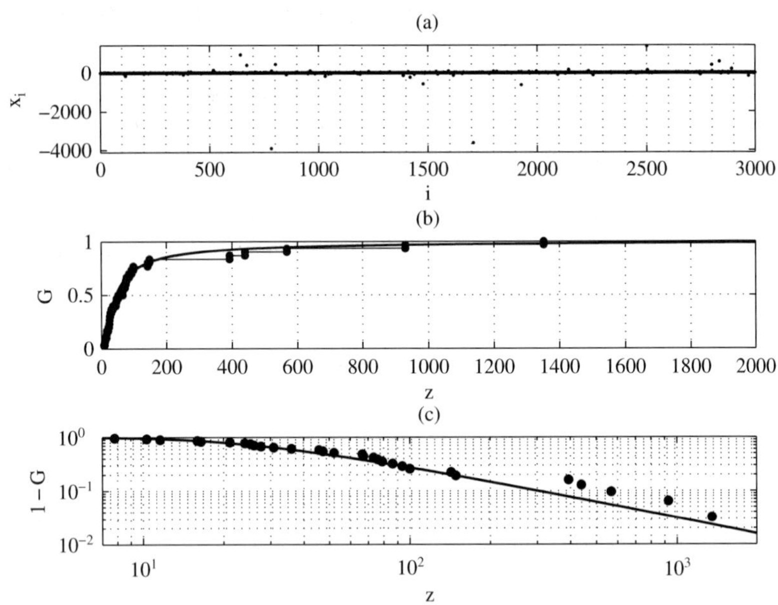

Figure 1.8. Analysis of maxima using the "block" method. (*a*) Simulated sample of size 3000 extracted from the standard Cauchy distribution. (*b*) Comparison between the empirical c.d.f. of the maxima (*marked line*) and the Fréchet probability law (*line*). (*c*) Comparison between empirical (*markers*) and theoretical (*line*) *survival* functions of the maxima

ILLUSTRATION 1.11 (GEV law of maxima for Uniform variates). ▶
Let us consider a standard Uniform distribution with c.d.f.

$$F(x) = x\mathbf{1}\{0 \le x \le 1\} + \mathbf{1}\{x > 1\}.$$

The quantile function is $F^{-1}(t) = t$, with $t \in (0, 1)$. Substituting F^{-1} in Eq. (1.77) yields

$$\lim_{\epsilon \to 0} \frac{1-\epsilon-1+2\epsilon}{1-2\epsilon-1+4\epsilon} = 2^{-1},$$

and thus $c = -1 < 0$. Then, the Uniform distribution belongs to the domain of attraction of the Weibull family. In addition it is possible to calculate the parameter γ of Eq. (1.72). Since $\omega_F = 1$, then Eq. (1.72) yields

$$\lim_{t \to \infty} \frac{1-(1-1/(tx))}{1-(1-1/t)} = \lim_{t \to \infty} \frac{1/(tx)}{1/t} = x^{-1},$$

and thus $\gamma = 1$. Figure 1.9 shows the analysis of maxima of a sample of size $n \times k = 3000$, obtained using $k = 30$ independent simulations of size $n = 100$ extracted from the standard Uniform distribution.

The parameters a_n and b_n are given by Eq. (1.75): $a_n = \omega_F = 1$ and $b_n = \omega_F - (1 - 1/n) = 1/n$, where n is 100. Note that $\omega_G = 1 < +\infty$, as discussed in Note 1.7. ◀

The limit behavior of the law of maxima (minima) is of great importance in practical applications, for it characterizes the probability of occurrence of extreme events. In order to investigate the asymptotic behavior of a distribution, the following definition is needed.

DEFINITION 1.6 (Tail equivalence). *Two distributions F and H are called right tail equivalent if, and only if,*

$$\omega_F = \omega_H \quad \textit{and} \quad \lim_{x \to \omega_F} \frac{1-F(x)}{1-H(x)} = c, \tag{1.78a}$$

for some constant $c > 0$. Similarly, F and H are called left-tail equivalent if, and only if,

$$\alpha_F = \alpha_H \quad \textit{and} \quad \lim_{x \to \alpha_F} \frac{F(x)}{H(x)} = c, \tag{1.78b}$$

for some constant $c > 0$.

The result given below clarifies the link between right-tail equivalence and the max-domain of attraction.

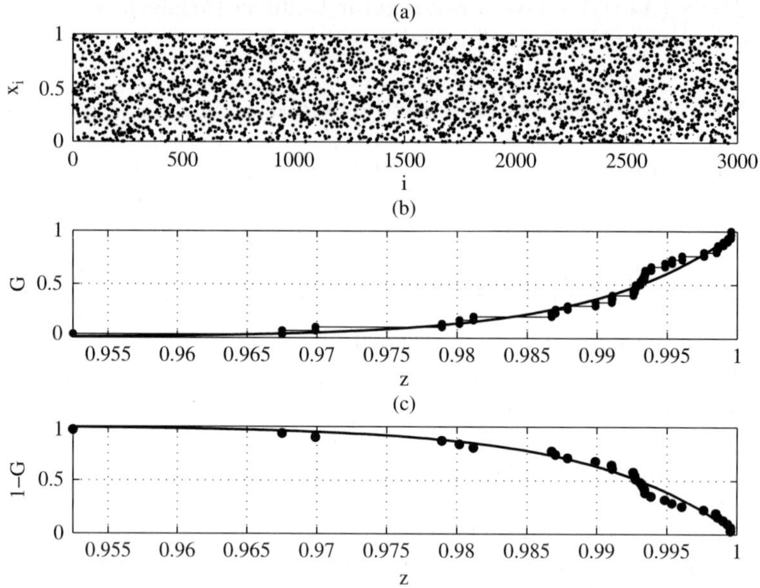

Figure 1.9. Analysis of maxima using the "block" method. (*a*) Simulated sample of size 3000 extracted from the standard Uniform distribution. (*b*) Comparison between the empirical c.d.f. of the maxima (*marked line*) and the Weibull probability law (*line*). (*c*) Comparison between empirical (*markers*) and theoretical (*line*) *survival* functions of the maxima

PROPOSITION 1.4. *If F and H are right-tail equivalent and*

$$\lim_{n\to\infty} F^n(a_n + b_n x) = G(x), \tag{1.79}$$

for all $x \in \mathbb{R}$, *then*

$$\lim_{n\to\infty} H^n(a_n + b_n x) = G(x), \tag{1.80}$$

for all $x \in \mathbb{R}$.

NOTE 1.12 (Tail equivalence). Proposition 1.4 yields the following results. (*a*) If two distributions are right-tail equivalent, and one belongs to a given max-domain of attraction, then the other also belongs to the same domain. (*b*) The norming constants are the same for both distributions. The practical consequences of Proposition 1.4 are important: in fact, if a distribution F can be asymptotically replaced by another tail-equivalent probability law, then the GEV distribution will suffice for the analysis of extremes.

1.2.2 "Threshold" Model

As anticipated in Note 1.11, sometimes the analysis of extremes via the "block" method provides a poor representation of the extremal behavior of a phenomenon. As an alternative, the maxima can be analyzed via the *Peaks-Over-Threshold (POT)* method.

Let us consider a sequence of i.i.d. r.v.'s $X_1, X_2, \ldots$, having a common c.d.f. F. The extremal behavior of these r.v.'s can be studied by considering events like $\{X > u\}$, where u is an arbitrary threshold, i.e. by investigating *exceedances* over a given threshold. In this case, the extremal behavior is usually described via the conditional probability

$$1 - H_u(x) = \mathbb{P}\{X > u + x \mid X > u\} = \frac{1 - F(u+x)}{1 - F(u)}, \qquad x > 0. \tag{1.81}$$

If the "parent" distribution F is known, then so is the probability in Eq. (1.81). However, F is generally not known in practical applications. Thus, a natural way to proceed is to look for an approximation of the conditional law *independent* of the "parent" distribution F. The following theorem is fundamental in the analysis of maxima using the POT method, and implicitly defines the *Generalized Pareto* (GP) distribution.

THEOREM 1.16 (Generalized Pareto distribution of maxima). *Let $X_1, \ldots, X_n$ be a sample of i.i.d. r.v.'s with "parent" distribution F. If F satisfies the conditions of Theorem 1.9, then, for $u \gg 1$, the conditional distribution of the exceedances $H_u(x)$ can be approximated as*

$$H_u(x) = \mathbb{P}\{X \le u + x \mid X > u\} \approx 1 - \left(1 + \gamma \frac{x}{b}\right)^{-1/\gamma}, \tag{1.82}$$

for $\{x : x > 0 \text{ and } 1 + \gamma x/b > 0\}$. Here $b = b' + \gamma(u - a')$ and $\gamma = \gamma' \in \mathbb{R}$ are, respectively, scale and shape parameters, and are functions of a', b', γ' given in Theorem 1.9. In the limit case $\gamma = 0$, H_u reduces to the Exponential distribution

$$H_u(x) \approx 1 - e^{-x/b}. \tag{1.83}$$

As a consequence of Theorem 1.16, we have the following definition.

DEFINITION 1.7 (Generalized Pareto distribution). *The distribution defined in Eq. (1.82) (and Eq. (1.83) in case $\gamma = 0$) is called the Generalized Pareto distribution (GP).*

NOTE 1.13 (Variant of the GP distribution of maxima). Sometimes Eq. (1.82) is also written by adding a suitable position parameter $a \in \mathbb{R}$:

$$H_u(x) = 1 - \left(1 + \gamma \frac{x - a}{b}\right)^{-1/\gamma}, \tag{1.84}$$

for $\{x: x > a$ and $1+\gamma(x-a)/b > 0\}$. Usually, a equals the threshold u (a critical level). Similarly, in the limit case $\gamma = 0$, H_u is written as a *shifted Exponential* distribution:

$$H_u(x) \approx 1 - e^{-(x-a)/b}. \tag{1.85}$$

Also these variants of the GP probability law are frequently used in practice.

The c.d.f. and p.d.f. of the GP distribution (using the canonical form) are shown, respectively, in Figures 1.10–1.11, considering different values of the shape parameter. In addition, in Figure 1.12 we show the quantile function x_H in the plane $(-\ln(1-H), x)$.

Note how the role taken by the GP distribution in the "Peaks-Over-Threshold" method is analogous to that assumed by the GEV probability law in the "block" method. Theorem 1.16 states that if "block" maxima obey (approximatively) a GEV probability law, then the exceedances over a given threshold u (assumed

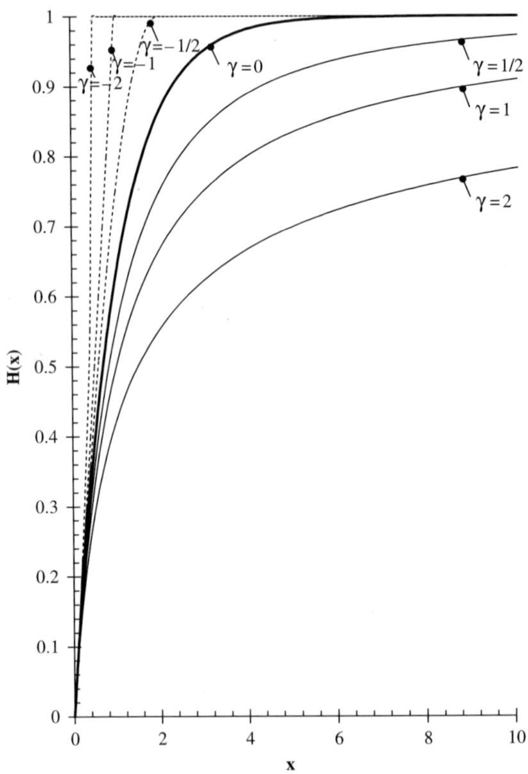

Figure 1.10. The c.d.f. of the GP law, in the canonical form ($a = 0$ and $b = 1$). The shape parameter takes the values $\gamma = -2, -1, -1/2, 0, 1/2, 1, 2$

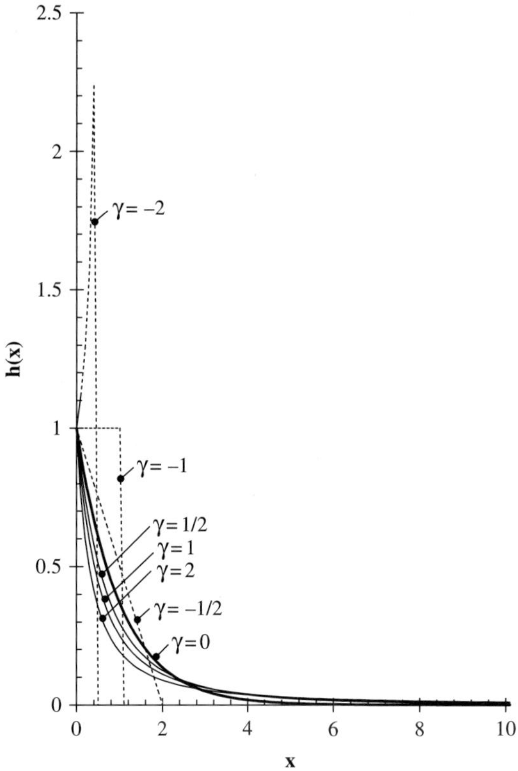

Figure 1.11. The p.d.f. of the GP law, in the canonical form ($a = 0$ and $b = 1$). The shape parameter takes the values $\gamma = -2, -1, -1/2, 0, 1/2, 1, 2$

sufficiently large, i.e. $u \gg 1$), follow (approximatively) a GP probability law. Interestingly enough, the expression of the GP law can be derived from that of the GEV distribution using the following formula:

$$H = 1 + \ln(\mathrm{G}). \tag{1.86}$$

In particular, the two shape parameters are equal: $\gamma = \gamma'$. This has important consequences on the finiteness of high order moments, as shown in Illustration 1.12 below.

ILLUSTRATION 1.12 (Distributions of exceedances). ►

Here we substitute in Eq. (1.86) the three canonical forms of the distributions of maxima, i.e. Eqs. (1.47)–(1.49), in order to derive the three types of distributions of exceedances.

1. **(Type I)** Let us consider as G the Type I distribution given by Eq. (1.47). According to Eq. (1.86), the corresponding distribution of exceedances is

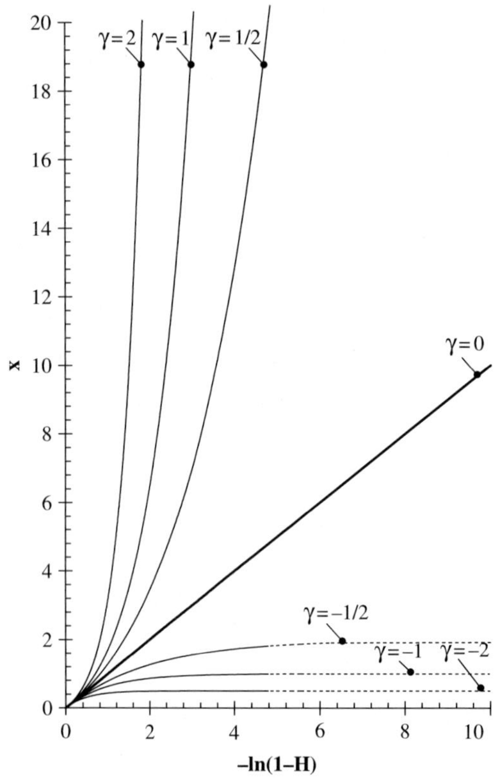

Figure 1.12. Quantiles of the GP distribution, in the canonical form ($a = 0$ and $b = 1$). The shape parameter takes the values $\gamma = -2, -1, -1/2, 0, 1/2, 1, 2$

$$H_u(x) = \begin{cases} 0, & x < 0 \\ 1 - \exp\left(-\frac{x}{b}\right), & x \geq 0 \end{cases}, \tag{1.87}$$

where $x = x' - a$, and $u = a$ is the threshold. Clearly, Eq. (1.87) is the c.d.f. of the *Exponential* distribution.

2. **(Type II)** Let us consider as G the Type II distribution given by Eq. (1.48). According to Eq. (1.86), the corresponding distribution of exceedances is

$$H_u(x) = \begin{cases} 0, & x \leq b \\ 1 - \left(\frac{x}{b}\right)^{-\gamma}, & x > b \end{cases}, \tag{1.88}$$

where $x = x' - a$ and $u = a$ is the threshold. Clearly, Eq. (1.88) is the c.d.f. of the *Pareto* distribution. For $x \gg 1$, the survival function $1 - H_u$ has an "algebraic" fall-off, and the statistical moments of order $m \geq \gamma > 0$ do not exist (see Note 1.7).

3. **(Type III)** Let us consider as G the Type III distribution given by Eq. (1.49). According to Eq. (1.86), the corresponding distribution of exceedances is

$$H_u(x) = 1 - \left(-\frac{x}{b}\right)^{\gamma}, \quad -b \le x \le 0, \tag{1.89}$$

where $x = x' - a$ and $u = a$ is the threshold. Clearly, Eq. (1.89) is the c.d.f. of a *Beta* distribution (properly reparametrized).

In all the three cases above, $b > 0$ is a scale parameter, and $\gamma > 0$ a shape parameter. ◄

We now reconsider the samples investigated in Illustrations 1.9–1.11, and provide a further, different, analysis by using the POT method.

ILLUSTRATION 1.13 (GP law of maxima for Exponential variates). ►

Let us consider the sample extracted from the standard Exponential distribution used in Illustration 1.9. Since the distribution of the "block" maxima is the Gumbel probability law, then the corresponding law of the "exceedances" is the GP distribution given by Eq. (1.83). Indeed, a direct calculation yields

$$1 - H_u(x) = \frac{1 - F(u+x)}{1 - F(u)} = \frac{e^{-(u+x)}}{e^{-u}} = e^{-x},$$

for $x > 0$, i.e. a GP distribution with $\gamma = 0$ and $b = 1$. Note that this result is *exact* for any threshold $u > 0$.

In Figure 1.13 we show the POT analysis. Here the threshold u is chosen in order to generate 30 exceedances, shown in Figure 1.13*bc* using the r.v. $X - u$.

Here the GP parameters are: $\gamma = 0$, $b = b_n + \gamma(u - a_n) = 1$, $n = 100$ is the number of "block" maxima, and a_n, b_n are as given in Illustration 1.9. Note that in Figure 1.13*c* the vertical axis is logarithmic, and the right tail exhibits an asymptotic linear behavior. This means that the distribution has an "exponential" fall-off, i.e. it is "light tailed" (see Note 1.7).

The analysis of Figure 1.13*a* shows how in the interval $2501 \le i \le 2600$ at least four "exceedances" occur. All these values contribute to the extremal behavior, and are reported in Figure 1.13*bc*. On the contrary, only the maximum of these values would be considered by the "block" method used in Illustration 1.9. ◄

ILLUSTRATION 1.14 (GP law of maxima for Cauchy variates). ►

Let us consider the sample extracted from the standard Cauchy distribution used in Illustration 1.10. Since the distribution of the "block" maxima is the Fréchet probability law, then the corresponding law of the "exceedances" is the GP distribution given by Eq. (1.82), with $\gamma > 0$. Indeed, using the fact that $\arctan(t) \approx \left(\frac{\pi}{2} - \frac{1}{t}\right)$ for $t \gg 1$, a direct calculation yields

$$1 - H_u(x) = \frac{1 - F(u+x)}{1 - F(u)} = \frac{\frac{1}{2} - \frac{\arctan(u+x)}{\pi}}{\frac{1}{2} - \frac{\arctan(u)}{\pi}} \approx \frac{\frac{1}{u+x}}{\frac{1}{u}} = \left(1 + \frac{x}{u}\right)^{-1},$$

for $x > 0$, i.e. a GP probability law with $\gamma = 1$ and $b = u > 0$.

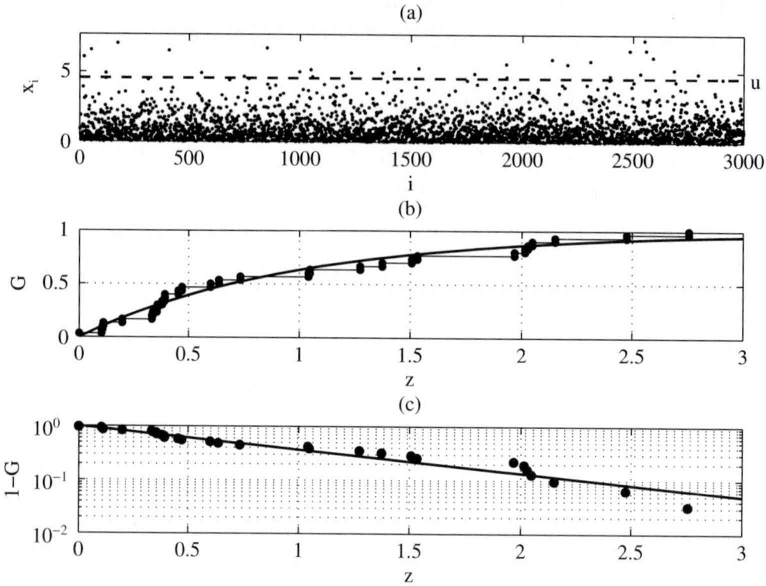

Figure 1.13. POT analysis of the sample used in Illustration 1.9. (*a*) Synthetic sample of size 3000 extracted from the standard Exponential distribution: the *dashed line* represents the threshold $u \approx 4.57$. (*b*) Comparison between the empirical c.d.f. of the POT maxima (*marked line*) and the GP probability law given by Eq. (1.83) (*line*). (*c*) Comparison between empirical (*markers*) and theoretical (*line*) *survival* functions of POT maxima

In Figure 1.14 we show the POT analysis. Here, the threshold u is chosen in order to generate 30 exceedances, shown in Figure 1.14*bc* using the r.v. $X - u$.

Here the GP parameters are: $\gamma = 1$, $b = b_n + \gamma(u - a_n) \approx 57.39$, $n = 100$ is the number of "block" maxima, and a_n, b_n are as given in Illustration 1.10. Note that in Figure 1.14*c* the axes are logarithmic, and the right tail exhibits an asymptotic linear behavior. This means that the distribution has an "algebraic" fall-off, as illustrated in Note 1.7.

The analysis of Figure 1.14*a* shows how in the intervals $601 \leq i \leq 700$ and $2801 \leq i \leq 2900$ several "exceedances" occur. All these values contribute to the extremal behavior, and are shown in Figure 1.14*bc*. On the contrary, only the corresponding "block" maxima are considered by the method used in Illustration 1.10. ◀

ILLUSTRATION 1.15 (GP law of maxima for Uniform variates). ▶

Let us consider the sample extracted from the standard Uniform distribution used in Illustration 1.11. Since the distribution of the "block" maxima is the Weibull probability law, then the corresponding law of the "exceedances" is the GP distribution given by Eq. (1.82), with $\gamma < 0$. Indeed, a direct calculation yields

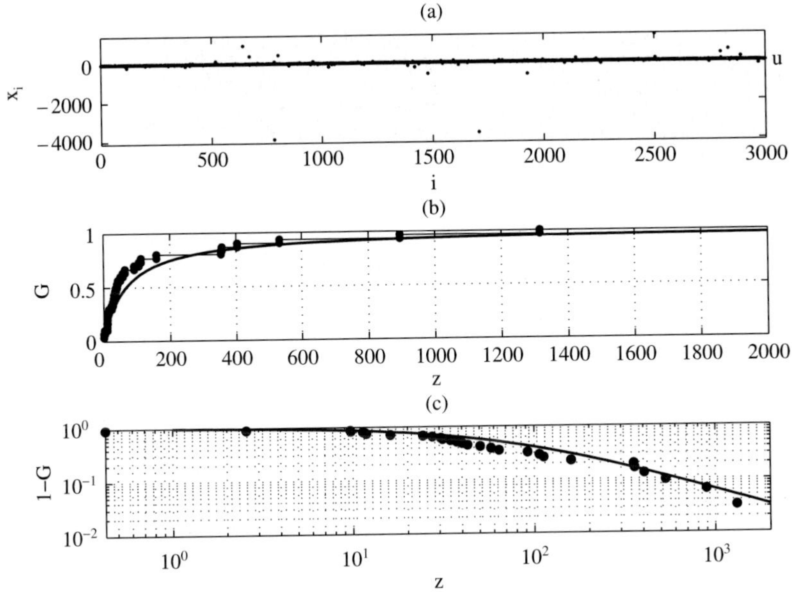

Figure 1.14. POT analysis of the sample used in Illustration 1.10. (*a*) Synthetic sample of size 3000 extracted from the standard Cauchy distribution: the *dashed line* represents the threshold $u \approx 36.04$. (*b*) Comparison between the empirical c.d.f. of the POT maxima (*marked line*) and the GP probability law given by Eq. (1.82) (*line*). (*c*) Comparison between empirical (*markers*) and theoretical (*line*) *survival* functions of POT maxima

$$1 - H_u(x) = \frac{1 - F(u + x)}{1 - F(u)} = \frac{1 - (u + x)}{1 - u} = 1 - \frac{x}{1 - u},$$

for $0 < x < 1 - u$, i.e. a GP probability law with $\gamma = -1$ and $b = 1 - u$.

In Figure 1.15 we show the POT analysis. Here the threshold u is chosen in order to generate 30 exceedances, shown in Figure 1.15*bc* using the r.v. $X - u$.

Here the GP parameters are: $\gamma = -1$, $b = b_n + \gamma(u - a_n) \approx 0.02$, $n = 100$ is the number of "block" maxima, and a_n, b_n are as given in Illustration 1.11. Note that $\omega_H = b < +\infty$. ◀

We now give a rationale for Eq. (1.82) in Theorem 1.16. Following Theorem 1.9, for $n \gg 1$,

$$F^n(z) \approx \exp\left\{ - \left[1 + \gamma' \frac{z - a'}{b'} \right]^{-1/\gamma'} \right\},$$

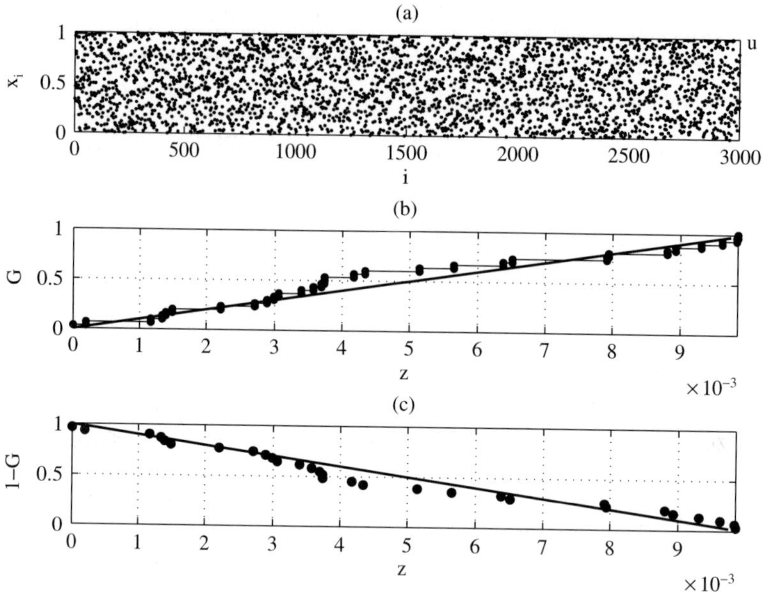

Figure 1.15. POT analysis of the sample used in Illustration 1.11. (*a*) Synthetic sample of size 3000 extracted from the standard Uniform distribution: the *dashed line* represents the threshold $u \approx 0.99$. (*b*) Comparison between the empirical c.d.f. of the POT maxima (*marked line*) and the GP probability law given by Eq. (1.82) (*line*). (*c*) Comparison between empirical (*markers*) and theoretical (*line*) *survival* functions of POT maxima

and thus

$$n \ln F(z) \approx -\left[1 + \gamma' \frac{z - a'}{b'}\right]^{-1/\gamma'}.$$

However, since $\ln F(z) = \ln(1 - (1 - F(z)))$, then $\ln F(z) \approx -(1 - F(z))$ for $z \gg 1$. Considering a threshold $u \gg 1$ we have

$$1 - F(u) \approx \frac{1}{n}\left[1 + \gamma' \frac{u - a'}{b'}\right]^{-1/\gamma'},$$

$$1 - F(u + x) \approx \frac{1}{n}\left[1 + \gamma' \frac{u + x - a'}{b'}\right]^{-1/\gamma'}.$$

Finally, the calculation of the conditional probality in Eq. (1.81) yields

$$\mathbb{P}\{X > u+x \mid X > u\} \approx \frac{n^{-1}\left[1+\gamma' \frac{u+x-a'}{b'}\right]^{-1/\gamma'}}{n^{-1}\left[1+\gamma' \frac{u-a'}{b'}\right]^{-1/\gamma'}} = \left(1+\gamma \frac{x}{b}\right)^{-1/\gamma},$$

where $b = b' + \gamma(u - a')$ and $\gamma = \gamma'$.

1.2.3 Scaling of Extremes

Fundamental concepts such as those of *scaling* (or *scale-free behavior*) have only recently been introduced and applied with success in many fields: geophysics, hydrology, meteorology, turbulence, ecology, biology, science of networks, and so on (see, e.g., [184, 8, 291, 237, 9]). The notion of scaling provides a useful tool for characterizing in a synthetic way the probabilistic structure of the phenomenon under investigation.

Systems involving scale-free behavior generally lead to probability distributions with a *power-law* analytical expression, and a (asymptotic) survival probability with an *algebraic* fall-off. This, in turn, yields heavy tailed distributions, that can generate extreme values with non-negligible probability, a fact that takes into account the actual features of extreme natural phenomena. The survival probability of the r.v. X representing a scaling phenomenon is (asymptotically) a power-law, i.e.

$$1 - F_X(x) = \mathbb{P}\{X > x\} \propto x^{-\delta}, \quad x \gg 1, \tag{1.90}$$

for a suitable *scaling exponent* $\delta > 0$. Examples of this type include earthquakes [130, 291], rock fragmentation [291], landslides [292], volcanic eruptions [225], hotspot seamount volumes [32, 283], tsumani runup heights [33], floods [183], river networks [237], forest fires [182], and asteroid impacts [41, 40].

The scaling features may considerably simplify the mathematical tractability of the phenomena under investigation. At the same time, they provide a synthesis of the mechanisms underlying the physical dynamics. In addition, the scaling approach may also offer a flexible tool for making inferences at different scales (temporal and/or spatial) without changing the model adopted.

Following [252], we now outline some properties of the Generalized Pareto distribution. In particular, we show how such a distribution may feature simple scaling, by assuming proper power-law expressions for both the position and scale parameters.

ILLUSTRATION 1.16 (Scaling of the GP distribution). ►

Let Δ be a given temporal scale, which specifies the time-scale of reference (e.g., one hour, one year, ...). Let us denote by X_Δ the intensity of the process, as observed at the scale Δ, and assume that X_Δ has a GP distribution:

$$F_{X_\Delta}(x) = 1 - \left(1 + \frac{\gamma_\Delta}{b_\Delta}(x - a_\Delta)\right)^{-1/\gamma_\Delta}, \tag{1.91}$$

where $a_\Delta \in \mathbb{R}$ is a *position* parameter, $b_\Delta > 0$ is a *scale* parameter, and $\gamma_\Delta \in \mathbb{R}$ is a *shape* parameter. To deal with *non-negative upper unbounded* variables, we only consider the case $\gamma_\Delta > 0$, and $x > a_\Delta \geq 0$.

Let $\Delta' = r\Delta$ denote a generic temporal scale, where $r > 0$ represents the *scale ratio*, and let $X_{\Delta'}$ be the intensity of the process as observed at the scale Δ'. If the following power-law relations hold

$$\begin{cases} a_{\Delta'} = a_\Delta \, r^{\delta_{\mathrm{GP}}}, \\ b_{\Delta'} = b_\Delta \, r^{\delta_{\mathrm{GP}}}, \\ \gamma_{\Delta'} = \gamma_\Delta, \end{cases} \tag{1.92}$$

where $\delta_{\mathrm{GP}} \in \mathbb{R}$ is called *scaling exponent*, then

$$F_{X_{\Delta'}}(x) = F_{X_\Delta}\left(r^{-\delta_{\mathrm{GP}}} x\right), \tag{1.93}$$

or, equivalently, $\mathbb{P}\{X_{\Delta'} \leq x\} = \mathbb{P}\left\{X_\Delta \leq r^{-\delta_{\mathrm{GP}}} x\right\}$. Then the process is *(strict sense) simple scaling*:

$$X_{\Delta'} \sim r^{\delta_{\mathrm{GP}}} X_\Delta, \tag{1.94}$$

where "$\sim$" means equality in probability distribution.

It must be stressed that, in practical applications, the *scaling regime* (when present) usually holds only between an *inner* cutoff $r_{\min}$ and an *outer* cutoff $r_{\max}$: that is, for $r_{\min} \leq r \leq r_{\max}$. Then, knowing the parameters $\{a_\Delta, b_\Delta, \gamma_\Delta\}$ of X_Δ, and the scaling exponent δ_{GP}, in principle it would be possible to calculate the distribution of $X_{\Delta'}$ for any given time scale Δ'.

Asymptotically, the survival probability of the GP law given in Eq. (1.91) has an algebraic fall-off:

$$1 - F_X(x) \approx x^{-\frac{1}{\gamma}}, \qquad x \gg 1, \tag{1.95}$$

which explains why only the moments of the order of less than $1/\gamma$ exist. Such a tail behavior is typical of *Lévy-stable* r.v.'s [92, 256], an important class of *stable* variables that play a fundamental role in modeling extreme events and (multiscaling) *multifractal* processes [259]. A straightforward calculation yields

$$\begin{aligned} \mathbb{E}(X_\Delta) &= a_\Delta + \frac{b_\Delta}{1-\gamma_\Delta}, \\ \mathbb{V}(X_\Delta) &= \frac{b_\Delta^2}{(1-\gamma_\Delta)^2 (1-2\gamma_\Delta)}, \end{aligned} \tag{1.96}$$

where, respectively, $0 < \gamma_\Delta < 1$ and $0 < \gamma_\Delta < 1/2$. Thus the shape parameter plays a fundamental role in modeling the physical process, since it tunes the order of divergence of the statistical moments. We observe that Eq. (1.92) yields similar relationships in terms of the expectation and the variance (if they exist):

$$\begin{aligned}\mathbb{E}(X_{\Delta'}) &= r^{\delta_{GP}}\,\mathbb{E}(X_{\Delta}),\\ \mathbb{V}(X_{\Delta'}) &= r^{2\delta_{GP}}\,\mathbb{V}(X_{\Delta}).\end{aligned} \tag{1.97}$$

also called (*wide sense*) *simple scaling* [128]. ◄

By analogy with the GP case, we now show how the GEV distribution may feature scaling properties if assuming suitable power-law expressions for both the position and scale parameters.

ILLUSTRATION 1.17 (Scaling of the GEV distribution). ►
Let $\tau > \Delta$ denote a reference time period, and let N be the number of Δ subperiods in τ reporting non-zero values of the process X_Δ, where X_Δ is as in Illustration 1.16. Then we define $Z_\Delta = \max\{X_\Delta\}$ in τ.

To calculate the asymptotic distribution of Z_Δ, we may either consider N to be equal to some non-random large value n, or condition upon N (see also Subsection 1.2.4). In particular, if N has a *Poisson* distribution (with parameter ν), and is independent of the process $\{X_\Delta\}$, the asymptotic laws will be of the same type in both cases. Note that, in practical applications, this latter Poissonian approach is more realistic, since it may account for the natural variability of the occurrences of X_Δ in non-overlapping reference time periods. In addition, the parameter ν is easily calculated, since it simply corresponds to the average of N.

According to the derivation method adopted, and introducing the parameters

$$\begin{cases} a'_\Delta = a_\Delta + \frac{b_\Delta}{\gamma_\Delta}\left(n^{\gamma_\Delta} - 1\right)\\ b'_\Delta = b_\Delta\, n^{\gamma_\Delta}\\ \gamma'_\Delta = \gamma_\Delta \end{cases}, \tag{1.98}$$

or

$$\begin{cases} a'_\Delta = a_\Delta + \frac{b_\Delta}{\gamma_\Delta}\left(\nu^{\gamma_\Delta} - 1\right)\\ b'_\Delta = b_\Delta\, \nu^{\gamma_\Delta}\\ \gamma'_\Delta = \gamma_\Delta \end{cases}, \tag{1.99}$$

it turns out that the distribution of Z_Δ is given by (see also Illustration 1.20)

$$G_{Z_\Delta}(z) = e^{-\left(1+\frac{\gamma'_\Delta}{b'_\Delta}\left(z-a'_\Delta\right)\right)^{-\frac{1}{\gamma'_\Delta}}}, \tag{1.100}$$

for $z > a'_\Delta - b'_\Delta/\gamma'_\Delta$. Evidently, this represents an upper unbounded GEV probability law with *position* parameter $a'_\Delta \in \mathbb{R}$, *scale* parameter $b'_\Delta > 0$, and *shape* parameter $\gamma'_\Delta > 0$.

In passing we observe that, assuming $a_\Delta = 0$ and using Eq. (1.99), in principle it would be possible to calculate the parameters ν and $\{b_\Delta, \gamma_\Delta\}$ simply through the estimate of the parameters $\{a'_\Delta, b'_\Delta, \gamma'_\Delta\}$. Therefore, from the knowledge of the distribution of the maxima Z_Δ's, it might be possible to make inferences about that of the "parent" process X_Δ's. In all cases, the shape parameters γ and γ' will always be the same (see below for a connection with the order of divergence of moments).

Using the same notation as before, if the following power-law relations hold,

$$\begin{cases} a'_{\Delta'} = a'_\Delta \, r^{\delta_{\mathrm{GEV}}} \\ b'_{\Delta'} = a'_\Delta \, r^{\delta_{\mathrm{GEV}}}, \\ \gamma'_{\Delta'} = \gamma'_\Delta \end{cases} \tag{1.101}$$

then also the maximum is *(strict sense) simple scaling*:

$$\mathrm{G}_{Z_{\Delta'}}(z) = \mathrm{G}_{Z_\Delta}\left(r^{-\delta_{\mathrm{GEV}}} z\right), \tag{1.102}$$

or equivalently, $\mathbb{P}\{Z_{\Delta'} \leq z\} = \mathbb{P}\left\{Z_\Delta \leq r^{-\delta_{\mathrm{GEV}}} z\right\}$, where $Z_{\Delta'} \sim r^{\delta_{\mathrm{GEV}}} Z_\Delta$ represents the maximum associated with the timescale Δ'.

There are other important facts. First, if Eq. (1.92) holds, then Eq. (1.101) also holds, and $\delta_{\mathrm{GEV}} = \delta_{\mathrm{GP}}$. Reasoning backwards, the converse is also true, i.e., from the scaling of the maxima it is possible to derive that of the "parent" process. Thus, the scaling of the "parent" distribution turns into that of the derived law of the maxima and vice versa. Second, as in the GP case, knowing the parameters $\{a'_\Delta, b'_\Delta, \gamma'_\Delta\}$ of Z_Δ and the scaling exponent δ_{GEV}, in principle it would be possible to calculate the distribution of $Z_{\Delta'}$ for any given timescale Δ'.

Incidentally we note that, as for the GP distribution, the survival probability of a GEV variable Z has an algebraic fall-off:

$$1 - \mathrm{G}_Z(z) \approx z^{-\frac{1}{\gamma'}}, \quad z \gg 1. \tag{1.103}$$

Thus the shape parameter (here γ') plays a fundamental role, since it tunes the order of divergence of the statistical moments. Note that the shape parameters γ (GP) and γ' (GEV) have the same values, and thus the processes modeled by such distributions would feature the same behavior of the high-order moments (as stated by Eq. (1.95) and Eq. (1.103)). Assuming, respectively, $0 < \gamma'_\Delta < 1$ and $0 < \gamma'_\Delta < 1/2$, a straightforward calculation yields

$$\begin{aligned} \mathbb{E}(Z_\Delta) &= a'_\Delta + \frac{b'_\Delta}{\gamma'_\Delta}\left(\Gamma(1-\gamma'_\Delta) - 1\right), \\ \mathbb{V}(Z_\Delta) &= \frac{b'^2_\Delta}{\gamma'^2_\Delta}\left(\Gamma(1-2\gamma'_\Delta) - \Gamma^2(1-\gamma'_\Delta)\right). \end{aligned} \tag{1.104}$$

Again, if Eq. (1.101) holds, we obtain the *(wide sense) simple scaling*:

$$\begin{aligned} \mathbb{E}(Z_{\Delta'}) &= r^{\delta_{\mathrm{GEV}}} \, \mathbb{E}(Z_\Delta), \\ \mathbb{V}(Z_{\Delta'}) &= r^{2\delta_{\mathrm{GEV}}} \, \mathbb{V}(Z_\Delta). \end{aligned} \tag{1.105}$$

Thus, Eq. (1.105) shows that, provided the expectation (or the variance) exists, to estimate δ_{GEV} it suffices to apply linear regressions to $\mathbb{E}(Z)$ (or $\mathbb{V}(Z)$) vs. r on a log-log plane for different durations Δ's, and then calculate the slope of the fit. Alternatively, δ_{GEV} can be estimated using Eq. (1.101), and exploiting the scaling of the parameters a' and b'. ◄

We now present an application to rainfall data [252].

ILLUSTRATION 1.18 (Scaling properties of rainfall). ►

Here we present the "scaling" analysis of the Bisagno drainage basin, located in Thyrrhenian Liguria (northwestern Italy). Hourly recorded rainfall data collected by five gauges are available for a period of seven years, from 1990 to 1996. Assuming homogenous climatic conditions within the basin, we can consider the data collected by these gauges as samples extracted from the same statistical population. Thus, this is essentially equivalent to analyzing a data set of 35 years of measurements. The size of the database is limited for the purposes of statistical analysis and inference, but this is the actual experimental situation.

In the analysis the fundamental rainfall duration Δ is equal to 1 hour, and the reference time period τ is taken as 1 year. We use four increasing levels of temporal aggregation: namely, $\Delta_1 = 1$ hour, $\Delta_2 = 2$ hours, $\Delta_3 = 3$ hours, and $\Delta_4 = 6$ hours. Correspondingly, the scale ratio r takes on the values: $r_1 = 1$, $r_2 = 2$, $r_3 = 3$, and $r_4 = 6$.

The rainfall data X_Δ, at the temporal scale Δ, are in effect calculated as the maximum rainfall depth in a time interval Δ within any single storm. Thus, N will represent the number of independent storms in the reference time period τ, which is assigned a Poisson distribution — with parameter ν to be estimated — as a working hypothesis.

Since the digital sampling of rainfall has introduced long series of identical data — i.e., rainfall measurements in different Δ subperiods may show the same values — then the (within storm) maximum rainfall depth data — calculated as outlined above — also show sets of identical values. Thus, these data are analyzed considering "classes" of arbitrary size of 10 mm, and calculating the frequency of each class, as shown shortly. In all cases we shall show that an upper unbounded GP probability law is compatible with the available data. In turn, the resulting distribution of maximum annual depth values approximates an upper unbounded GEV probability law, with a shape parameter practically identical to that found in the GP case.

In Figures 1.16–1.17 we plot, respectively, $\ln(1-\hat{F}_i)$ vs. $\ln x_{(i)}$ for the available rainfall depth data, and $\ln(-\ln \hat{F}_i)$ vs. $\ln z_{(i)}$ for the corresponding maximum annual depth measurements. Here the Weibull plotting position is used for $\hat{F}_i$'s (see Illustration 1.6). Actually, this is the QQ-plot test [87, 13] for verifying whether or not a GP (GEV) distribution could be used as a model of the processes investigated: GP (GEV) distributed data should show a linear behavior for $x \gg 1$ ($z \gg 1$), respectively, in Figures 1.16–1.17.

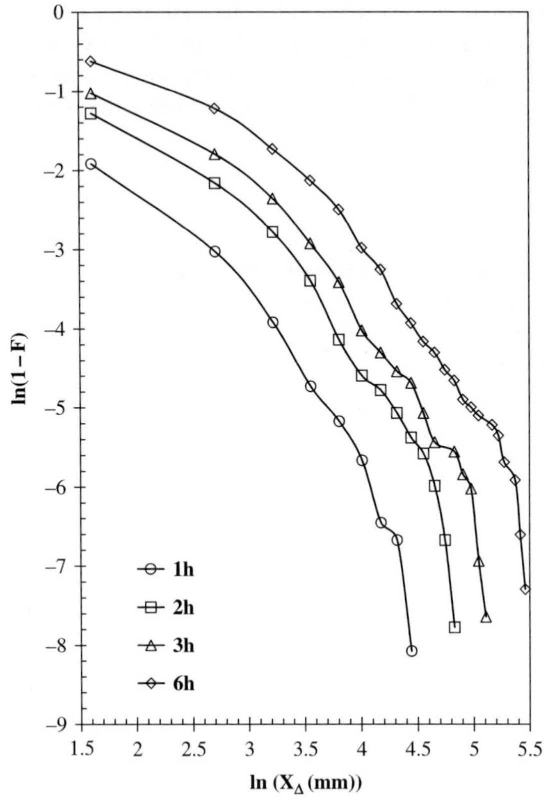

Figure 1.16. Plot of $\ln(1-\hat{F}_i)$ vs. $\ln x_{(i)}$ for the *maximum rainfall depth* data (on a storm basis) for different temporal aggregations

Indeed, in both cases a linear trend is evident asymptotically, supporting the hypothesis that X has a GP distribution and that Z has a GEV distribution. Also, the steep asymptotic behavior indicates that both the shape parameters γ and γ' are expected to be close to zero.

Further relevant features emerge from a careful analysis of Figures 1.16–1.17, and deserve discussion. On the one hand, both in the GP and in the GEV cases the same asymptotic behavior is evident for all the four series presented. Thus, the GP and the GEV laws may be considered as general models of the respective processes, for any given temporal duration Δ. On the other hand, the asymptotic linear trends look roughly the same, independently of the level of temporal aggregation. Hence *unique* shape parameters γ and γ' could be taken for all the four GP and GEV data series. In addition, we expect such common values of γ and γ' to be close to one another.

The estimation of these shape parameters is not straightforward. Quite a few methods exist to calculate them: for instance, besides standard techniques such as

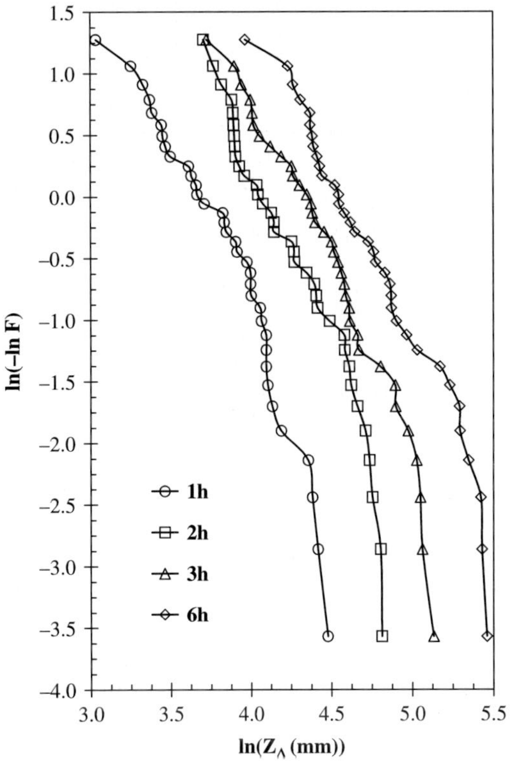

Figure 1.17. Plot of $\ln(-\ln \hat{F}_i)$ vs. $\ln z_{(i)}$ for the *maximum annual rainfall depth* data for different temporal aggregations

the *method of moments*, one could use the *L-moments* [139] or the LH-moments [297]. Also Hill's [135] and Pickands's [216] estimators are frequently used in practice (see [87, 13] for a thorough review and discussion). These methods exploit the asymptotic behavior of the heavy upper tail of Pareto-like distributions, without any specific assumption about the existence of low-order moments.

In Table 1.3 we show the estimates of γ and γ', for different levels of temporal aggregation. Roughly, all the values are quite close to one another, and lie within one sample standard deviation (*s.d.*) from the respective sample means (*av.*).

The estimate of γ and γ' as a function of r provides an empirical evidence that a common value $\bar{\gamma}$ for these parameters can be assumed. Thus we may take $\overline{\gamma} \approx 0.078$, i.e. the average of all the values of γ and γ'. As anticipated, such a value is close to zero.

The estimate of the remaining position and scale parameters is made using different techniques for, respectively, the GP and the GEV case (for a review of methods, see [168] and also [249, 250]). Since the rainfall depth data are organized in classes, the parameters a and b are estimated by minimizing *Pearson* χ^2, whereas

Table 1.3. Estimates of GP and GEV parameters. Subscripts "M" and "L" refer to the use of the *method of moments* and *L-moments* technique as estimation procedures. The subscript "GP" refers to the calculation of the GEV parameters using the estimates of the GP parameters $\{a, b\}$ and Eq. (1.99)

Param.	**Unit**	**Estimates**				**Statistics**	
r	[-]	$1h$	$2h$	$3h$	$6h$	av.	s.d.
γ	[-]	0.089	0.076	0.080	0.065	0.078	0.010
a	[mm]	0.117	0.277	0.498	0.661		
b	[mm]	5.826	9.379	12.447	19.459		
γ'	[-]	0.084	0.065	0.074	0.089	0.078	0.011
b'_M	[mm]	12.160	18.167	24.537	36.701		
a'_M	[mm]	41.343	62.593	77.369	101.304		
b'_L	[mm]	14.335	21.449	28.646	41.905		
a'_L	[mm]	40.917	61.938	76.670	100.804		
b'_{GP}	[mm]	8.051	12.961	17.200	26.890		
a'_{GP}	[mm]	28.281	45.616	60.668	94.726		
$\overline{\gamma}$	[-]	0.087	0.071	0.077	0.077	0.078	0.007

the parameters a' and b' are calculated both by the *method of moments* and by the *L-moments* technique. In all cases, a common value $\overline{\gamma}$ for γ and γ' is used. The results are shown in Table 1.3.

Evidently, from a practical point of view, the parameter a (corresponding to the lower bound of X) can always be taken equal to zero, and the values of a' and b' are essentially the same, either they are estimated via the *method of moments* or via the *L-moments* technique. Thus, using Eq. (1.99), we calculate a' and b' using the estimates of a and b (or vice versa). Here we use a dry period $D = 7$ hours to separate different storms. Such an ad hoc choice is motivated by the known behavior of the meteorology in the region under investigation. In turn, the annual storm rate ν turns out to be $\nu \approx 60$. The estimates a'_{GP} and b'_{GP} of the parameters a' and b' obtained via Eq. (1.99) are shown in Table 1.3. Indeed, these values are empirically consistent with the other estimates of a' and b'. It must be recalled that, on the one hand, the estimates of a and b are obtained using a technique completely different from that adopted for a' and b', and in addition they may depend upon both the size of the classes and the length of the dry period. On the other hand, the database is of limited size (only 35 measurements are available for each level of temporal aggregation), and both the *method of moments* and the *L-moments* technique may be biased, when the sample size is small.

Given the fact that $\overline{\gamma}$ is effectively smaller than $1/2$, we may test whether or not a scaling behavior, as described by Eq. (1.101) and Eq. (1.105), is present. For this purpose we plot $\mathbb{V}(Z_\Delta)$, $\mathbb{E}(Z_\Delta)$, and the different estimates of b'_Δ and a'_Δ versus r on a log-log plane, as shown, respectively, in Figures 1.18–1.19.

The scaling of the variables of interest is always well evident, and thus it is possible to try and fit a unique parameter δ_{GEV}. The results are presented in Table 1.4.

Apparently, the estimate of δ_{GEV} using the variance and the parameter b' seems slightly different from that obtained using the expectation and the parameter a'.

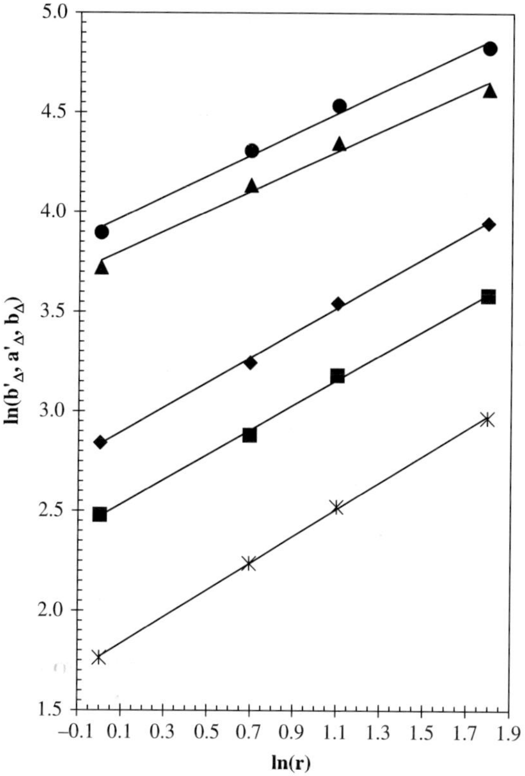

Figure 1.18. Plot of b'_Δ (*squares*) and a'_Δ (*triangles*) versus r on a log-log plane. The parameters are estimated using the *method of moments* (see Table 1.3 and Table 1.4). For the sake of comparison, we also show the scaling of the following quantities: the sample standard deviation (*diamonds*), the sample average (*circles*), and the GP parameter b_Δ (*asterisks*). The straight lines represent linear regressions

However, the corresponding confidence intervals indicate that the two sets of estimates are statistically the same, at least in the range of temporal aggregations investigated. In addition, for the sake of comparison, in Figures 1.18–1.19, we also show the scaling of the GP parameter b_Δ (using Eq. (1.92)), and calculate the corresponding estimate of the parameter δ_{GP} (see Table 1.4). Note that, since the GP parameter a_Δ is taken as zero for any level of temporal aggregation Δ, then Eq. (1.92) still holds. Given the previous discussion about the different techniques used to calculate the GP-GEV parameters (and noting that only four levels of temporal aggregation are used to estimate δ), although δ_{GP} appears to be slightly larger than δ_{GEV}, we may conclude that both values are empirically consistent.

Finally, given the estimates $\{a'_\Delta, b'_\Delta, \overline{\gamma}\}$ for a series of rainfall durations Δ, we may compare the empirical c.d.f.'s estimated on the available data with the corresponding "theoretical" distribution functions, as shown in Figure 1.20. Note that, for any level of temporal aggregation Δ, the values a'_Δ and b'_Δ considered are

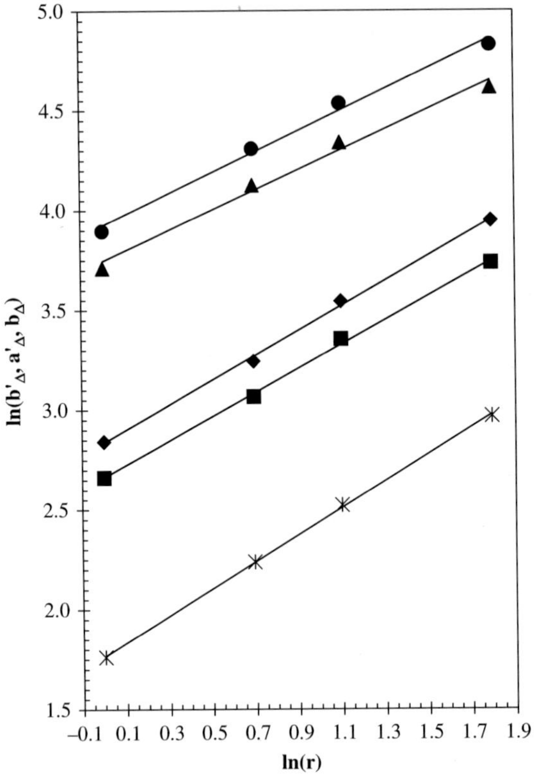

Figure 1.19. Plot of b'_Δ (*squares*) and a'_Δ (*triangles*) versus r on a log-log plane. The parameters are estimated using the *L-moments* technique (see Table 1.3 and Table 1.4). For the sake of comparison, we also show the scaling of the following quantities: the sample standard deviation (*diamonds*), the sample average (*circles*), and the GP parameter b_Δ (*asterisks*). The straight lines represent linear regressions

Table 1.4. Estimates of the parameter δ_{GEV} and corresponding 99% Confidence Interval (CI). In the first row the scaling of $\mathbb{V}(Z)$ and $\mathbb{E}(Z)$ is used. In the second row the scaling of b'_M and a'_M is used. In the third row the scaling of b'_L and a'_L is used. Subscripts "M" and "L" refer to the use of the *method of moments* and of the *L-moments* technique, respectively, as estimation algorithms. For the sake of comparison, the last row reports the estimate of the parameter $\delta_{\text{GP},c}$ and an approximate 99% CI

	δ_{GEV}	**99% CI**		δ_{GEV}	**99% CI**
$\mathbb{V}(Z)$	0.62	(0.43, 0.81)	$\mathbb{E}(Z)$	0.52	(0.20, 0.85)
b'_M	0.62	(0.43, 0.81)	a'_M	0.50	(0.11, 0.90)
b'_L	0.60	(0.42, 0.79)	a'_L	0.50	(0.12, 0.89)
c	0.67	(0.59, 0.76)	c	0.67	(0.59, 0.76)

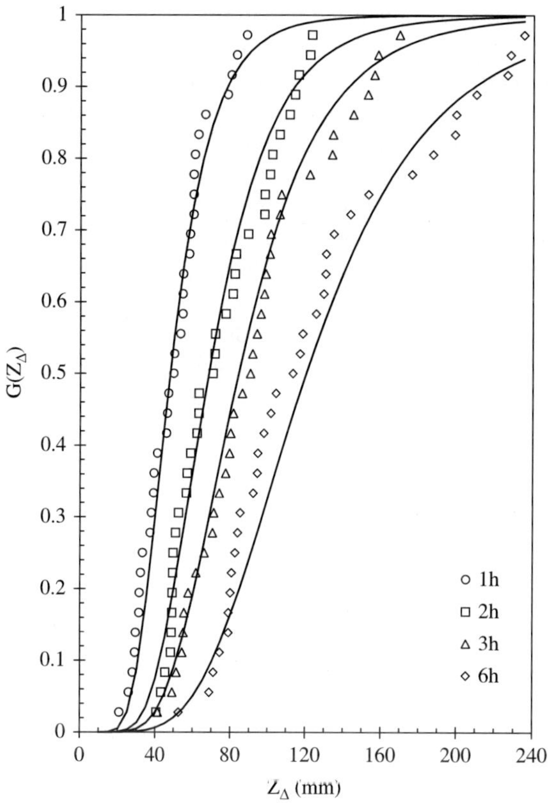

Figure 1.20. Plot of the empirical c.d.f.'s (*markers*) calculated for all the four levels of temporal aggregation mentioned in the text, and the corresponding GEV distributions (*lines*) fitted using the *scaling estimates* of the parameters b'_Δ and a'_Δ

obtained via Eq. (1.101) using b'_1 and a'_1 as basic estimates (i.e., those obtained for $\Delta_1 = 1$ hour calculated on the original non-aggregated raw data through the *L-moments* technique).

As a general comment, we see that the agreement between the empirical and the theoretical c.d.f.'s seems to be satisfactory, especially considering the limited size of the available database. Lastly, because of the scaling relations just found, in principle it would be possible to draw theoretical c.d.f.'s for any desired level of temporal aggregation. ◀

Alternatively, other distributions can used to describe the scaling properties of geophysical processes and their extremal properties. For example, in [27] both the Gumbel and the Lognormal distributions are applied to model the scaling of annual maximum storm intensity with temporal duration at a point in space.

The concept of scaling is also used to explain the variability of r.v.'s in space (where Δ is a spatial scale). For example, in [172, 243] the concept of random scaling is shown to explain the *Horton's law* of drainage network composition [138]. In [3] this concept is further applied to topologically random network. The scaling is used to represent the variability of the maximum annual flood peak in a river, as parameterized by its basin area [128, 126, 127], and the maximum annual rainfall depth as parameterized by its temporal duration [27]. In addition, the probability distribution of catastrophic floods is investigated in [61] by combining the scaling properties of maximum annual flood peaks with those of river network composition.

The concept of scaling can be extended from one-dimensional to multi-dimensional problems, where more than one coordinate is involved. The concepts of *self-similarity* and *self-affinity* are used when the transformation from a scale to another is, respectively, *isotropic* and *anisotropic*. For instance, in [62, 63] the concept of self-affinity is used to investigate the scaling properties of extreme storm precipitation with temporal duration and area.

1.2.4 Contagious Extreme Value Distributions

Contagious Extreme Value distributions play an important role in the analysis of extreme events. As will be shown shortly, they make it possible to investigate samples of random size, to account for the full sequence of extreme events associated with a given r.v. (possibly produced by different physical mechanisms), or to deal with distribution parameters that are themselves random variables. Practical examples are the annual maximum flood in a river, that can be generated by storm runoff and snowmelt runoff, or the maximum sea wave height, that may be due to the combined effect of astronomical tides, local storm surges, and large oceanic water-mass movements (like *el Niño* and *la Niña*). Note that, to define such extreme events, a threshold is sometimes introduced in applications: for instance, a flood event may be identified by the peak flow exceeding a given level.

Consider a sample $X_1, X_2, \ldots, X_N$ of i.i.d. r.v.'s with a common distribution F, where N is the random number of occurrences in a "block" (say, one year). Some examples are floods or droughts occurring at a site in a year, or earthquakes or hurricanes affecting a region in a year.

If p_N denotes the p.m.f. of N, then the c.d.f. G of the maximum of the N variables X_i's can be calculated as

$$G(x) = \sum_{n=0}^{+\infty} F(x)^n p_N(n). \tag{1.106}$$

Let us assume that N has *Poisson* distribution as follows:

$$p_N(n) = \frac{\nu^n e^{-\nu}}{n!}, \tag{1.107}$$

where $n \in \mathbb{N}$, and the parameter $\nu > 0$ represents the mean of N. Then

$$\begin{aligned} G(x) &= \sum_{n=0}^{+\infty} (F(x))^n \frac{\nu^n e^{-\nu}}{n!} \\ &= \sum_{n=0}^{+\infty} \frac{(\nu F(x))^n e^{-\nu}}{n!} \\ &= e^{-\nu(1-F(x))} \sum_{n=0}^{+\infty} \frac{(\nu F(x))^n e^{-\nu F(x)}}{n!}. \end{aligned} \tag{1.108}$$

Since the series on the right-hand side sums to one, G is then given by

$$G(x) = e^{-\nu(1-F(x))}, \tag{1.109}$$

for all $x \in \mathbb{R}$. The distribution G is known as a (Poisson) *Compound Contagious Extreme Value* probability law.

ILLUSTRATION 1.19 (Shifted Exponential). ►

Let F be a *shifted Exponential* distribution, i.e.

$$F(x) = 1 - \exp\left(-(x-a)/b\right), \tag{1.110}$$

for $x \geq a$, where $a \in \mathbb{R}$ is a position parameter and $b > 0$ is a scale parameter. Then G is given by

$$G(x) = \exp\left(-\nu e^{-(x-a)/b}\right) = \exp\left[-\exp\left(-(x-a-b\ln\nu)/b\right)\right].$$

Thus, G is a *Gumbel* distribution, with position parameter $a' = a + b\ln\nu$ and scale parameter $b' = b$. ◄

ILLUSTRATION 1.20 (Generalized Pareto). ►

Let F be a GP distribution (see Eq. (1.82)). Then G is given by

$$\begin{aligned} G(x) &= \exp\left[-\nu\left(1+\gamma\frac{x}{b}\right)^{-1/\gamma}\right] \\ &= \exp\left[-\left(1+\frac{\gamma}{b/\nu^{-\gamma}}\left(x-\frac{1-\nu^{-\gamma}}{\gamma}b/\nu^{-\gamma}\right)\right)^{-1/\gamma}\right]. \end{aligned}$$

Thus, G is a GEV distribution, with position parameter $a' = \frac{\nu^{\gamma}-1}{\gamma} b$, scale parameter $b' = b\nu^{\gamma}$, and shape parameter $\gamma' = \gamma$. Clearly, this reduces to the Gumbel distribution derived in Illustration 1.19 in the limit $\gamma \to 0$. ◄

Let us now investigate the case of a phenomenon generated by several mechanisms characterized by different probability distributions. In this context, *mixtures* are needed. Let us consider m independent sequences of r.v.'s, each having common c.d.f. F_i and occurring according to a *Poisson* chronology with parameter $\nu_i > 0$. Then, the c.d.f. G of the maximum of these variables can be calculated as

$$G(x) = \prod_{i=1}^{m} \exp\left(-\nu_i(1 - F_i(x))\right). \tag{1.111}$$

ILLUSTRATION 1.21 (Shifted Exponential). ►

Let $m = 2$ and let F_i be *shifted Exponential* distributions (see Eq. (1.110)). Then G is given by

$$\begin{aligned} G(x) &= \exp\left(-\nu_1 e^{-(x-a_1)/b_1} - \nu_2 e^{-(x-a_2)/b_2}\right) \\ &= \exp\left[-\exp\left(-\frac{x - a_1 - b_1 \ln \nu_1}{b_1}\right) - \exp\left(-\frac{x - a_2 - b_2 \ln \nu_2}{b_2}\right)\right]. \end{aligned}$$

Putting $b_1' = b_1$, $b_2' = b_2$, $a_1' = a_1 + b_1 \ln \nu_1$, and $a_2' = a_2 + b_2 \ln \nu_2$, yields

$$G(x) = \exp\left[-\exp\left(-\frac{x - a_1'}{b_1'}\right) - \exp\left(-\frac{x - a_2'}{b_2'}\right)\right], \tag{1.112}$$

which is sometimes referred to as the *Two-Component Extreme Value (TCEV)* distribution [242]. ◄

Another type of Contagious Extreme Value distribution is derived considering r.v.'s characterized by a c.d.f. whose parameters are themselves r.v.'s. Let Θ be the (vector of the) parameter(s) of the family of distributions $H(\cdot; \Theta)$ associated with X, and assume that Θ is a r.v. with distribution Λ_Θ over $\mathbb{R}^d$. Then, using a simple notation, the actual distribution F of X can be calculated as

$$F(x) = \int_{\mathbb{R}^d} H(x; t) \, \mathrm{d}\Lambda_\Theta(t), \tag{1.113}$$

where the Lebesgue-Stieltjes integral is used in order to deal with continuous, discrete, or mixed r.v.'s.

The interesting point is that contagious distributions can be used to deal with situations where, e.g., the parameters, as well as the sample size, are random, as shown below.

ILLUSTRATION 1.22 (Converse Weibull). ►

Let us consider N i.i.d. r.v.'s X_i's having a *Converse Weibull* distribution given by Eq. (1.56), with scale parameter b and shape parameter γ (the position parameter is set equal zero). Suppose now that b is a r.v. with Exponential p.d.f. $h(t) = \eta e^{-\eta t}$ for $t > 0$, with $\eta > 0$. Then, the actual c.d.f. F of the X_i's is given by

$$F(x) = \int_0^\infty \left[1 - \exp\left(-\frac{x}{b}\right)^\gamma\right] \eta e^{-\eta b} \, \mathrm{d}b. \tag{1.114}$$

Assuming a Poissonian chronology for the number N of occurrences, the distribution G of the maximum of the X_i's is

$$G(x) = \exp\left(-\nu \int_0^\infty \left[\eta \exp\left(-\frac{x^\gamma + \eta b^{\gamma+1}}{b^\gamma}\right)\right] db\right), \tag{1.115}$$

which must be solved numerically to compute the desired probabilities of extreme events. ◀

1.3. HAZARD, RETURN PERIOD, AND RISK

In the analysis of extremes it is of great importance to quantify the occurrences of particular (rare and catastrophic) events, and determine the consequences of such events. For this purpose, three important concepts are illustrated shortly: the *hazard*, the *return period*, and the *risk*.

Let us consider a sequence $E_1, E_2, \ldots$ of independent events. Also let us assume that such events happen at times $t_1 < t_2 < \cdots$ (i.e., we use a *temporal marked point process* as a model). Each event E_i is characterized by the behavior of a single r.v. $X \sim F$, and can be expressed as either $E_x^< = \{X \leq x\}$ or $E_x^> = \{X > x\}$. Let T_i be the *interarrival time* between E_i and E_{i+1}, $i = 1, 2, \ldots$. As is natural, we assume that $T_i > 0$ (*almost surely*), and that $\mu_{T_i} = \mathbb{E}(T_i)$ exists and is finite; therefore $\mu_{T_i} > 0$. If we let $N_x^<$ and $N_x^>$ denote, respectively, the number of events E_i between two successive realizations of $E_x^<$ and $E_x^>$, and defining $T_x^<$ and $T_x^>$ as, respectively, the interarrival time between two successive realizations of $E_x^<$ and $E_x^>$, it turns out that

$$T_x^< = \sum_{i=1}^{N_x^<} T_i, \tag{1.116a}$$

$$T_x^> = \sum_{i=1}^{N_x^>} T_i. \tag{1.116b}$$

Assuming that the interarrival times T_i are i.i.d. (and independent of X), via *Wald's Equation* [122, 241] it is easy to show that

$$\tau_x^< = \mathbb{E}(T_x^<) = \mathbb{E}(N_x^<)\,\mu_T, \tag{1.117a}$$

$$\tau_x^> = \mathbb{E}(T_x^>) = \mathbb{E}(N_x^>)\,\mu_T, \tag{1.117b}$$

where T denotes any of the i.i.d. r.v.'s T_i. Clearly, $N_x^<$ and $N_x^>$ have a Geometric distribution with parameters $p_x^<$ and $p_x^>$ given by, respectively,

$$p_x^< = \mathbb{P}\{E_x^<\} = \mathbb{P}\{X \leq x\}, \tag{1.118a}$$

$$p_x^> = \mathbb{P}\{E_x^>\} = \mathbb{P}\{X > x\}. \tag{1.118b}$$

The above results yield the following definition.

DEFINITION 1.8 (Hazard). *The probabilities* $p_x^<$ *and* $p_x^>$ *given by Eqs. (1.118) define, respectively, the hazard of the events* $E_x^<$ *and* $E_x^>$.

Then, using Eqs. (1.117), we obtain

$$\tau_x^< = \mu_T / p_x^<, \tag{1.119a}$$

$$\tau_x^> = \mu_T / p_x^>. \tag{1.119b}$$

The above results yield the following definition; in Section 3.3 a generalization to a multivariate context will be presented.

DEFINITION 1.9 (Return period). *The positive numbers* $\tau_x^<$ *and* $\tau_x^>$ *given by Eqs. (1.119) define, respectively, the return periods of the events* $E_x^<$ *and* $E_x^>$.

The return period of a given event is the average time elapsing between two successive realizations of the event itself. Note that $\tau_x^<$ and $\tau_x^>$ are decreasing functions of the corresponding hazards $p_x^<$ and $p_x^>$: this is obvious, since the inter-arrival time gets longer for less probable events. The concepts of hazard and return period provide simple and efficient tools for the analysis of extremes: one can use a single number to represent a large amount of information. A thorough review can be found in [273].

NOTE 1.14. In applications the "block" method is often used for the analysis of extremes, where the block size is one year when annual maxima (or minima) are considered. Thus, in this context, $\mu_T = 1$ year.

Rational decision-making and design require a clear and quantitative way of expressing *risk*, so that it can be used appropriately in the decision process. The notion of risk involves both uncertainty and some kind of loss or damage that might be received. For instance, the random occurrence of extreme events may cause disasters (injuries, deaths, or shutdown of facilities and services) depending on the presence of damagable objects. The concept of risk combines together the occurrence of a particular event with the *impact* (or consequences) that this event may cause. The hazard can be defined as the source of danger, while the risk includes the likelihood of conversion of that source into actual damage. A simple method for calculating the risk is given below.

ILLUSTRATION 1.23 (Risk matrix approach). ►

According to FEMA guidelines [91], the risk can be assessed by discretizing the domain of the variables of interest in a finite number of classes. Here the events are considered on a annual basis, and $p = \mathbb{P}\{E\}$ denotes the probability of the relevant event E.

Firstly, the hazard is ranked in four quantitative classes, according to the rate of occurrence of the events, as shown in Table 1.5.

Table 1.5. Classes of hazard according to [91]

Hazard class	Description
High	Events occur more frequently than once every 10 years ($p \geq 0.1$)
Moderate	Events occur once every 10–100 years ($0.01 \leq p \leq 0.1$)
Low	Events occur once every 100–1000 years ($0.001 \leq p \leq 0.01$)
Very Low	Events occur less frequently than once every 1000 years ($p \leq 0.001$)

Secondly, the impact is ranked in four qualitative classes, according to the consequences of the events, as shown in Table 1.6.

Finally, the hazard classes are merged with the impact classes. This yields the risk matrix shown in Table 1.7.

It is worth noting how the same risk condition can be associated with events having different combinations of hazard and impact. For instance, a moderate risk "C" is related both to a high hazard with a negligible impact, and a very low hazard with a catastrophic impact.

A similar approach is described in [158], where the risk analysis consists of an answer to the following three questions:

1. "What can happen?", corresponding to a *scenario* identification;
2. "How likely is it that it will happen?", corresponding to the hazard;
3. "If it happens, what are the consequences?", corresponding to the impact.

A list of scenarios, and corresponding likelihoods and consequences, is then organized in a suitable matrix. If this set of answers is exhaustive, then the whole matrix is defined as the risk. ◄

The occurrence of a potentially dangerous event is closely related to its hazard or, alternatively, to its return period. The impact of such an event is usually quantified through two variables: the *exposure*, representing the elements potentially damageable, and the *vulnerability*, quantifying the potential damages and losses. Note that the classification of the impact is not an easy task: the evaluation of

Table 1.6. Classes of impact according to [91]

Impact class	Description
Catastrophic	Multiple deaths, complete shutdown of facilities for 30 days or more, more than 50% of property severely damaged
Critical	Multiple severe injuries, complete shutdown of critical facilities for at least 2 weeks, more than 25% of property severely damaged
Limited	Some injuries, complete shutdown of critical facilities for more than one week, more than 10% of property severely damaged
Negligible	Minor injuries, minimal quality-of-life impact, shutdown of critical facilities and services for 24 hours or less, less than 10% of property severely damaged

Table 1.7. Risk matrix according to [91]. The entries represent the risk condition: "A" denotes extreme risk, "B" denotes high risk, "C" denotes moderate risk, and "D" denotes low risk

	Impact			
Hazard	**Negligible**	**Limited**	**Critical**	**Catastrophic**
High	C	B	A	A
Moderate	C	B	B	A
Low	D	C	B	B
Very Low	D	D	C	C

exposure and vulnerability often requires a multidisciplinary approach, combining qualitative and quantitative data.

The stochastic component of the risk is given by the hazard, while the impact is a function of known structural factors. The important point is that hazard and impact, as defined above, do not affect one another. For instance, the rate of occurrence of earthquakes does not depend on the number of buildings in a given area. Similarly, the impact is only a physical characteristic of the structures present in the region (how much they are susceptible to seismic shocks). However, the impact may depend on the intensity of the phenomenon via a deterministic function.

A widespread notion of risk defines it as "risk = hazard *times* impact". As pointed out in [158], this definition can sometimes be misleading, and it is suggested to use the more proper statement "risk = hazard *and* impact". Here the matter is more essential than a simple linguistic question: in order to define the risk, it is necessary to make explicit the link between hazard and impact via a suitable functional form (that may not be the simple product "hazard *times* impact"). Proceeding along this way, the definition of risk may become arbitrary: for instance, the function linking hazard and impact may result from a subjective evaluation — however, subjectivity can be made "objective" via specific directives and guidelines. As an advantage, this approach can be adapted to any particular situation of interest (see below).

First of all it must be noted that the risk is always associated with a well specified event: there does not exist the risk "per se". In turn, we consider the risk $\mathbf{R}$ as a function of a specific event E, where E is written in terms of the random variable(s) X generating the phenomenon under investigation. This approach is quite general, for X can be a multivariate r.v.: for instance, E can be defined in terms of storm duration and intensity, or wave height-frequency-direction. Then $\mathbf{R}$ can be considered as a function on the same probability space as of X, which can be explored by changing E: for example, if $E_1, E_2, \ldots$ represent earthquakes of increasing intensity, then $\mathbf{R}(E_i)$ measures the risk associated with rarer and rarer events. Actually, $\mathbf{R}$ is a genuine measure defined on the same σ-algebra of events as of X.

The second step consists in the construction of the *impact function* ζ, linking the stochastic source of danger X with its potential consequences. The idea is to provide a framework where the intensity of the phenomenon under investigation and the corresponding probability of occurrence yield a well specified impact.

As already mentioned above, the impact is a deterministic function of the intensity; randomness, however, is associated with intensity, since only a statistical estimate can be provided. In order to provide a way to quantify and compare the impact, we arbitrarily choose for ζ the range $\mathbb{I} = [0, 1]$: $\zeta \approx 0$ denotes a negligible impact, and $\zeta \approx 1$ a catastrophic impact. We define the impact function as follows.

DEFINITION 1.10 (Impact function). *A measurable integrable function* $\zeta : \mathbb{R}^d \to \mathbb{I}$ *is called impact function.*

Note that Definition 1.10 gives a large freedom of choice for ζ. For instance, suppose that some structures collapse only when subjected to specific frequencies (e.g., due to a resonance phenomenon). Then ζ can be easily constructed in order to accentuate the impact of dangerous frequencies and ignore the others. More generally, ζ is the key function to transform a hazard into damage.

Finally, let us assume that X has distribution F_X over $\mathbb{R}^d$. We calculate the risk as a suitable integral over the (measurable) event $E \subseteq \mathbb{R}^d$ of interest:

$$\mathbf{R}(E) = \frac{1}{\mathbb{E}(\zeta(X))} \int_E \zeta(x)\, \mathrm{d}F_X(x). \tag{1.120}$$

Here $\mathbb{E}(\zeta(X))$ plays the role of a normalizing constant, which also makes $\mathbf{R}$ adimensional. Note that the use of the Lebesgue-Stieltjes integral in Eq. (1.120) gives the possibility to deal with the continuous, the discrete, and the mixed cases. In particular, if X is absolutely continuous with density f_X, then $\mathrm{d}F_X(x) = f_X(x)\,\mathrm{d}x$. This yields the following definition.

DEFINITION 1.11 (Risk). *The non-negative number* $\mathbf{R}(E)$ *given by Eq. (1.120) represents the risk of the event E.* $\mathbf{R}$ *is called risk function, and* $\mathbb{E}(\zeta(X))$ *is the expected risk.*

Evidently, the range of $\mathbf{R}$ is $\mathbb{I} = [0, 1]$: $\mathbf{R} \approx 0$ identifies a low risk, and $\mathbf{R} \approx 1$ an extreme risk. Thus, $\mathbf{R}$ is a probability measure on the same probability space as of X: indeed, a natural interpretation of the risk. Most importantly, the calculation of the risk of complex events (e.g., those obtained via the union and/or the intersection operators) is achieved using the standard rules of Measure Theory.

The definition of risk given above offers interesting perspectives. First of all, the "risk matrix method" discussed in Illustration 1.23 turns out to be a particular case of the present approach: Table 1.7 is no more than a discretized version of the integral in Eq. (1.120). Secondly, the integral in Eq. (1.120) acts as a sort of "filtering" operator: for instance, very unlikely events ($f_X \approx 0$), associated with very large impacts ($\zeta \approx 1$), may yield the same risk as likely (or characteristic) events associated with average impacts (see, e.g., the moderate risk condition "C" in Table 1.7).

If we assume that f_X can always be evaluated or estimated via statistical techniques, the fundamental problem of the present approach is to find an appropriate functional expression for the impact function ζ. This corresponds to set up a suitable "damage" function of the intensity X: it can be done either starting from first principles (e.g., when the dynamics of the phenomenon and the physics of the structures are known), or through a trial-and-error procedure.

NOTE 1.15. A further definition of risk was proposed by UNESCO [295] as follows:

$$\mathbf{R} = X * E * V, \tag{1.121}$$

where "$*$" indicates the convolution operator, and E and V represent, respectively, the exposure and the vulnerability (see also [159] and references therein).

1.4. NATURAL HAZARDS

In this section we deal directly with various *natural hazards*. Their magnitudes and effects vary in time and space. As discussed previously, they cause loss of human life sometimes running into hundreds of thousands and tend to destroy national, economic and social infrastructures. During the past 15 years there has been, in some types, an increase in their occurrence and severity, affecting more than two billion people.

We show how to estimate the risks involved using the theory developed in the previous sections. Initially the geological hazards of earthquakes, volcanic eruptions and tsunamis are considered. The related subjects of landslides and avalanches follow. These have links with the weather. We then focus on climatic hazards. Windstorms, extreme sea levels and high waves, droughts, and wildfires are included. In each case, we provide substantial discussion of the nature of the hazard and its physical aspects, give details of past events and demonstrate appropriate procedures of analysis.

Hazards caused by storm rainfall and floods are investigated in Subsection 1.2.3 and elsewhere in the book. More importantly, later Chapters and Appendices contain numerous examples based on Copulas.

1.4.1 Earthquakes

Earthquakes pose a severe threat to the built environment. In some years there are more than 50 potentially destructive earthquakes. When they occur close to urban areas, the consequences can be catastrophic. Many types of buildings collapse, and there is a destructive effect on dams, bridges, and transport systems. With the steep rise in economic development during the past century, there have been high increases in the threats imposed. For example, the disaster potential in California has grown at least ten-fold from the time of the calamitous 1906 San Francisco earthquake. From the 1990's, however, there has been a better awareness worldwide, and a willingness to cope, or minimize, the impacts of earthquakes. At least 38 countries susceptible to earthquakes, from Mexico to India, are revising their seismic codes for designing safer structures [213, 148]. Nevertheless, innumerable poorly constructed buildings exist in earthquake-prone areas.

The cause of the most severe earthquakes can be explained by means of plate tectonic theory. Basically, the surface of the earth consists of many huge pieces of flat rocks, called *tectonic plates*, that have relative motions and interactions at their boundaries. Deep earthquakes occur at convergent plate boundaries at depths of more than 300 kilometers below the surface. However, many destructive earthquakes occur in the 20–50 kilometer depth range. This often happens when one plate descends very slowly beneath the other. Alternative motions, that cause earthquakes, are when plates collide or slide past each other horizontally, or vertically, or separately from one another. Regardless of boundary actions, seismic waves are generated by the sudden fracturing caused when elastic strain energy, accumulated over a very long period, exceeds the crushing strength of rock. These activities take place along fault lines at plate boundaries of which there are 13 major sites worldwide (for example, the North American or Indian plates). Other destructive earthquakes are in the 20–50 kilometers depth range.

Immediately after an event, the earthquake is located by means of a seismograph, an instrument initiated nearly 100 years ago. This shows records of the shaking of the ground caused by vibrating seismic waves, with wide frequency, and amplitude ranges, travelling along the surface, or through the earth. The earthquake originates at a hypocenter, or focus, vertically above which, at ground level, is the epicenter. The primary longitudinal waves, called *P* waves, travel at the highest speeds in the direction of propagation, and are transmitted through both solid and liquid media. The slower secondary, or *S* waves, vibrate at right angles to the direction of propagation, and pass through rock. Subsequently, *Love and Rayleigh* waves move on the surface with very high amplitudes. At large distances, they are the cause of ground displacements, or distortions, and much of the shaking felt. In engineering seismology applied to structural safety, the late Bruce Bolt [22] pioneered the use of data from sensors along fault lines, records of previous earthquakes, and the analysis of subterranean rock formations. Bolt showed, for instance, how some locations in active seismic zones, regardless of the distance from the epicenter, can be more prone to this type of hazard.

It is a common practice to classify earthquakes according to the mode of generation. The most common are of the tectonic type, which poses the greatest threat, as just discussed. They occur when geological forces cause rocks to break suddenly. Volcanic earthquakes constitute a second category. However, volcanic eruptions can occur independently of earthquakes. More follows in Subsection 1.4.2. Some of the most damaging earthquakes occur where ground water conditions are appropriate to produce significant liquefaction. This has also been studied across active faults in China and Japan, for example, but there are other exceptional circumstances.

Amongst past events worldwide, the earthquake on November 1st, 1755 in Lisbon, Portugal, had a death toll of 60,000, and the most devastating one in Tianjin, China in 1965 killed 255,000 inhabitants. Earthquakes that have caused tsunamis are described in Subsection 1.4.3. In North America, an unprecedented earthquake occurred on April 18th, 1906 at San Francisco in California, as already mentioned.

It was caused by a rapture along one large section of the San Andreas fault which runs roughly parallel with the coastline. Specifically, the dates given of paleoearthquakes in the region are 1857 (the Fort Tejon, California event of January 9th), and going further backwards in time, in 1745, 1470, 1245, 1190, 965, 860, 665 and 545. This shows an average of one major earthquake per 160 years, but with a large variation.

The *magnitude* of an earthquake is a quantitative measure of its size based on the amplitude of the ground measured by a seismograph at a specified distance from the rupture of the crust. A well-known measure of the strength of an earthquake is the *Richter magnitude scale*. It was originated in 1931 in Japan by K. Waldati, and extended practically by C. Richter in 1935 in California, USA. The Richter magnitude M_L is related to the distance in kilometers from the point of rupture and the amplitude in millimeters, see, for example, [22, p. 153]. $M_L \in \mathbb{R}$, has a logarithmic basis. Generally it assumes values in the range 0 to 10, even if negative values can occur for very small events (such as rock falls) and can theoretically exceed 10, although this has not been observed. Thus, an earthquake of magnitude 9 is 1,000 times as large as one of magnitude 6. In addition, if the size of an earthquake is expressed by its release energy, then it is proportional to $10^{1.5M_L}$, so a magnitude 9 event has about 32,000 times more energy than a magnitude 6 event. Worldwide, an earthquake of magnitude not less than 7.5 is generally considered to be a major event. The highest recorded so far is the Valdivia earthquake in Chile, of May 22nd, 1960, which had a magnitude of 9.5.

Global seismic networks can record earthquakes anywhere with magnitudes exceeding four, but in some parts of the world, with intensive seismic networks, earthquakes with magnitudes as low as 1 and 2 can be monitored. As already mentioned, the first signals to arrive at a distant recording station, after an earthquake, are through seismic *P* waves, see [195]. Thus, it is possible to determine the hypocenter within 15 minutes. At a station in the upper Tiber Valley in Tuscany, Italy, *P* waves arrived 740 seconds after the Sumatra-Andaman Islands earthquake of December 26th, 2004 [235]. However, at least, several hours may be required to estimate the magnitude reliably; it may sometimes take months or longer as in case of the earthquake just cited.

Predictions of earthquakes, in time and space, have not been sufficiently reliable. Some of the reasons are the varied trigger mechanisms and insufficient instrumentation. Bolt et al. [23] classify two types of credible earthquake predictions in California, which has a recent 180-year record. One is a general forecasting method, which gives probabilities of occurrences over a long period. The second attempts to be specific by stating the time interval, region and range of magnitude. The Bulletin of the Seismographic Stations of the University of California lists 3638 earthquakes of magnitude M_L, in the range $3.0 \leq M_L \leq 7.0$, observed during the period 1949 to 1983 over an area of 280,000 square kilometers in northern and central California.

For instance, near the town of Parkfield, California, on the line of the San Andreas fault, seismographic records have shown that the area has been struck by moderate-sized earthquakes ($5.5 \leq M_L \leq 6.0$) in the years 1857, 1881, 1901, 1922, 1934 and

1966, with a nearly constant return period of 22 years. Another event "predicted" to occur in the 1980's happened with delay in September 2004. In some cases a cluster of foreshocks occur over a period of 6 months within about 3 kilometers of the epicenter, over an area called the preparation zone, implying a release of strain energy in this zone prior to a major rupture some distance away.

In contrast to magnitude, an *intensity* scale provides a qualitative description of earthquake effects such as human perception, and effects on buildings and the surrounding landscape. For example, in Italy, and elsewhere in Southern Europe, one uses the *Mercalli-Cancani-Sieberg* (MCS) intensity scale, whereas in central and eastern Europe, the *Medvedev-Sponheuer-Karnik* (MSK) intensity scale is used. The European Macroseismic Scale (EMS), developed after significant contributions by seismologists in Italy, is probably a better tool for describing intensity. We commence with a simple example, and introduce some theoretical concepts in the next.

ILLUSTRATION 1.24 (Catalog of Italian Earthquakes). ►

Catalogo dei Terremoti Italiani dall'anno 1000 al 1980 ("Catalog of Italian Earthquakes from Year 1000 to 1980") has historical information on Italian Earthquakes. It was edited by D. Postpischl in 1985 and published by the National Research Council of Italy. Rome has experienced 329 earthquakes. Table 1.8 is a summary of earthquakes in Rome during the stated period according to the Century of occurrence.

We provide answers to some basic questions. What is the probability of more than 2 earthquakes in a century? 4/10. Also, the events can be divided according to their MCS intensities as shown in Table 1.9.

What is the mean intensity of earthquakes? Mean intensity $= (2\times 113+3\times 132+4\times 56+5\times 22+6\times 4+7\times 2)/329 = 3.02$. What is the probability of occurrence of an earthquake of intensity greater than 5? 6/329. ◄

For applications in the United States a modified Mercalli intensity (MMI) scale is adopted; see [22, pp. 311–314] for an abridged description with intensities dependent on average peak velocity or acceleration. The range is from I to XII. For example, an earthquake that frightens everyone, and makes them run outside after slight damage to property and movement of heavy furniture indoors has intensity value VI, with an average peak velocity of 5 to 8 cm/s. At the top end of the scale, if all masonry structures and bridges are destroyed, and objects are thrown in the air with waves seen on the ground, it has an intensity of XII.

There is an approximate linear relationship between the epicentral intensity and magnitude. For example, epicentral intensities of 6 and 9 as measured using MMI units are associated with magnitudes of 5 and 7, respectively. There are of course field variations.

Table 1.8. Timing of the 329 earthquakes that occurred in Rome

Century	XI	XII	XIII	XIV	XV	XVI	XVII	XVIII	XIX	XX
Total	2	1	1	0	3	0	1	15	301	5

Table 1.9. MCS intensity of the 329 earthquakes that occurred in Rome

MCS	2	3	4	5	6	7
Total	113	132	56	22	4	2

As regards the frequency of occurrence of earthquakes and its relationship with magnitude, the Gutenberg-Richter law [129]

$$\nu = a\,10^{-bx} \tag{1.122}$$

is widely used. Here ν denotes the mean number of earthquakes in unit time (say, 1 year) with *magnitude* greater than x, and the parameters a and b vary from one region to another. Turcotte [291] gives $a = 10^8$ and $b = 1$ in worldwide data analysis of surface-wave magnitudes based on surface waves with a period of 20 seconds (b is modified in $b' = b\ln(10) = 2.3b$ considering the natural logarithm instead of the logarithm in base 10). This gives globally an average of ten earthquakes of magnitude 7, or greater, each year. Note that regional earthquake occurrence characteristics can cause significant deviations from default parameter values of $a = 10^8$ and $b = 1$.

A lower bound of magnitude, x_{min}, can be used to represent the minimum level of earthquake of any consequence, and an upper bound, x_{max}, to represent the largest possible earthquake considered in a particular zone. Then, the modified Gutenberg-Richter law has the truncated Exponential relationship

$$\nu = \nu_0 \frac{e^{-b'(x-x_{\text{min}})} - e^{-b'(x-x_{\text{max}})}}{1 - e^{-b'(x_{\text{max}}-x_{\text{min}})}}, \tag{1.123}$$

where ν_0 is the number of earthquakes equal to the lower bound or larger. Note that the normalization and truncation is usually done by modifying the probability density function.

During an earthquake crystal deformation occurs at the boundaries between major surface plates, and relative displacements take place on well defined faults, which are considered to have memory. Cornell and Winterstein [53] suggest that a Poissonian chronology can be applied to the annual number of earthquakes in an area, even if fault memory exists, unless the elapsed time between a significant event with memory exceeds the average recurrence time between such events.

Thus, if the annual number of earthquakes in an area is Poisson distributed with mean ν, the probability that no earthquake occurs in a year, with magnitude greater than, say, x is $e^{-\nu}$. This is also the probability of nonexceedance of x by the annual maximum magnitude of an earthquake. Thus, the c.d.f. of the magnitude of the annual maximum earthquake is given by

$$G(x) = e^{-\nu} = \exp(-a\,e^{-b'x}). \tag{1.124}$$

Eq. (1.124) is the Gumbel distribution (see Eq. (1.47)), with scale parameter $1/b'$ and location parameter $b'^{-1}\ln(a)$. Note that regional earthquake occurrence characteristics cause significant deviations from the default parameter values of $a = 10^8$ and $b = 1$.

For worldwide data, $b = b'/2.3 = 1$ and $a = 10^8$, as already mentioned; thus the scale and location parameters of the Gumbel distribution are 0.43 and 8, respectively, see Figure 1.21. This distribution is widely used to predict the annual maximum magnitude of earthquakes in a region. For example, values of $b = 0.90$ and $a = 7.73 \cdot 10^4$ are estimated from the catalogue of earthquakes exceeding magnitude 6 in southern California for the period 1850–1994 as reported by [305], see Figure 1.21.

Figure 1.22 gives the probability distribution of maximum annual earthquake magnitude, where on the x-axis $-\ln(-\ln(G))$ represents the probability level, and on the y-axis the magnitude is plotted. Note that the dotted lines show the effect of truncation with $x_{min} = 6$ and $x_{max} = 10$.

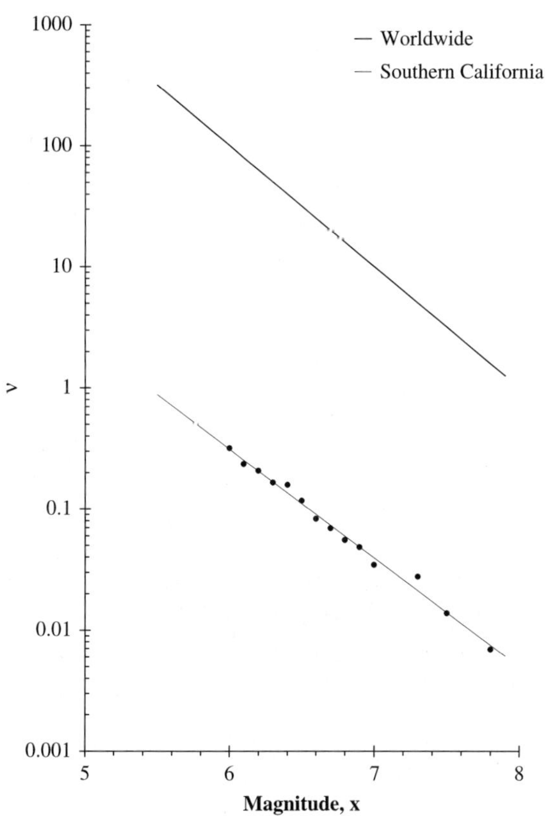

Figure 1.21. The Gutenberg-Richter law

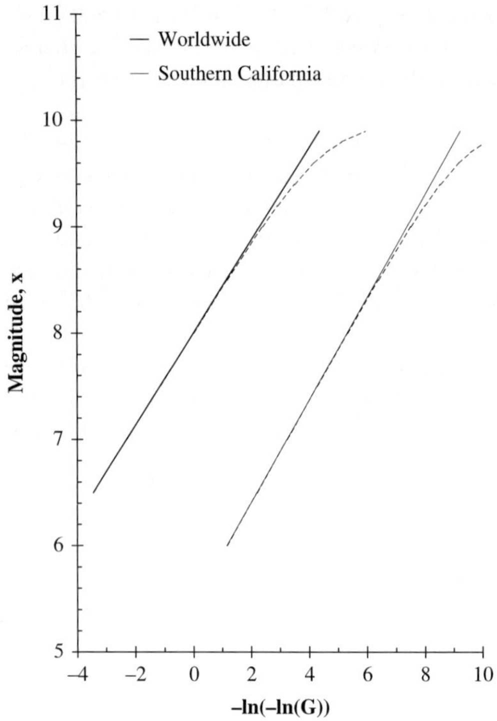

Figure 1.22. Probability distribution of maximum annual earthquake magnitude

ILLUSTRATION 1.25 (Earthquake Intensity in Rome). ►

The data of the Mercalli-Cancani-Sieber (MCS) index from the second part of Illustration 1.24 is used to show the Gumbel fit. Let X represent the MCS intensity for the metropolitan area of Rome, and let ν denote the observed number of earthquakes with intensity not less than x subdivided by the number of years of observation, that is, 980. One can estimate the values of b' and a in Eq. (1.122) by a simple linear regression of $\ln \nu$ against x. By considering intensities greater than 2, it is found that $b' = 1.10$, and $a = 4.21$. Thus the Gumbel distribution for the annual maximum earthquake MCS intensity of Eq. (1.124) has parameters $1/b' = 0.91$, and $b'^{-1} \ln(a) = 1.3$ for this area. The plot is shown in Figure 1.23.

Because of linearity between intensity and magnitude, the corresponding c.d.f. of annual maximum magnitude can be easily determined. ◄

The total energy E, measured in Joules, in the seismic waves generated by an earthquake can be related to its magnitude X by a log-linear relationship:

$$\ln(E) = 1.44\,X + 5.24. \tag{1.125}$$

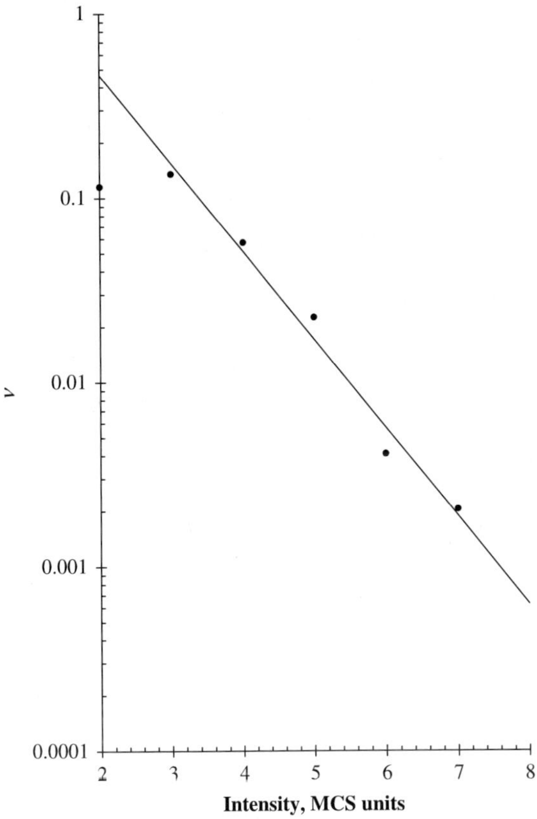

Figure 1.23. Variability of the mean annual number of earthquakes in Rome with MCS intensity greater than a specified value

The strain release associated with an earthquake is proportional to the moment of the earthquake, which can be related accordingly to its magnitude by using either a heuristic linear relationship or a theoretical log-linear law. The area of the rupture is also related to the moment by a power-law.

The assessment of seismic hazard, at a given site, requires the evaluation of ground motion acceleration at that site. This can be determined by combining intensity, or magnitude of earthquakes in the region, with the attenuation of epicentral magnitude, or intensity for a specified probability distribution of the distance from the epicentre. Therefore, one needs to estimate the spatial distribution of epicentral distance for earthquakes in the region, which depends on active faults or point sources. In many areas seismic risk maps are available for planning purposes. The expected intensity of ground shaking is shown by the *effective peak acceleration* (EPA).

For reviews of probabilistic seismic analysis, see, for example, [130, 52, 179, 230, 291, 193]. Wang and Ormsbee [298] make a limited comparison between probabilistic seismic hazard analysis, and flood frequency analysis.

1.4.2 Volcanic Eruptions

Volcanoes are the manifestation of a thermal process rooted deep inside the Earth from which heat is not readily emitted by conduction and radiation. A volcano is formed, as part of the heat eviction process, where the earth's crust opens and magma, a molten rock material that forms igneous rocks upon cooling, reaches out from huge pressure chambers. The magma pours out as lava, generally accompanied by a glowing plume and an avalanche of hot gases, steam, ash and rock debris. Some volcanoes seem to have erupted once only, whereas others have had several eruptions. The phenomenon is studied by scientists from various disciplines such as physics, geology, biology and meteorology.

The eruption of volcanoes occur simultaneously with earthquakes in some tectonic regions, where subterranean forces cause deformation near the earth's surface. Thus, tectonic plate theory can explain the location of most volcanoes, for example, the island areas, and the mountains alongside the Pacific. Volcanic eruptions, however, can occur far from plate boundaries as in Iceland, and in Wyoming, USA.

The hazard due to volcanoes is comparable to the one due to earthquakes, but there are some benefits from volcanic activities. The lava that spreads can subsequently aid some forms of agriculture. Furthermore, after a prehistoric eruption in the area of the present Yellowstone National Park, USA, the lava spread as far as eastern Nebraska, and is still a source of scouring powder for kitchens. Incidentally, it may be of interest to note that this park has many geysers, or hot springs, that sprout a column of hot water, and steam into the air. The most regular one is named Old Faithful because it has been performing once every 40 to 80 minutes, with heights of around 40 meters, for more than a century. Deep below, the pressure in a column of water makes the boiling point to be as high as 150°C. When bubbles of steam form, after boiling starts in the cyclic action and hot water spills from the vent, the pressure becomes less in the column down below. A lower boiling point is reached and, consequently, there is a sudden gush of water until the supply is depleted. Then, the conduit is replenished from ground water, and the process continues.

Worldwide, there are around 650 potentially active volcanoes on land. Many volcanoes seem to have erupted only a few times, whereas others have had several eruptions. Most of the volcanic activity is around the periphery of the Pacific Ocean. Some of the other notable volcanoes are in Hawaii, the Canary Islands and along the Mediterranean Sea. Going back historically, Mount Mazama, in Oregon, USA, erupted around 5,700 BC. Also, there had been a cataclysmic eruption of the Santorin Volcano about 100 kilometers north of Crete in the Mediterranean around 1,500 BC. Near Naples, Italy, the eruption of Vesuvius in 79 AD was the next major event recorded. In more recent times, the Mount Pelé event on the island of Martinique in the Caribbean in 1902 and the Krakatoa volcanic eruption on August 27th,1983 in Indonesia were two of the most violent. Another major explosion occurred on May 18th, 1980 at Mount Saint Helens, Washington, USA, causing a billion dollars of damage to the timber industry.

There are various hazards associated with volcanic eruptions. Some of the main concerns are pyroclastic flows of low viscosity and high density with temperatures possibly exceeding 600°C and speeds that can reach a few kilometers per second, toxic gas clouds containing hydron sulphide, carbon monoxide and carbon dioxide and ash falls causing structural and agricultural damages. Furthermore, volcanoes can cause avalanches, tsunamis and mudflows.

Risk evaluation from volcanic activity, depends on the past behavior of a volcano. For this purpose one should have a recorded history of eruptions and geological knowledge of the composition and structural formation of the cone. Prediction of the type of eruption also depends on past records but prediction of the time of occurrence is subject to considerable uncertainty, as in the case of earthquakes.

There is much variation in the sizes of volcanoes, and a single volcano can erupt in many ways. Some eruptions produce mainly ash or tephra, and others yield primarily liquid rock or magma as discussed initially. The conditions that produce given amounts of tephra or magma during an eruption are not well understood. Different types of volcanic eruption require different approaches. Thus it is much more difficult to quantify a volcanic eruption than an earthquake.

ILLUSTRATION 1.26 (Distribution of volcanic eruptions). ►

McLelland et al. [194] used the volume of tephra as a measure of the size of an eruption. They found a frequency-volume law for volcanic eruptions using data from 1975 to 1985 and also historic information on eruptions during the past 200 years. From this data [291] shows that the mean number ν of eruptions per year with a volume of tephra larger than x varies according to a power-law, that is,

$$\nu = c\,x^{-d}, \tag{1.126}$$

where $d = 0.71$, and $c = 0.14$ for the volume of tephra measured in cubic kilometers and the given data, see Figure 1.24. The frequency-volume law for volcanic eruptions is similar to the which frequency-magnitude Gutenberg-Richter law for earthquakes (see Subsection 1.4.1).

As in the case of earthquakes, one may assume that the number of eruptions in a year is a Poisson variate with mean ν. Then the probability that no eruptions occur in a year with tephra volume larger than x is $e^{-\nu}$. This is the same probability that the maximum tephra volume of a volcanic eruption in a year does not exceed x. Thus, the c.d.f. of annual maximum tephra volume is as follows:

$$\mathrm{G}(x) = e^{-\nu} = \exp(-c\,x^{-d}). \tag{1.127}$$

Note that this is the Fréchet distribution (see Eq. (1.48)) with shape parameter d and scale parameter $c^{1/d}$. ◄

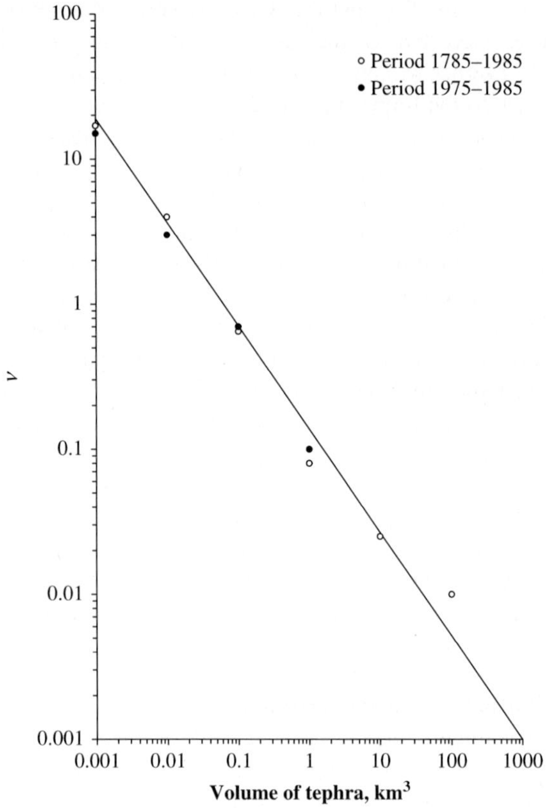

Figure 1.24. Variability of the mean annual number of volcanic eruptions with a volume of tephra greater than a specified value

Warning and evacuation of inhabitants are more effective than after earthquakes which are too sudden. Furthermore, inhabitants of valleys below volcanoes are warned to be alert to the low-frequency roar of an impending eruption. A guide to active volcanoes is given by [180, 98].

1.4.3 Tsunamis

Tsunamis are extraordinary occurrences of sea waves generally caused by seismic, volcanic, or other geological activities. The waves are long, and are originated by sudden near-vertical displacements, or distortions of the seabed under water. In the most severe cases, a tsunami arises from an earthquake along a submerged fault line. Associated with a massive earthquake, two plates of the earth's crust suddenly grind against each other after stress develops as one plate pulls down on the other. It is this instantaneous non-horizontal shifting of the plates, and the sudden uplifting of the seabed that causes the water to move up and down. The waves can move in all directions, outwards from the focus of the earthquake, and in this respect the waves are similar to those caused by a pebble falling into a shallow pond.

On average, there are more than one fatal tsunami per year. The Pacific Ocean accounts for 58% annually, whereas the Mediterranean Sea, Atlantic and Indian Oceans contribute 25, 12 and 5% respectively. The word is derived from Japanese and means "harbor waves" according to events on the Japanese coasts in which damage was caused in harbors. We note that on average Japan has around 20% of the world's major earthquakes (exceeding magnitude 6). In particular, Japan has had more than ten disastrous tsunamis during the past 400 years including the 1707 event which sank more than 1000 ships in Osaka Bay.

Other historical tsunamis, with associated earthquake magnitudes (in Richter scale), and causalities given, respectively, in parentheses, occurred in Lisbon, Portugal (9.0; 100,000) in 1755, Messina, Italy (7.5; 100,000) in 1908, and Chile (9.5; 2,000) in 1960. Along the western Coast of the Americas from Alaska to Chile numerous events have also been recorded. However, along the previously mentioned San Andreas fault, near the Californian Coast, some plate movements are close to horizontal and any resulting tsunami is much less effective.

Most recently, unprecedented death and destruction mainly in Indonesia, Sri Lanka, India and Thailand were caused by the Indian Ocean tsunami of December 26th, 2004. It is estimated that around 230,000 lives were lost, and more than ten countries were affected. This event began with an earthquake exceeding 9.3 in magnitude, the most severe in 40 years. It originated from a sudden movement along the Burma micro-plate, a part of the Indian plate, about 200 kilometers off the northwestern coast of the island of Sumatra in Indonesia, and was propagated 1250 kilometers northwards to the Andaman Islands. On this day two associated plates slid past each other over a width of about 25 meters in less than 15 minutes, about 30 kilometers below the seabed. This happened when strain energy accumulated over a very long period was suddenly released. The seabed, 10 kilometers below sea level in some areas, was suddenly distorted over an estimated area 100 kilometers wide and 400 kilometers long. Much of the seabed was probably deformed in this area. The resulting massive displacement of water was the likely cause for the waves to move at speeds over 800 kilometers per hour, taken as inversely proportional to the square root of the sea depth. The waves that ran westwards across the Indian Ocean reached East Africa within ten hours. On reaching shallow water, wave heights increased drastically to a maximum of 30 meters or more in some countries and inundation of coastal areas was extensive.

About 75% of tsunamis are caused by earthquakes below sea level as in the cases just cited, hence the alternative term seismic sea waves. Tsunamis can also be caused by other factors such as submarine landslides (8% of all cases) as, for example, in Valdez, Alaska in 1964 and in Sugami Bay, Japan in 1933. Elsewhere, a quake magnitude of 7.1, less severe than other devastating earthquakes, occurred in Papua New Guinea during 1998 and apparently caused mudslides. These led to a severe tsunami in the vicinity resulting in a death toll exceeding 1,500 inhabitants.

Another less frequent cause of tsunamis is an avalanche into a sea, such as the one that happened during 1958 in Alaska. A volcanic eruption can also initiate a tsunami (5% of all cases) as on Krakatoa Island near Java, Indonesia following an eruption in 1883, with the death toll around 40,000. Other less frequent causes are meteor strikes and underwater sediment slides.

The three main aspects of tsunamis to consider are their impact at generation, the deep sea water type of propagation, and the effects on approaching coastal areas through shallow waters. In the open seas the extraordinary waves of a tsunami are hardly visible, rarely exceeding one meter in height, but they reach great heights when shallow waters are reached. The wavelengths of ordinary sea waves do not generally exceed 100 meters in open seas. The same characteristic of a tsunami can be in the range 100 to 200 kilometers, much greater than the sea depth. The times between the passage of consecutive troughs are measured in a few minutes, or less. A tsunami propagates in sea as a gravity wave obeying the classical laws of hydrodynamics. The low amplitude waves (generally 0.3 to 0.6 meters) cause the system to conserve much energy, minimizing work done against gravity. Consequently very long distances can be traversed, provided there is sufficient initial force, until it reaches land. The time interval between tsunami crests at a coast line is generally around 10 to 15 minutes. Besides, if the tide level is high when a tsunami wave reaches a coast the further the water travels inland.

As already stated, wave heights increase drastically when shallow water is reached near a shoreline and the speed of propagation is sharply reduced with the conversion of kinetic energy to potential energy. The topography of the bottom of the sea affects the wave height; a long and shallow approach to the seashore gives rise to higher waves. An undamaged coral reef can act as a breakwater and reduce the effect of a tsunami, as known from the experience along part of the southwestern Coast of Sri Lanka during the December 26th, 2004 event. Thus there are geographical variations in coastal configurations of the sea level.

It is not well understood how the coastal shelf waters begin to oscillate after a rise in sea level. Most of the destruction is caused by the first 3 to 5 major oscillations but the movements may continue for more than a day. The first arrival may be a peak or it may be a trough, which draws people to view the sea bottom, followed by a peak.

The speed of propagation at sea is given by $\sqrt{gD}$ (cm/s), where $g = 981$ cm/s^2 is the acceleration due to gravity, and D (cm) is the ocean depth. Along a shoreline, tsunami waves can reach heights above 10 meters; crest heights of 25 meters were noted on the Sanriku Coast in Japan in 1933. The magnitude or severity of a tsunami is generally related to the maximum height of the wave. For a coastal maximum wave height of h meters, the *magnitude* of a tsunami can be defined as $M = 3.32\ \ln(h)$.

A basic type of risk analysis can be done by considering the maximum wave height reached by each tsunami in a particular risk-prone area and the numbers of occurrences of tsunamis greater than various heights over a time period. For example, on a particular Japanese coastal region the chance of a wave height greater than 5 meters during a 25 year time span is 60 percent. This may not seem as alarming as a tornado or a hurricane but the effects of a tsunami can be devastating as observed in the Pacific and Indian Oceans.

In the open seas, crest heights of tsunami waves are hardly noticeable as already stated, and are separated by distances exceeding 100 kilometers. Therefore, some special methods of detection, and warning need to be devised. In an advanced system, networks of seismographs and tidal gauges are linked to radio and tidal stations with high speed computers that simulate the propagation of tsunami waves. Then, warnings are issued to nations at risk, if a system is in operation. This may be in the form of an assessment of the hazard and risk involved.

It is the prerogative of the respective governments to warn the public. They should prepare communities for disasters. The people should know, for example, that when the sea is drawn down rapidly, it is a clear sign that a tsunami is approaching. They should be lead to higher ground, or other appropriate places, away from the sea, which may seem counter-intuitive to some people. What happened along the coastal areas in Sri Lanka, and in many other countries during the 2004 event, is a tragedy that could have been easily avoided; there was sufficient time for warnings in contrast to the situations in countries much closer to the earthquake.

A tsunami generally takes time to traverse an ocean. Therefore a warning system can be put in place, after scientific analysis of seismic and pressure sensor data. This makes it possible to evacuate coastal areas. From Chile to Hawaii, for example, a tsunami takes about 15 hours to travel. Pacific travel times are expressed as 10 hours on average. Note that there will be sufficient energy left to cause possible death and destruction after reaching a shoreline. During the 1990's there were more than 4000 deaths in this zone consequent to ten tsunami events. Thus the incentive for warning systems is higher in the Pacific.

However, at present there is a high probability of a false alarm arising from a warning system; this may be as high as 75 percent in some cases. One of the problems is that not all earthquakes under the sea cause tsunamis. Some seismologists take the critical magnitude as 6.5. However, less severe earthquakes have sometimes caused a higher loss of life. It is important to have faster seismological techniques for making accurate estimation of the magnitudes and characteristics of earthquakes liable to cause tsunamis. For instance, the magnitude of the Sumatra-Andaman Islands earthquake of 2004 was estimated in the initial hours as around 8; subsequently it has been revised to 9.3. Besides, tsunamis can be caused by other natural events as already cited.

Note also that signatures associated with the 2004 tsunami that originated near Sumatra were recorded via infrasound arrays in the Pacific and Indian Oceans. These are part of the International Monitoring System of the Comprehensive Nuclear Test Ban Treaty. The sound may be radiated from the ocean surface during the tsunami initiation and propagation or generated by the vibration of landmasses caused by an earthquake. This is a potential new source of help.

It is reported that the 40-year old Pacific Ocean warning system, for which 26 countries are responsible, will be applied to cover the Indian Ocean by 2007.

Currently the Pacific network has 130 remotely reporting sea level gauges with several deep-ocean pressure sensors.

The new types of warning systems planned may have links to satellites. These are likely to incorporate well-located buoy-anchored detectors placed deep below sea levels. Thus warnings can be based on sea level data. Those, that depend only on seismic data, are more liable to be false eventually. Following the practice in the Pacific Ocean warnings can be cancelled within 2 to 3 hours if the subsequent evidence is contrary. Thus there will be a much higher chance of success in forecasting the passage of a tsunami.

1.4.4 Landslides

Landslides occur mainly in mountainous, or hilly terrains. A downward mass movement of earth and rock on unstable slopes is involved. There are variations with differences in types of materials, and degree of slope. In the most common cases, the preconditions are that there is a critical increase of pore water pressures in a sufficiently thick, and inclined soil layer, which consists of various geological materials (such as clays, sands, and gravels) below a steep, or undermined surface with sparse vegetative cover. Landslides can be initiated by rainfall, earthquakes, volcanic activity, changes in groundwater, disturbance and change of a slope by man-made construction activities, or any combination of these factors. Landslides can also occur underwater, causing tsunami waves and damage to coastal areas. These landslides are called submarine landslides. Landslides are widespread throughout the world and can cause severe loss of property and lives and damage to buildings, dams, bridges, services and communication systems. The influence of rain on landslides is more when high intensity rainfall occurs over a long period of hours. The approach to risk evaluation, in this geotechnical problem, is similar, and sometimes identical, with that applied to debris flow (a rapid downslope flow of debris over a considerable distance) with its associated hazard mapping. Moreover, landslides occur over short distances, usually less than a kilometer, whereas debris flow occurs over distances exceeding one kilometer. The subjects are generally interconnected. In the classification of landslides, the material type and velocity, which is linked to soil behavior, are important criteria. Also the distance traveled, in relation to the velocity, gives a measure of the effect of a landslide.

Seismic effects can also be a major cause. For instance, the devastating 1910 landslide in the Lake Taupo area of New Zealand had debris flow with a mean velocity of 8 m/s over a travel path of 1.5 kilometers and caused a large loss of life. This was originated either by a seismic event or by geothermal cooling in the Hipua Thermal Area as described by [134]. The largest landslide recorded during the past century occurred during the 1980 eruption of Mount St. Helens, a volcano in the Cascade Mountain Range in the State of Washington, USA. It was estimated that 2.8 cubic kilometers of soil and shattered rock were involved. Elsewhere, a slide of about 27 million cubic meters of material occurred during an earthquake in Madison County, Montana, USA in August 1959.

Human activities, such as the construction of highways and embankments, can also lead to landslides as in undercutting and steepening of slopes. For instance, severe landslides occurred in the excavation of the Panama canal, one of which involved 400,000 cubic meters of material. Further, in northern Italy, an impounding reservoir had been constructed by means of a thin dome or cupola dam between the steep slopes of the Vajont Valley of the Piave River. It appeared that no proper investigations had been made previously for possible landslides. In November 1960, shortly after completion, there was a partial movement of the side slope above the dam. The movements continued until there was a large slide in October 1963, and an estimated 250 cubic kilometers of soil, and rock fell into the reservoir. Consequently, the dam overtopped by 100 meters. The structure survived, but the sudden release of a very large quantity of water killed 3,000 inhabitants in the valley below. Subsequent geological surveys revealed that the sliding failure occurred in highly fractured oolitic limestone. Artesian pressures along the shear surface also affected the instability of the slope. Similarly, the coal-mining town of Aberfan, Wales in the United Kingdom was partly engulfed in October 1966 by a slope failure and resulted in a loss of 144 lives. Another cause of landslides is deforestation; in Indonesia, for example, heavy tropical downpours cause dozens of landslides annually in highly populated areas.

The risk to life, and property caused by avalanches, discussed in Subsection 1.4.5, is a closely related subject. Many of the conditioning factors of landslides have strong random elements such as the non-homogeneity of the soil strata, variations in the water contents, and inadequate knowledge and inconsistencies of the physical and topographical characteristics. Hence, there is justification for the use of statistical and probabilistic methods in the assessments of landslides, and the threats imposed. Calculations such as the number of landslides that occur during a specified time in a particular area, may be made assuming a Poisson distribution, particularly those affecting large areas.

ILLUSTRATION 1.27 (Occurrence of landslides). ►

Suppose the objective is to calculate the probability of the occurrence of at least one landslide, during a fixed time Δt. Let N be the number of landslides, that occur during time Δt in the given area. Assuming a Poissonian chronology for the number of the occurrences, its p.m.f. is given by Eq. (1.107), where ν denotes the mean number of landslides in a time interval Δt. Hence,

$$\mathbb{P}\{N \geq 1\} = 1 - \mathbb{P}\{N = 0\} = 1 - e^{-\nu}. \tag{1.128}$$

For the estimation of the Poisson parameter ν, suppose that n landslides were observed historically during an interval Δt (say, 200 years); then the maximum likelihood estimate is $\widehat{\nu} = n/\Delta t$. ◄

The above procedure is applicable when there is sufficient data on landslides from a particular area. Alternatively, a physically based method can be used to calculate the failure over an infinite plane with a certain slope (called *infinite slope analysis*), as given, for example, by [39] following [279]. However other mechanisms of failure like rockfalls, rock avalanches, rotational slides, debris flow, earthflows, sagging are treated differently. For the terminology on landslides we follow widely accepted classifications, for instance, by Cruden and Varnes [54].

ILLUSTRATION 1.28 (A physical approach to landslides). ►

The resistance stress τ_R at the bottom of a soil layer, that moves during a landslide over an infinite plane with a certain slope, is given by

$$\tau_R = c + N_S \tan\phi, \tag{1.129}$$

where c is the cohesion, ϕ is the internal angle of friction, and N_S is the effective normal stress.

Consider a soil layer of unlimited length with specific gravity G_s, thickness H, and porosity n_p that is inclined at an angle α. Let the water depth be h. For a given degree of saturation s of the soil in the unsaturated zone, specific weight of water γ_w and a relative water depth $r = h/H$, the resistance stress can be expressed as

$$\tau_R = c + \left[(1-n_p)(G_s - d) + sn_p(1-d)\right]\gamma_w H \cos\alpha \tan\phi, \tag{1.130}$$

in which $d = r$ if $r < 1$, and $d = 1$ if $r \geq 1$.

At the point of equilibrium, or given a limiting or critical state, that is before a landslide begins, the resistance stress τ_R is equal to the driving stress τ_D. This acts on the bottom of the soil layer and is equivalent to the sum of the weight components of water and solid particles along the inclined plane. The driving stress can be written, using similar notation, as

$$\tau_D = \left[(1-n_p)(G_s - d) + sn_p(1-d) + r\right]\gamma_w H \sin\alpha. \tag{1.131}$$

Let the difference between the resistance stress and the driving stress be the state function W, also known as *safety margin*. Thus

$$W = \tau_R - \tau_D. \tag{1.132}$$

A failure occurs when $W < 0$.

We note that except for the specific gravity G_s of the soil, and the specific weight γ_w of water, the other 7 parameters can have high variances. These are the degree of soil saturation s, porosity n_p, internal angle of friction ϕ, water depth h, thickness of the soil layer H, angle of inclination of the soil layer α, and cohesion c. Usually, these are not interrelated. Let us therefore treat these as independent random variables.

By using the *first order second moment* (FOSM) method [168, pp. 583–584], the nonlinear function of random variables W can be approximated by using a Taylor series expansion about the respective mean values:

$$\begin{aligned} W(x_1, x_2, \ldots, x_7) \approx\; & W(\mathbb{E}(X_1), \mathbb{E}(X_2), \ldots, \mathbb{E}(X_7)) + \\ & \sum_{i=1}^{7}(x_i - \mathbb{E}(X_i)) \left[\frac{\partial W}{\partial x_i}\right]_{x_i = \mathbb{E}(X_i)}, \end{aligned} \tag{1.133}$$

in which the partial derivatives are obtained at the respective means $\mathbb{E}(X_i)$ of the variables. It follows from the foregoing equations that the mean value of the state function W is obtained as

$$\mathbb{E}(W) = \mathbb{E}(\tau_R) - \mathbb{E}(\tau_D), \tag{1.134}$$

in which

$$\begin{aligned}\mathbb{E}(\tau_R) = \mathbb{E}(c) + \big[(1-\mathbb{E}(n_p))(G_s - \mathbb{E}(d)) + \\ \mathbb{E}(s)\,\mathbb{E}(n_p)\,(1-\mathbb{E}(d))\big]\cdot \\ \gamma_w \mathbb{E}(H) \cos \mathbb{E}(\alpha) \tan \mathbb{E}(\phi),\end{aligned} \tag{1.135}$$

and likewise,

$$\begin{aligned}\mathbb{E}(\tau_D) = \big[(1-\mathbb{E}(n_p))(G_s - \mathbb{E}(d)) + \\ \mathbb{E}(s)\,\mathbb{E}(n_p)\,(1-\mathbb{E}(d)) + \mathbb{E}(r)\big]\cdot \\ \gamma_w \mathbb{E}(H) \sin \mathbb{E}(\alpha).\end{aligned} \tag{1.136}$$

Because the relative water depth r is given by h/H, we take $\mathbb{E}(r) = \mathbb{E}(h)/\mathbb{E}(H)$ as the relative depth. When $\mathbb{E}(r) < 1$, $\mathbb{E}(d) = \mathbb{E}(r)$, and when $\mathbb{E}(r) \geq 1$, $\mathbb{E}(d) = 1$. Also, the standard deviation of the state function W, $\mathbb{S}(W)$, is obtained from the variances $\mathbb{V}(X_i)$'s as

$$\mathbb{S}(W) = \sqrt{\sum_{i=1}^{7} \mathbb{V}(X_i) \left(\left. \frac{\partial W}{\partial x_i} \right|_{x_i = \mathbb{E}(X_i)} \right)^2}. \tag{1.137}$$

A reliability index β can be defined as

$$\beta = \mathbb{E}(W)/\mathbb{S}(W). \tag{1.138}$$

Assuming the variables are Normally distributed, the probability p_{LS} of occurrence of a landslide can be expressed as

$$p_{LS} = 1 - \Phi(\beta). \tag{1.139}$$

where Φ is the standard Normal c.d.f.. As expected, p_{LS} increases with the relative depth $\mathbb{E}(r)$. ◀

Soil mass movements are the major landform shaping processes in mountainous and steep terrain. Shallow landslides result from infrequent meteorological, or seismic, events that induce unstable conditions on otherwise stable slopes, or accelerate movements on unstable slopes. Thus, the delicate equilibrium between the resistance of the soil to failure and the gravitational forces tending to move the soil downslope can be easily upset by external factors, such as rainstorms, snowmelt, and vegetation management. The major triggering mechanism for slope failures is the build-up of soil pore water pressure. This can occur at the contact between the soil mantle and the bedrock, or at the discontinuity surface determined by the wetting front during heavy rainfall events. The control factors of landslide susceptibility in a given area may be subdivided into two categories: quasi-static and dynamic. The quasi-static variables deal with geology, soil geotechnical properties, slope gradient, aspect and long term drainage patterns. The dynamic variables deal with hydrological processes and human activities, which trigger mass movement in an area of given susceptibility. Shallow landslides hazard assessment is based on a variety of approaches and models. Most rely on either multivariate correlation between mapped (observed) landslides and landscape attributes, or general associations of landslides hazard from rankings based on slope lithology, land form or geological structure. Antecedent precipitation amounts, and daily rainfall rate, are further triggering factors of shallow landsliding.

The statistical approach can provide an insight of multifaced processes involved in shallow landsliding occurrence, and useful assessments of susceptibility to shallow landslide hazard in large areas. But the results are very sensitive to the dataset used in the analysis, and it is not straightforward to derive the hazard (i.e. probability of occurrence) from susceptibility. As an alternative to the use of probabilistic concepts, a fuzzy-approach is possible. But the results are very sensitive to the dataset used in the analysis, and it is not straightforward to derive the hazard (i.e. probability of occurrence) from susceptibility. The intensity and duration of rainfalls that trigger landslides can be analysed using critical rainfall threshold curves, defined as envelope curves of all rainfall triggering landslides events for a certain geographic area. Due to the lack of a process-based analysis, this method is unable to assess the stability of a particular slope with respect to certain storm characteristics and it does not predict the return period of the landslide-triggering precipitation.

Another approach deals with models coupling slope stability equation with a hillslope hydrological model. This can provide an insight of triggering processes of shallow landslides at the basin scale, also accounting for the spatial variability of the involved parameters. For example, [197] developed a simple model for the topographic influence on shallow landslides initiation by coupling digital terrain data with near-surface through flow and slope stability models. Iverson [149] provided an insight of physical mechanism underlying landslide triggering by rain infiltration by solving the Richard's equation. D'Odorico et al. [73] coupled the short term infiltration model by [149] and the long term steady state topography driven subsurface flow by [197] and analyzed the return period of landslide triggering precipitation using hyetograph at different shapes. Iida [146] presented a hydrogeomorphological model considering both the stochastic character of rainfall intensity and duration and the deterministic aspects controlling slope stability using a simplified conceptual model. Rosso et al. [246] improved the modelling approach by [197] to investigate the hydrological control of shallow landsliding, and coupled this model with the simple scaling model for the frequency of storm precipitation by [27] to predict the return period of the landslide-triggering precipitation. This can help understanding the temporal scales of climate control on landscape evolution associated with the occurrence of shallow landslides, see Figure 1.25.

1.4.5 Avalanches

An avalanche is a large mass of snow that moves on a mountain slope causing destruction in its wake. Landslides and avalanches of snow have similar causes. As in the case of soil, snow has some complex interrelated properties such as density, cohesion and angle of internal friction. Furthermore, after snow falls, and accumulates over a long slope, it will remain stationary if the shearing strength at

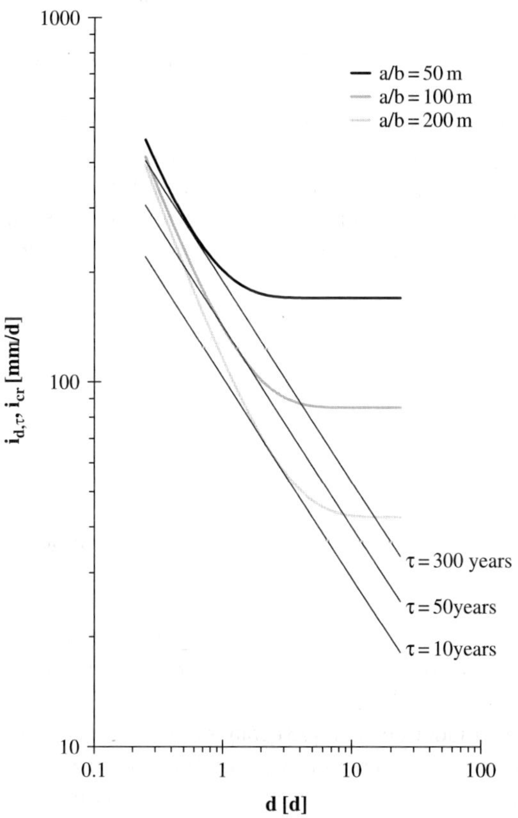

Figure 1.25. Coupling of the relationship between critical rainfall rate i_{cr} and duration d of the precipitation triggering shallow landslides (*thin lines*), with the Intensity-Duration-Frequency curves $i_{d,\tau}$, for the failure return period τ (*thick lines*), under specified hillslope and climate conditions. Different values of topographic index a/b (i.e. the ratio between drainage area to the contour length) indicate the fundamental role of hillslope topography (modifed after [246])

all depths is in excess of the shearing stress caused by the weight of snow and the angle of repose. Subsequently, at some critical depth of snow, the frictional resistance of the sloping surface will be overcome and movement of the snow mass will commence.

The trigger mechanism may be spring rains that loosen the foundation or the rapid melting caused by a warm dry wind (Föhn); other possible causes are thunder or blasting or artillery fire that induce vibrations. Some avalanches commence while snow is still falling. Avalanches of wet snow are particularly dangerous because of the huge weight involved, its heavy texture, and the tendency to solidify when movement stops. Dry avalanches also cause danger because large amounts of air are trapped and this induces fluid motion. Avalanches may include large quantities of rock debris; they can move long distances apparently on thin cushions of compressed air.

As regards soil, the complexities of the material notwithstanding, some aspects of its strength are known and the shearing strength, for instance, can be estimated. On the other hand, there are some physical processes that affect the mechanical properties of snow that are not well known.

The structure, type and interrelationships of the snow particles change over time with the effects of pressure, temperature and migration of water vapor, for instance. The shearing strength of new snow is similar to that of a dry soil but as time passes the density and cohesion properties will change and vary with the depth of snow. As already mentioned, there may come a time when the snow layer becomes unstable and begins to move as an avalanche. Apart from the depth of snow the commencement of an avalanche is affected largely by the permanent features of the topography and the transience of the weather.

Because the occurrence, frequency, and type are affected by meteorological factors, a variety of possible avalanches may develop after snow falls in winter giving rise to a classification system as in the case of landslides. In Japan, for instance, avalanches are classified in type according to the weight of material moved. The logarithm of the mass of snow in tons, a MM scale, is used; this is similar to that of the Richter magnitude scale for earthquakes as discussed in Subsection 1.4.1. Such a MM scale for avalanches varies from less than 1 (small) to greater than 5 (very large). Another measure is the velocity of travel of an avalanche. This depends on the angle of slope, the density and the shearing strength of the snow and the distance traveled. It can vary from 1 kilometer per hour to 300 kilometers per hour.

Avalanches impose a continuous threat to life and property in mountainous areas subject to heavy snowfalls particularly in temperate regions. We note that many villages, settlements and buildings have been destroyed by avalanches in the Alps and other regions. For instance, on January 10th, 1962, on the Andes mountains of South America a mass estimated to contain 3 million cubic meters of snow and ice broke from the main glacier of Mount Huascaran, Peru and fell a vertical distance of 1 kilometer initially and eventually 3.5 kilometers in height, destroying a town, villages, bridges and highways in its wake. A similar event occurred 8 years later but it was initiated by an earthquake. This area has had a long history of avalanches. In the Canadian Rocky Mountains zone of British Columbia a very large avalanche occurred on February 18th, 1965. A whole camp, named Camp Leduc, was destroyed with a loss of many lives. Because it would have been possible to predict the avalanche occurrence and path in the area, this tragedy could have been avoided with proper location and construction of appropriate structures.

By using the Icelandic procedure [161], in Italy, Switzerland, and some other countries, the reinforcement of structures reduces the risk of destruction over the whole area, as expected. However, the risk levels are unacceptably high in inhabited areas. This may be reduced by constructing retaining works of steel and concrete to protect at least part of the release or ideally to provide balancing forces to avert ruptures. In the rapture zone, snow fences and wind baffles can be used to check snow settlement on leeward slopes. For diverting flowing snow, vulnerable buildings in some towns of Switzerland are constructed like prows of ships. The Swiss have long relied on forests to stop or retard small avalanches affecting mountain villages; however, because of acid rains nearly half the trees have been destroyed or damaged in many areas. The best protection for highways and railroads is of course to construct tunnels.

With an increasing number of people using Swiss ski resorts, avalanche studies are an important aspect of Alpine climatology. During early spring and other periods of threatening weather, the research station at Davos publishes daily bulletins as warnings to villages and tourists of avalanches; these average 10,000 annually. Similarly, in the United States, the Forestry Service of the Department of Agriculture is responsible for forecasting, monitoring and control of avalanches.

As in the treatment of landslides of the previous section, statistical methods are used in the evaluation of the risk involved (see Section 1.3). Let us assume that one has knowledge of the p.d.f. f_U of the velocity U of avalanches at a given site. This is generally termed the *hazard* component of risk. Also suppose that an associated *impact* relationship $u \mapsto \zeta(u)$ is known. This refers to the possibility of consequential death inside a building subject to avalanches: it depends on the situation, construction and foundation of the building. Suppose that the event of interest is $E = \{u_1 \le U \le u_2\}$, where $0 < u_1 < u_2 < \infty$. Then, according to Eq. (1.120), the risk of death applied to working in a building situated in the flow path of the avalanche can be written as

$$\mathbf{R}(E) = \frac{1}{\int_0^\infty \zeta(x) f_U(x)\, \mathrm{d}x} \int_{u_1}^{u_2} \zeta(x) f_U(x)\, \mathrm{d}x. \tag{1.140}$$

Different types of impact functions ζ can be used. In Italy and elsewhere these have been often based on the experience of Icelandic avalanches where, for instance, data have been collected from the 1995 events of Sudavik and Flateyri. The relationship for reinforced structures in Iceland given by [161] seems to provide conservative estimates.

ILLUSTRATION 1.29 (A physical approach to avalanches). ►
For the estimation of the hazard component of risk, f_U, [11] proposed a method following the work by [93] in debris flow hazard mapping. This is based on the calculation of the survival function of U as follows:

$$1 - F_U(u) = \int_0^{\infty} \mathbb{P}\{U \geq u \mid H = h\}\, f_H(h)\, \mathrm{d}h, \tag{1.141}$$

where H indicates the *avalanche release depth* and f_H is the corresponding p.d.f.. For this assessment the Swiss assumption of three days of prior snow depth is used. The conditional probability in Eq. (1.141) is evaluated using an avalanche dynamic model for velocity, with $U < 24$ m/s, and f_H is estimated by statistical inference from snowfall data. The risk estimation procedure was applied to the Val Nigolaia region of the Italian Alps in the province of Trento. The risk was evaluated along the main flow direction of the Val Nigolaia, with and without velocity thresholds considering the typical runout distance, that is the furthest point of reach of the debris. This aspect is further discussed later. ◄

It can be assumed that in high mountainous regions avalanches occur as Poisson events, with a mean number ν in a given time interval. Then the size, that is the volume of snow moved by the avalanche, is an exponentially distributed variate X with scale parameter b. This parameter depends on local topography and other factors as discussed. Using the average density of the snow, the weight of material moved, as discussed previously, can be assessed from the size.

ILLUSTRATION 1.30 (The size of an avalanche). ►
Let us assume that the exponential parameter b, used to model the size of the avalanche, varies uniformly in a certain area from, say, from b_1 to b_2. Using Eq. (1.113) it follows that

$$F(x) = \int_{b_1}^{b_2} \frac{1 - e^{-x/b}}{b_2 - b_1}\, \mathrm{d}b = 1 - \frac{b_1 b_2}{x} \frac{e^{-x/b_2} - e^{-x/b_1}}{b_2 - b_1}. \tag{1.142}$$

Then, the associated extreme value c.d.f. of the size, for an average number ν of avalanches occurring in a year within the area, is given by

$$\mathrm{G}(x) = \exp\left[-\nu \frac{b_1 b_2}{x} \frac{e^{-x/b_2} - e^{-x/b_1}}{b_2 - b_1}\right], \tag{1.143}$$

which is obtained as a contagious extreme value distribution (see Subsection 1.2.4). It is found the following estimate of the parameters: $\widehat{\nu} = 9$, $\widehat{b}_2 = 1000$ m^3, and $\widehat{b}_1 = 100$ m^3. Thus

$$\mathrm{G}(x) = \exp\left[-1000 \frac{e^{-0.001x} - e^{-0.01x}}{x}\right]. \tag{1.144}$$

This is shown in Figure 1.26.
◄

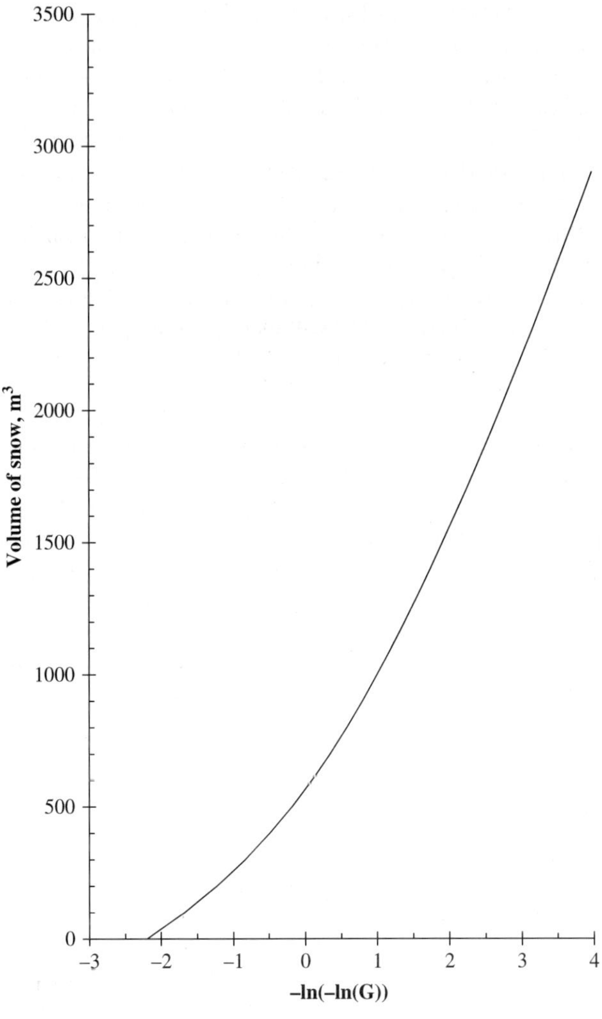

Figure 1.26. Probability distribution of avalanche volume

Further studies are required to estimate the vulnerability relationship of Eq. (1.140) as applied to typical buildings in the Alps and other specific areas. The vulnerability component of risk is often inadequately specified. Knowledge on how the impact of an avalanche damages structures and leads to deaths is still limited. The effects may be different in some countries from those in Iceland.

As regards the runout distance, that is the furthest reach of the avalanche, [177] gave useful empirical calculations of the maximum distance from topographic parameters using data from 423 avalanches. Additional regression analysis of a nonlinear type using topographic variables such as the ground slope are reported by [191, 192].

In the Italian Alpine areas of Lombardia, Piemonte and Veneto regions, and the autonomous provinces of Bolzano and Trento, risk maps have been made to safeguard lives and property in avalanche-prone areas. These maps (scale 1:25,000) of probable localizations of avalanches for land-use planning have been prepared by Piemonte region in 2002. In northeastern Italy, the Forestry Corporation had previously collected long series of data regarding damages caused to the wood from avalanches. These have been used in the calculations with other historical information. In addition, daily records of new snow and snow depth are available at numerous sites in series of 17 to 44 years. Within the runout distances, three zones are demarcated on the maps following the Swiss practice, but with thresholds modified according to local specific situations. The Gumbel distribution, Eq. (1.47), has been used in calculating the return periods [5].

The risk zones are:

1. **Red or high risk zone.** Expected avalanches have either for a return period of 30 years an impact pressure $P = 3$ kPa, or for a return period of 100 years, $P = 15$ kPa. Land use is restricted here and no new constructions are allowed.
2. **Blue or moderate risk zone.** Likewise, for a return period of 30 years, $P = 3$ kPa, or for a return period of 100 years, $P = 15$ kPa. New buildings to be constructed here should be adequately reinforced; also, low buildings are specified.
3. **Yellow or low risk zone.** Likewise, for a return period of 100 years, $P < 3$ kPa. New constructions are allowed here with minor restrictions.

Monitoring and evaluation plans are prepared for the safety of people in the red, blue and yellow zones. In Piemonte region, historical information on avalanches in the province of Torino is available at the internet site:

`http://gis.csi.it/meteo/valanghe/index.html`

The τ-year return period avalanche at a given site is computed by deriving the avalanche run out and magnitude from the statistics of snow cover, mostly related to avalanche event magnitudes, i.e. snow fall depth in the days before the event. The snow depth in the avalanche release zone is often assumed to coincide with the snow depth precipitation in the three days before the event, or three days snow fall depth, H_{72} [25, 4]. This is first evaluated for a flat surface, and then properly modified to account for local slope and snow drift overloads [12, 10]. Accordingly, avalanche hazard mapping based on these criteria requires to input the τ-year quantile of H_{72} (e.g., for $\tau = 30$ and $\tau = 300$ years, as a standard) for each avalanche site. The estimation of the τ-year quantile of H_{72} is often carried out by fitting data observed at a gauged site with an extrene value distribution using the "block" method, i.e. the maximum annual observed values of H_{72} for the available years of observation at a given snow gauging station. Both the Gumbel and the GEV distributions are adopted for the purpose. In Italian Alps, except for a very few cases, only short series of observed snow depth are available, covering a period of about 20 years [19]. This is also applicable to other countries. However, in the Swiss Alps, daily snow data series are generally available for periods of about 60 to 70 years [175]. Regionalization methods (similar to those adopted for flood frequency analysis, see Subsection 1.4.9) can be used to overcome the lack of observed data. These include the index value approach by [18]. The advantage of using this approach stems with the reduction of uncertainty of quantile estimates because of unadequate length of site samples available.

1.4.6 Windstorms

Windstorms account for 70 percent of insurers' total claims from natural catastrophes. For example, insurance losses of \$17 billions were sustained in the 1992 Hurricane Andrew on the Gulf Coast of USA, a historic record. Regarding economic losses, the percentages for windstorms and earthquakes are 28 and 35 respectively. Where human fatalities are concerned earthquakes at 47 percent are the main cause followed very closely by wind storms at 45 percent. In the design of tall buildings (long span bridges and several other wind-sensitive structures) engineers need to provide for resistance to counter the effects of high wind speeds. Quite frequently, the Weibull distribution is used to model wind speeds, following the European practice.

ILLUSTRATION 1.31 (Probabilities of wind speeds). ►

The Converse Weibull distribution (see Eq. (1.56)) is often used as a model for wind speeds. The p.d.f. is

$$f(x) = \frac{\gamma}{b}\left(\frac{x}{b}\right)^{\gamma-1} \exp\left[-\left(\frac{x}{b}\right)^{\gamma}\right], \tag{1.145}$$

for $x > 0$, with scale and shape parameters b and γ. In Northern Europe, estimates of γ are close to 2. In such situations the *Rayleigh* distribution can be used, with p.d.f.

$$f(x) = \frac{2x}{b^2} \exp\left[-\left(\frac{x}{b}\right)^{2}\right]. \tag{1.146}$$

The theory is also applicable to calculations of available wind power. This is an environment friendly and inexpensive energy resource. ◄

ILLUSTRATION 1.32 (Hurricane winds). ►

Over a 30-year period, maximum wind speeds of 12 hurricanes have been recorded. These have a mean of $\overline{X} = 40$ m/s, and a coefficient of variation $\widehat{V} = 0.32$. In order to determine the wind speed with a 100-year return period, one may proceed by assuming that the number of hurricanes per year is Poisson distributed. The mean rate of occurrence is thus $\widehat{\nu} = 12/30 = 0.4$. Also one may assume that the wind speed is Pareto distributed. Its c.d.f. is given in Eq. (1.88). Using the method of moments, the shape parameter γ is estimated as

$$\widehat{\gamma} = 1 + \sqrt{(1 + 1/\widehat{V}^2)} = 1 + \sqrt{(1 + 1/0.32^2)} = 4.281.$$

The scale parameter b is estimated as $\widehat{b} = \overline{X}(\widehat{\gamma}-1)/\widehat{\gamma} = 40 \times 3.281/4.281 = 30.65$ m/s. One also assumes that the annual maximum wind speeds have a Fréchet distribution given by Eq. (1.48). An estimate of Fréchet parameters, b' and γ, is obtained from the one of the Pareto parameters, b and γ, as $\widehat{b'} = \widehat{b}\widehat{\nu}^{1/\widehat{\gamma}} = 30.65 \times 0.4^{1/4.281} = 24.75$ m/s, and $\widehat{\gamma} = 4.281$. For a return period τ_x of 100 years, the Gumbel reduced variate $y = -\ln[-\ln(1-1/\tau_x)]$ is 4.600. Hence the 100-year quantile of the annual maximum wind speed is $\widehat{x}_{100} = \widehat{b'}\exp(y/\widehat{\gamma}) = 24.75\exp(4.600/4.281) = 72.48$ m/s. ◀

Structural engineers frequently adopt the highest recorded of wind speeds, or values with return period of 50 years, for most permanent structures. The return period is modified to 25 years for structures having no human occupants, or where there is a negligible risk to human life, and 100 years for structures with an unusually high degree of hazard to life, and property in case of failure.

The probability distribution describing extreme wind speed applies to homogeneous micrometeorological conditions. Thus, one should consider initially the averaging time, the height above ground, and the roughness of the surrounding terrain. If different sampling intervals are used, when observations are made, the entire sample must be adjusted to a standard averaging time, say, a period of 10 minutes. Also, if there is a change in the anemometer elevation, during the recording period, the data must be standardized to a common value, such as 10 m above ground, using a logarithmic law to represent the vertical profile of wind speed. With regard to roughness, wind data from different nearby locations must be adjusted to a common uniform roughness over a distance of about 100 times the elevation of the instrument by using a suitable relationship. In addition, one must consider sheltering effects, and small wind obstacles. Besides, in modeling extreme wind speeds, one must also distinguish cyclonic winds from hurricane and tornado winds, because they follow different probability laws [178, 272].

If one assumes that the occurrence of extreme winds is stationary, the annual maximum wind speed X can be represented by the Gumbel distribution [6]. Thus, calculations of design wind speeds for various return intervals are based on the estimated mean, and standard deviation. For stations with very short records, the maximum wind, in each month, can be used instead of annual maxima. The design wind speed is thus given by

$$x_\tau = \mathbb{E}(X)_m - \mathbb{S}(X)_m \left\{ \frac{\sqrt{6}}{\pi} \left[n_e + \ln\left(\ln \frac{12\tau}{12\tau - 1} \right) \right] \right\}, \tag{1.147}$$

where n_e is Euler's constant, τ the return period, and $\mathbb{E}(X)_m$ and $\mathbb{S}(X)_m$ represent the mean and standard deviation of the sample of monthly maxima, respectively for month m. The Fréchet distribution is an alternative to the Gumbel distribution, although appreciable differences are found only for large return periods $\tau > 100$ years. Also, the Weibull distribution is found to fit wind speed data for Europe, as already mentioned [290].

Then, one determines the extreme value distribution using either the contagious distribution approach, see Subsection 1.2.4, with a specified threshold, or the probability distribution of continuous wind speeds. Such data are generally given as a sequence of time averages, say, for example, as 10-minute average wind speeds, and the c.d.f. of the average wind speed X is a mixed Weibull distribution

$$F(x) = p_0 + (1 - p_0)\{1 - \exp[-(x/b)^\gamma]\}, \tag{1.148}$$

for $x \geq 0$, where b and γ are the scale, and shape parameters of the Weibull distribution used for $x > 0$, and p_0 is the probability of zero values. Solari [276] suggests the following two methods to obtain the c.d.f. of annual maximum wind speed. In the simple case, as shown in Subsection 1.1.2, one can consider the distribution of the largest value with a sample of n independent data in a year, so the c.d.f. of annual maximum wind speed can be obtained through Eq. (1.16). Alternatively, one can use the crossing properties of a series of mutually i.i.d. r.v.'s. In this case, the c.d.f. is obtained through Eq. (1.109). Following Lagomarsino et al. [173], the values of n in Eq. (1.16), and ν in Eq. (1.109), are found by combining the observed frequencies of annual maxima with the values of F computed for the annual maxima. Recently, the application of POT method (see Subsection 1.2.2) to the frequency analysis of wind speed is supported [137, 43] and discussed [50, 131, 132].

In hurricane regions, the data are a mixture of hurricane and cyclonic winds. Gomes and Vickery [119] find that a single probability law cannot be assumed in these situations. The resulting extreme value distribution is a mixture of the two underlying distributions. A possible approach is to use the Two-Component Extreme Value distribution (see Eq. (1.112)) for this purpose. More generally, one can use Eq. (1.111) to model extreme wind mixture of hurricanes and cyclones; the number of occurrences of hurricanes is taken as a Poisson distributed variate, and the Pareto distribution is used to represent the correspondent wind speed data. A shifted Exponential distribution is adopted to fit cyclonic wind speed data. The resulting extreme value distribution of annual maximum winds is

$$\mathrm{G}(x) = \exp\left[-e^{-\frac{x-a_1}{b_1}} - \nu_2\left(\frac{x}{b_2}\right)^{-\gamma}\right], \tag{1.149}$$

where b_1 and a_1 denote the scale and location parameters of Gumbel distributed cyclonic wind speed, as estimated from the annual maxima of cyclonic winds, b_2 and γ are the scale and shape parameters of Pareto distributed hurricane wind speed, and ν_2 is the mean number of annual occurrences of hurricanes. An application is given by [168, pp. 483–484].

The probability distribution of the annual maximum tornado wind speed is affected by large uncertainties. There is a lack of records of tornado wind speeds. Besides, instrument damages during a tornado can add to data shortages. Observations are currently derived from scales of speed based on structural damage to the area, see [301, 103].

A standard approach is to consider the occurrence of a tornado in the location of interest as a Poisson event with parameter $\nu = \nu_0 a/A_0$, where a is the average damage area of a tornado, A_0 is a reference area taken as a one-degree longitude-latitude square, and ν_0 is the average annual number of tornadoes in the area A_0. Thus,

$$G(x) = \exp\{-\nu_0 a/A_0 [1 - F(x)]\}, \tag{1.150}$$

where F is the c.d.f. of tornado wind speed, and ν_0, a, and A_0 are regional values. As regards tornado risk analysis, reference may be made to [107, 293].

In the book by Murname and Liu [200] the variability of tropical cyclones is examined, mainly in the North Atlantic Ocean, from pre-historic times on various time scales. Reference may be made to [107, 293] for further aspects of tornado wind hazard analysis.

1.4.7 Extreme Sea Levels and High Waves

Sea levels change continuously. Variations take on a variety of scales. Waves cause instantaneous quasi-cyclical changes in level with amplitudes exceeding 10 meters on average. Wave action intensifies with storms. Then, there are tides, the periodic motions arising from the gravitational pull of the moon, and to a partial extent by the sun. Tides have generally periods of 12 hours, but there can be small-amplitude 24 hour tides, depending on the locality. Additionally, storm surges occur such as the one that affected the Mississippi Coast and New Orleans on August 29th, 2005 after hurricane Katrina, the highest recorded in the United States. Besides, there are small variations called seiches.

Visible changes in sea levels are rapid but sometimes there appear to be trends. For example, St. Mark's Square in Venice, Italy, was flooded on average 7 times a year in the early 1900's but currently inundations have surpassed 40 times a year. Measurements of sea levels have been made at many fixed sites from historic times. These refer to a mean sea level at a particular point in time and space. For example, at Brest, France, observations have been recorded from 1810. In the United Kingdom, the Newlyn gauge in southwest England is used as a bench mark for the mean sea level. Not surprisingly, there is high correlation between the French and English readings. There are many other such gauges, in the United States and elsewhere.

Many physical forces and processes affect sea levels, apart from the deterministic tidal force. The behavior of sea surfaces is such that it is justifiable to treat sea levels as stochastic variables. After measurements one has a time series of sea levels at a particular location over a period of years. For purposes of applications, one needs to abstract important information pertaining to sea levels and waves. One such characteristic is the wave height. This is clearly a random variable. More illustrations about the sea state dynamics are given in subsequent Chapters dealing with copulas.

ILLUSTRATION 1.33 (Model of the highest sea waves). ►

Suppose one takes the mean of the highest one-third of the waves at a site; this is designated as h_{sig}, that is the significant wave height. The simplest probabilistic model, which provides a good fit to the wave height, is the Rayleigh distribution, in which a parameter such as h_{sig} can be incorporated. Thus the survival probability is

$$\mathbb{P}\{H > h\} = \exp\left[-2(h/h_{sig})^2\right]. \quad (1.151)$$

The probability of exceedance of h_{sig} is $\exp(-2) = 0.1353$.

As an alternative to h_{sig}, one can use the square root of the sum of squares of the wave height in Eq. (1.151), which for an observed n waves is

$$h_{rms} = \sqrt{\sum_{i=1}^{n} h_i^2}. \quad (1.152)$$

◄

The selection of an appropriate wave height is important in offshore and coastal engineering. From statistical analyses made on observed measurements the Lognormal [74], and Weibull distributions are also suitable candidates. The statistical procedure for extreme wave heights involves the following steps: (1) selecting of appropriate data, (2) fitting of a suitable probability distribution to the observed data, (3) computing extreme values from the distribution, and (4) calculating confidence intervals [190, 282, 199]. The largest storms that occur in a particular area are usually simulated, and parts of the wave heights attributed to a storm are estimated using meteorological data. The POT method (see Subsection 1.2.2) is generally used [215].

Extreme wave heights occurring in different seasons, and from different causes are analyzed separately. This is because of the enforced nonstationary behavior of wave heights. For example, in midlatitude areas high waves may arise from tropical, extra-tropical, or other causes. Also, differences of fetch concerning maximum distances from land play an important role. These effects require the use of mixed distributions.

The number of annual maxima is most often inadequate to incorporate in a model of extremes. For the POT method, one selects the sample from the full data set of wave heights. The threshold is usually chosen on physical or meteorological considerations. For example, by using weather charts one can determine the number of high storm events per year. Where there is a significant seasonal variation in storms, the threshold of wave height is generally determined such that an average, say, between one and two storms per season are considered. For this procedure, the three-parameter Weibull distribution given by Eq. (1.49) is found to provide an acceptable fit to significant wave heights for most maritime surfaces. The truncated Weibull distribution for storm peaks above the threshold x_0 is

$$\begin{aligned} F_{X|X>x_0}(x) &= \frac{F(x) - F(x_0)}{1 - F(x_0)} \\ &= 1 - \exp\left[-\left(\frac{x-a}{b}\right)^{\gamma} + \left(\frac{x_0-a}{b}\right)^{\gamma}\right], \end{aligned} \tag{1.153}$$

where a, b, and γ are parameters to be estimated. The values of γ, most frequently found from data analysis, range from 1 to 2. Quite often γ is close to 1. In this case, a truncated Exponential distribution can suffice. The wave height with return period $\tau_x^>$ is then computed as the value x satisfying

$$F_{X|X>x_0}(x) = 1 - \frac{\mu_T}{\tau_x^>}, \tag{1.154}$$

where μ_T is the average interval between two subsequent storms.

Alternatively, one must consider the annual number of storm occurrences, N, to be a random variable. If N is a Poisson distributed variate with mean ν, then the c.d.f. of annual maximum wave heights takes the form

$$\begin{aligned} G(x) &= \exp\left\{-\nu\left[1 - F_{X|X>x_0}(x)\right]\right\} \\ &= \exp\left\{-\nu \exp\left[-\left(\frac{x-a}{b}\right)^{\gamma} + \left(\frac{x_0-a}{b}\right)^{\gamma}\right]\right\}. \end{aligned} \tag{1.155}$$

This method has been applied to the highest sea waves in the Adriatic by [168, pp. 486–487].

In general, one can consider the wave height X at a given time to have three additive components: mean sea level U, tidal level W, and surge level S. The mean sea level, which is taken from the variability in the data of frequencies longer than a year, varies as a result of changes in land and global water levels. For example, 100 years of data show that the mean sea level increases with a rate of 1 to 2 millimeters per year on the global scale. Also, the presence of inter-annual variations due to the Southern Oscillation means that nonstationarity (of the mean) can no longer be modeled by a simple linear trend in the Pacific Ocean. The deterministic astronomical tidal component, generated by changing forces on the ocean produced by planetary motion, can be predicted from a cyclic equation including global and local constants. The random surge component, generated by short-term climatic behavior, is identified as the $X - U - W$ residual.

Woodworth [304] finds that around the Coast of Great Britain, there is an apparent linear trend in the sea level. The dominant tide has a cycle of 12 hours and 26 minutes. Also, Tawn [282] observes that extreme sea levels typically arise in storms which produce large surges at or around the time of a high tide. Therefore, the probability distribution of the annual maximum sea wave height must account for nonstationarity. Also, the extreme values of S may cluster around the highest values of W, because extreme sea levels typically arise in storms that happen to produce large surges at or around the time of a high tide. However, it is often assumed that the astronomical tide does not affect the magnitude of a storm surge (as in the following example taken from [168, p. 488]). It is then unlikely that the highest values of S coincide with the highest values of W.

How sea levels change and the effects of tides are discussed in detail by [224]. For the complex interactions between ocean waves and wind see [150]. More illustrations about the sea state dynamics are given in subsequent chapters dealing with copulas.

1.4.8 Low Flows and Droughts

A drought is the consequence of a climatic fluctuation in which, as commonly conceived, rainfall is unusually low over an extended period and hence the entire precipitation cycle is affected. Accompanying high temperatures lead to excessive evaporation and transpiration with depleted soil moisture in storage. However, droughts can sometimes occur when surface-air temperatures are not higher than normal, as in the period 1962 to 1965 in northeastern USA. Thus a drought is associated with drastic reductions in reservoir or natural lake storages, the lowering of groundwater levels and decrease in river discharges. It may be spread over a year, or longer period, and can affect a large area: a whole country or even a continent. Droughts and associated water shortages play a fundamental role in human life.

Associated with such an event is the resource implications of the availability of water for domestic and other uses. With regard to agriculture, drought is the most serious hazard in most countries. Accordingly, its severity can be measured or defined; the other characteristics are the duration and frequency. In this way one can define three indicators of a drought: (1) vulnerability, a measure of the water shortage, or drought deficit, (2) resilience, an indication of its length, and (3) reliability, a probability measure.

It follows that a drought may be meteorological, agricultural, hydrological or simply associated with water management. Thus it is a complex phenomenon that can defined in different ways. Invariably, a water deficit is involved. Lack of precipitation is the usual cause although minimal precipitation can sometimes lead to a severe drought as experienced in the US Gulf and East coasts in 1986. A drought is associated with persistent atmospheric circulation patterns that extends beyond the affected area. Attempts to minimize the impact of a drought have been made from historical times. However, cloud seedings for controlling rainfalls as in the United States have not had much success.

Some parts of the world, such as the Sahel region of Africa, have permanent drought characteristics with sparse vegetation and thus agriculture is wholly dependent on irrigation. These regions have endured historical famines, as in Senegal, Mauritania, Mali, Upper Volta, Nigeria, Niger and Chad during the period from 1968 to 1974. The 1973 drought in Ethiopia took 100,000 human lives; some regions of India, China and the former Soviet Union have endured more severe tragedies caused by droughts in previous centuries. In the 1990s droughts of severity greater than in the previous 40 years have been experienced worldwide. Millions in Africa are now living under acute drought conditions. In Brazil, the livelihoods of millions are affected from April 2006 by the driest period of the past 50 years. Across the Pacific Ocean in Australia droughts have also intensified.

Low flows refer to river flows in the dry period of the year, or the flow of water in a river during prolonged dry weather. Such flows are solely dependent on ground water discharges or surface water outflows from lakes and marshes or melting of glaciers. The occurrence of low flows is considered to be a seasonal phenomenon. On the other hand, a drought is a more general phenomenon that includes other characteristics as just stated, in addition to a prolonged low flow period.

Low flow statistics are needed for many purposes. They are used in water supply planning to determine allowable water transfers and withdrawals, and are required in allocating waste loads, and in siting treatment plants and sanitary landfills. Furthermore, frequency analysis of low flows is necessary to determine minimum downstream release requirements from hydropower, water supply, cooling plants, and other facilities.

In this section, we study the statistics associated with low river discharges so as to provide measures of probability. For the purpose of demonstration at a simple level, let us consider an annual minimum daily flow series in the following example. It seems reasonable to fit an appropriate two-parameter distribution here because the length of the data series is only 22 years. Subsequently, as we move towards a drought index, an annual minimum d-day series is considered.

ILLUSTRATION 1.34 (Annual minimum daily flows). ►

The set {2.78, 2.47, 1.64, 3.91, 1.95, 1.61, 2.72, 3.48, 0.85, 2.29, 1.72, 2.41, 1.84, 2.52, 4.45, 1.93, 5.32, 2.55, 1.36, 1.47, 1.02, 1.73} gives the annual minimum mean daily flows, in m^3/s, recorded in a sub-basin of the Mahanadi river in central India (a tropical zone) during a 22-year period.

Assuming that a Converse Weibull distribution provides a good fit, determine the probability that the annual minimum low flow does not exceed 2 m^3/s over a period of two years. The Converse Weibull c.d.f. is given in Eq. (1.56), where $\tilde{\gamma}$ is the shape parameter, and $\tilde{b}$ is the scale parameter. The location parameter $\tilde{a}$ is assumed to be equal to zero. Let $z = \ln(x)$, and $y = \ln(-\ln(1 - \tilde{G}(x)))$. For the sample data, $z_i = \ln(x_i)$, and $y_i = \ln(-\ln(1 - \hat{F}_i))$, for $i = 1, 2, \ldots, n$. We use the *APL* plotting positions to calculate $\hat{F}_i$ (see Table 1.1). Hence $z_i = (y_i/\tilde{\gamma}) + \ln(\tilde{b})$. The plot of z vs. y for the sample data is shown in Figure 1.27.

One has for $n = 22$, $\bar{z} = 0.764$, and if we substitute the *APL* plotting position $\bar{y} = -0.520$. The shape parameter is estimated by Least Squares as $\widehat{\tilde{\gamma}} = 2.773$. Hence, the scale parameter has estimate $\widehat{\tilde{b}} = \exp(\bar{z} - \bar{y}/\widehat{\tilde{\gamma}}) = 2.591$. Substituting $x = 2$ m^3/s, and the estimated parameter values,

$$\tilde{G}(2) = 1 - \exp\left[-\left(\frac{2}{2.591}\right)^{2.773}\right] = 0.386. \tag{1.156}$$

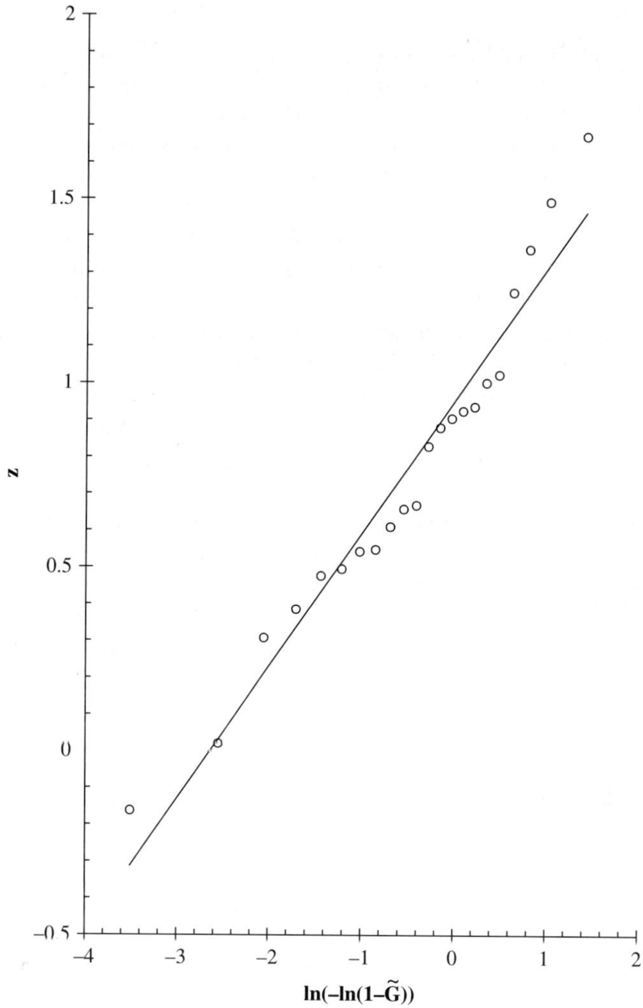

Figure 1.27. Plot of annual minimum mean daily flows from Central India. Shown is z vs. y, comparing the sample data (*markers*) and the theoretical distribution (*line*)

Assuming independence in the low flow series, the probability that the annual minimum low flow will be less than 2 m^3/s over a two year period is $\tilde{G}(2)^2 = 0.149$. Confidence limits can be placed on the shape parameter $\tilde{\gamma}$, knowing that $X^2 = \frac{2n\tilde{\gamma}}{\hat{\tilde{\gamma}}}$ has an approximately Chi-squared distribution with $2n$ degrees of freedom, where n is the sample size. Thus, if the confidence limits are 99%,

$$\frac{\hat{\tilde{\gamma}}}{2n} X^2_{2n,0.995} < \tilde{\gamma} < \frac{\hat{\tilde{\gamma}}}{2n} X^2_{2n,0.005}, \tag{1.157}$$

in which $X^2_{2n,\alpha}$ is a value exceeded with probability α by a Chi-squared variate with $2n$ degrees of freedom. Hence by substituting $\widehat{\tilde{\gamma}} = 2.773$, $n = 22$, and the two Chi-square limiting values, obtained from standard tables, the 99 percent confidence limits for $\tilde{\gamma}$ are $1.57 < \tilde{\gamma} < 4.73$. This provides justification for the use of the Weibull distribution in lieu of the Exponential for which the shape parameter $\tilde{\gamma} = 1$. However, this does not preclude the use of other distributions. ◄

Whereas a single variable may be sufficient to characterize maximum flood flows, the definition of drought, and low flow in rivers, often involve more than one variable, such as the minimum flow level, the duration of flows which do not exceed that level, and the cumulated water deficit. One can use a low flow index, in order to circumvent the problem of evaluating the joint probability of mutually related variates, such as the annual minimum d-day consecutive average discharge with probability of nonexceedance q, say, $x_q(d)$. For instance, the 10-year 7-day average low flow, $x_{0.1}(7)$ is widely used as a drought index in the United States.

An essential first step to low flow frequency analyses is the "deregulation" of the low flow series to obtain "natural" stream flows. Also, trend analysis should be made so that any identified trends can be reflected in the frequency analyses. This includes accounting for the impact of large withdrawals and diversions from water and wastewater treatment facilities, as well as lake regulation, urbanization, and other factors modifying the flow regime. To estimate the quantile $x_q(d)$ from a stream flow record, one generally fits a parametric probability distribution to the annual minimum mean d-day low flow series. The Converse Gumbel distribution, see Eq. (1.54), and the Converse Weibull distribution, see Eq. (1.56), are theoretically plausible for low flows.

Studies in the United States and Canada have recommended the shifted Weibull, the Log-Pearson Type-III, Lognormal and shifted Lognormal distributions based on apparent goodness-of-fit. The following example which pertains to a 7-day mean low flow series is a modification from [168, pp. 474–475].

ILLUSTRATION 1.35 (Annual minimum 7-day flow). ►

The mean and standard deviation of the 7-day minimum annual flow in the Po river at Pontelagoscuro station, Italy, obtained from the record from 1918 to 1978 are 579.2 m^3/s and 196.0 m^3/s, respectively, and the skewness coefficient is 0.338. Let us consider the Converse Gumbel and Converse Weibull distributions of the smallest value given in Eq. (1.54) and Eq. (1.56), respectively.

The values of the parameters estimated via the method of moments are $\widehat{\tilde{b}} = 152.8$ m^3/s, $\widehat{\tilde{a}} = 667.4$ m^3/s for the Gumbel distribution, and $\widehat{\tilde{\gamma}} = 3.26$, $\widehat{\tilde{b}} = 646.1$ m^3/s, for the Weibull distribution (equating the location parameter to zero). These c.d.f.'s are shown in Figure 1.28. On the x-axis $\ln[-\ln(1-\tilde{G})]$ is plotted, and the y-axis gives the r.v. of interest.

The Weibull distribution provides a good approximation to the observed c.d.f.. This is because of the large drainage area of more than 70×10^3 km^2, but the distributions may be quite divergent for small areas, say, less than 10^3 km^2. The estimated

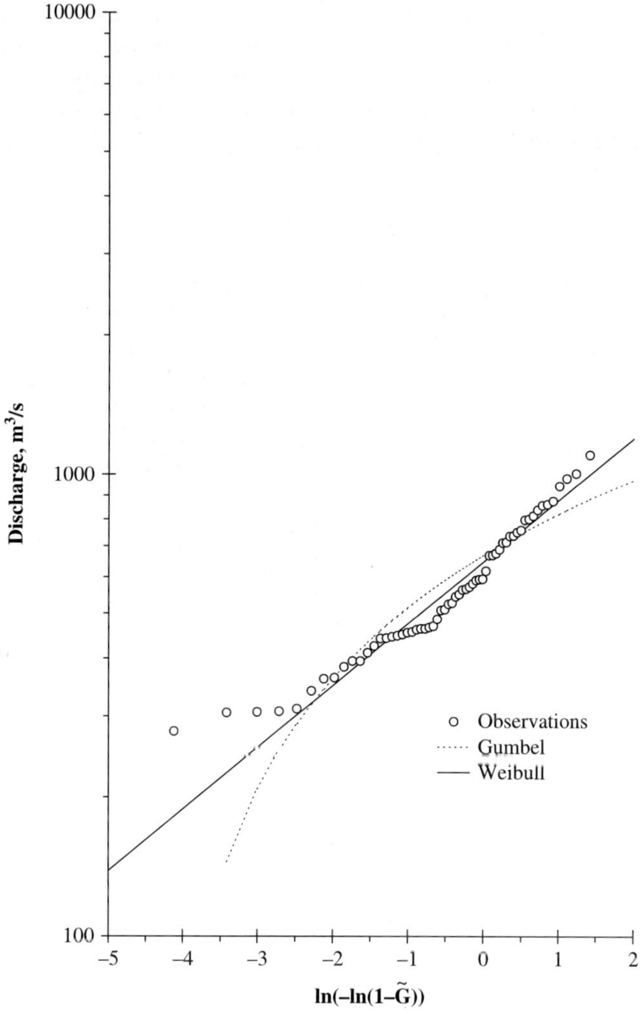

Figure 1.28. Plot of extreme value distributions of annual 7-day minimum flow in the Po river at Pontelagoscuro, Italy

10-year 7-day average low flow, $x_{0.1}(7)$, is given by $x_{0.1}(7) = \ln[-\ln(1-0.1)] \times 152.8 + 667.4 = 323.5$ m^3/s, for the Gumbel distribution, which has a poor fit. For the Weibull distribution it becomes $x_{0.1}(7) = 323.3$ m^3/s. ◄

An alternative to d-day averages is the flow duration curve, a form of cumulative frequency diagram with specific time scales. This gives the proportions of the time over the whole record of observations, or percentages of the duration, in which different daily flow levels are exceeded. However, unlike $x_q(d)$, it cannot be interpreted on a annual event basis.

Moreover, low flow data can contain zero values, which is common in small basins of arid areas, where zero flows seem to be recorded more often than nonzero flows. Accordingly, the c.d.f. of a low flow index, say, X, is a mixed distribution. It has a probability mass at the origin, p_0, and a continuous distribution for nonzero values of X, which can be interpreted as the conditional c.d.f. of nonzero values, say, $F(x|x > 0)$. Thus,

$$F(x) = p_0 + (1 - p_0)F_X(x|x > 0). \tag{1.158}$$

The parameters of $F(x|x > 0)$ of Eq. (1.158) can be estimated by any procedure appropriate for complete samples using only nonzero data, while the special parameter p_0 represents the probability that an observation is zero. If r nonzero values are observed in a sample of n values of data, the natural estimator of the exceedance probability $q_0 = (1 - p_0)$ of the zero value or perception threshold is r/n, and $p_0 = 1 - r/n$. With regard to ungaged sites, regional regression procedures can be used to estimate low flow statistics by using physical and climatic characteristics of the catchment. If one uses only the drainage area, for instance, the low flow quantile for an ungaged river site draining an area of A is $(A/A_x)x_q(d)$, where $x_q(d)$ is the corresponding d-day low flow quantile for a gauging station in the vicinity which drains an area of A_x. This may be modified by a scaling factor $(A/A_x)^b$, $b < 1$, which is estimated by regional regression of quantiles for several gauged sites.

When records are short there are also other record augmentation methods and these are compared by [171]. Besides, with regard to droughts note that they can be measured with respect to recharge and groundwater discharge; in this way the performance of ground water systems can be evaluated [214].

1.4.9 Floods

A flood consists in high water levels overtopping natural, or artificial, banks of a stream, or a river. The flood is the consequence of a meteorological condition in which rainfall precipitation is high over an extended period of time and a portion of space. The study of floods has a long history. After ancient agricultural nations, which depended heavily on water flows, realised the economic significance of floods, the importance of this natural phenomenon has increased in modern industrialized countries. Water has become a permanent and inexpensive source of energy. Impounded in reservoirs, or diverted from streams, it is essential for irrigating field crops. Also, one must have a sufficient knowledge of the quantity of water flow to control erosion. It is widely accepted that life, and property, need to be protected against the effects of floods. In most societies a high price is paid to reduce the possibilities of damages arising from future floods. Indeed, the failure of a dam caused by overtopping is a serious national disaster. To safeguard against such an event, engineers cater for the safe passages of rare floods with estimated return periods from 1,000 to 1,000,000 years depending on the height of the dam.

The recurrent problem is that the observed record of flow data at the site of interest does not extend over an adequately long period, or is not available at all. Note that flood frequency analysis for a given river site usually relies on *annual maximum flood series* (AFS) of 30–60 years of observations, thus it is difficult to obtain reliable estimates of the quantiles with small exceedance probabilities of, say, 0.01 or less. As first indicated by [14], reliable quantile estimates can be obtained only for return periods less than $2n$, where n denotes the length of the AFS. Also, Hosking et al. [141] showed that reliable quantile estimates are obtained only for non-exceedance frequencies less than $1 - 1/n$, which correspond to a return period equal to n.

In this situation, it is possible to calculate the flood frequency distribution using flood records observed in a group of gauged river sites, defining a region. It is denominated *regionalization method.* The regionalization method provides a way of extrapolating spatial information; in other words, it substitues the time information, not available, in the considered site, with the spatial information available in a region, which includes the site of interest. As a regionalization technique, we now illustrate the *index-flood* method.

ILLUSTRATION 1.36 (Regionalization method). ►

The *index-flood* method was originally proposed by [57]. Initially, this method identifies a homogeneous region, given that a number of historical data series are available. This operation is traditionally carried out using statistical methods of parametric and non-parametric type [49], with large uncertainties, and a certain degree of subjectiveness, mainly due to the lack of a rigorous physical basis. Alternatively, this operation is done using the concept of scaling of maximum annual flood peaks with basin area [128, 126, 127, 233], or through seasonality indexes [30, 31, 218, 147]. Successively, the data collected at each river site are normalized with respect to a local statistic known as *index-flood* (e.g., the mean [57] or the median [147] annual flood). Thus, the variables of interest are

$$X_i = Q_i / \mathcal{I}_i, \tag{1.159}$$

i.e. the ratio between the maximum annual flood peak Q_i at a river site i, and the corresponding index-flood $\mathcal{I}_i$.

Assuming statistical independence among the data collected within the homogeneous region, the normalized data are pooled together in a unique sample denominated *regional normalized sample*. Then, a parametric distribution is fitted to the normalized sample. This distribution, sometimes referred to as the *growth curve* [202], is used as a regional model to evaluate flood quantiles for any river site in the region.

The index-flood method calculates the τ-quantile of flood peak $q_{i,\tau}$ at the i-th site as

$$q_{i,\tau} = x_\tau \mathcal{I}_i, \tag{1.160}$$

where x_τ denotes the τ-year quantile of normalized flood flows in the region, and $\mathcal{J}_i$ is the index-flood for the site considered. In this way, the flood quantile at a particular site is the product of two terms: one is the normalized flood quantile, common to all river sites within the region, and the other the index-flood, which characterizes the river site under analysis and the corresponding river basin. The index-flood $\mathcal{J}$ incorporates river basin characteristics like geomorphology, land use, lithology, and climate.

If the GEV distribution is considered as growth curve, then the quantile x_τ is written as

$$x_\tau = a'_R + \frac{b'_R}{\gamma'_R}(e^{\gamma'_R y_\tau} - 1), \tag{1.161}$$

where a'_R, b'_R and γ'_R are the parameters of the regional model, and $y_\tau = -\ln(-\ln((\tau-1)/\tau))$ is the Gumbel reduced variate.

The index-flood $\mathcal{J}$ can be estimated through several methods [17]. However the applicability of these methods is related to data availability. Due to their semplicity, empirical formulas are frequently used to evaluate the index-flood. Empirical formulas link $\mathcal{J}$ to river basin characteristics C like climatic indexes, geolithologic and geopedologic parameters, land coverage, geomorphic parameters, and anthropic forcings. Thus $\mathcal{J}$ is written as

$$\mathcal{J} = \beta_0 \prod_{i=1}^{m} C_i^{\beta_i}, \tag{1.162}$$

where C_i is the i-th characteristic, and β_0 and β_i are constants, with $i = 1, \ldots, m$. Note that Eq. (1.162) corresponds to a multiple linear regression in the log-log plane, and the parameters can be estimated using the Least Squares technique. Substituting Eq. (1.161) and Eq. (1.162) in Eq. (1.160) yields

$$q_\tau = [a'_R + \frac{b'_R}{\gamma'_R}(e^{\gamma'_R y_\tau} - 1)] \cdot \beta_0 \prod_{i=1}^{m} C_i^{\beta_i}. \tag{1.163}$$

◄

The following example is an application of the index-flood method [64].

ILLUSTRATION 1.37 (Regionalization method (cont.)). ►

Climate is highly variable under the multifaceted controls by the atmospheric fluxes from the Mediterranean sea, and the complex relief including the southern range of western and central Alps and the northern Apennines range. In northwestern Italy, 80 gauging stations provide consistent records of AFS, with drainage areas ranging from 6 to 2500 km^2. Some additional AFS, available for river sites with larger drainage areas, are not included in the analysis, because of major influence of river regulation.

Both physical and statistical criteria are used to cluster homogeneous regions, namely seasonality measures (Pardé and Burn indexes), and the scale invariance of flood peaks with area are used to identify the homogeneous regions. Homogeneity tests (L-moment ratio plots and Wiltshire test), are further applied to check the robustness of the resulting regions, see for details [64]. This yields four homogeneous regions (also referred to as A, B, C, and D). Two of them are located in the Alps, and the other two in the Apennines. In particular, region A, or central Alps and Prealps, includes Po sub-basins from Chiese to Sesia river basin; region B, or western Alps and Prealps, includes basins from Dora Baltea river to Rio Grana; region C, or northwestern Apennines and Thyrrhenian basins, includes Ligurian basins with outlet to the Thyrrhenian sea and Po sub-basins from Scrivia river basin to Taro river basin; region D, or northeastern Apennines, includes basins from Parma to Panaro river basin (including Adriatic basins from Reno to Conca river basin).

In addition, also a transition zone (referred to as TZ) is identified. This deals with one or more river basins, generally located on the boundaries of the homogeneous regions that cannot be effectively attributed to any group, due to anomalous behavior. Anomalies could be ascribed to either local micro-climatic disturbances or superimposition of the different patterns characterizing the neighboring regions. For example, in the identified transition, correspondent to the Tanaro basin, some tributaries originate from the Alps, others from the Apennines, so there is a gradual passage from region B to region C.

For each homogeneous region, we assume the GEV distribution as growth curve, and evaluate the parameters from the regional normalized sample using the unbiased PWM method [278]. The data are normalized with respect to the empirical mean assumed as index-flood. Table 1.10 gives the values of regional GEV parameters, a'_R, b'_R, γ'_R, and the size of regional normalized sample n'_R. Note that all the estimates of the shape parameter, γ'_R, indicate that the normalized variable X is upper unbounded. The huge size of the normalized sample allows to obtain reliable estimates of the normalized quantiles with very small exceedance probabilities: 0.003 for regions A and B, 0.002 for region D, and 0.001 for region C.

The normalized quantile estimates for selected return periods are given in Table 1.11 for the four homogeneous regions.

Table 1.10. Estimates of the regional GEV parameters and the size of regional normalized sample, for the four homogeneous regions

Region	n'_R	a'_R	b'_R	γ'_R
A	316	0.745	0.365	0.110
B	347	0.635	0.352	0.320
C	753	0.643	0.377	0.276
D	439	0.775	0.334	0.089

Table 1.11. Estimates of the regional quantile x_τ for the four homogeneous regions in northwestern Italy, for selected return periods $\tau = 10, 20, 50, 100, 200, 500$ years

Region	$x_{\tau=10}$	$x_{\tau=20}$	$x_{\tau=50}$	$x_{\tau=100}$	$x_{\tau=200}$	$x_{\tau=500}$
A	1.68	2.03	2.52	2.93	3.37	4.00
B	1.80	2.38	3.37	4.33	5.52	7.57
C	1.82	2.38	3.29	4.14	5.17	6.87
D	1.61	1.91	2.33	2.67	3.03	3.55

The growth curve for the four homogeneous regions in northwestern Italy is reported in Figure 1.29. Table 1.11 and Figure 1.29 show how different is the flood behavior of the four regions. For instance, the quantile $x_{\tau=500}$ is 4.00 for region A, and close to the double for region B, contiguous to region A.

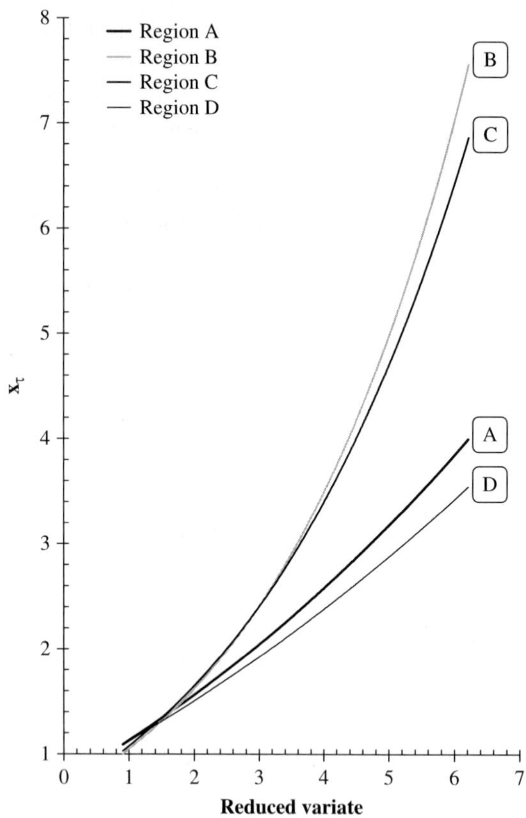

Figure 1.29. Growth curve for the four homogeneous regions in northwestern Italy

Then, the index-flood $\mathcal{J}$ can be computed using different approaches. These include empirical formulas and sophisticated hydrologic models of rainfall-runoff transformations. For instance, if one takes the basin area A as the only explanatory variable, the following empirical formulas are obtained for the four homogeneous regions: $\mathcal{J} = 2.1A^{0.799}$ for region A, $\mathcal{J} = 0.5A^{0.901}$ for region B, $\mathcal{J} = 5.2A^{0.750}$ for region C, and $\mathcal{J} = 2.5A^{0.772}$ for region D, with $\mathcal{J}$ in m^3/s and A in km^2. However, these simple formulas are inaccurate to describe the variability of $\mathcal{J}$ within a region, and more complex formulations involving other explanatory variables are needed. ◄

Another method is to derive the flood frequency distribution via a simplified representation of its generation processes, i.e. rainfall and runoff. This approach is called *derived distribution method*. It was used by [81], who first developed the idea of deriving flood statistics from a simplified schematization of storm and basin characterisation. Indeed, the mechanism of derived distribution is well established in Probability Theory, where a variable $Y = Y(\boldsymbol{X})$ is functionally related to a *random vector* $\boldsymbol{X}$, whose components are random variables with joint p.d.f. $f_{\boldsymbol{X}}$ and joint c.d.f. $F_{\boldsymbol{X}}$. Due to the randomness of $\boldsymbol{X}$, also Y is expected to be a random variable, with distribution function F_Y given by, for $y \in \mathbb{R}$,

$$F_Y(y) = \mathbb{P}\{Y(\boldsymbol{X}) \leq y\} = \int_{\{\boldsymbol{x}:\, Y(\boldsymbol{x}) \leq y\}} f_{\boldsymbol{X}}(\boldsymbol{x})\, \mathrm{d}\boldsymbol{x}. \tag{1.164}$$

For instance, Y may represent the peak flow rate, and the components of $\boldsymbol{X}$ may include, e.g., soil and vegetation characteristics parametrized via both deterministic and random variables.

The derived distribution method can be followed by either analytical function (see, among others, [81, 82, 302, 165, 51, 72, 303, 228]), or via the statistical moments using the *second-order second-moment approximation* (*SOSM*) of extreme flood flows [233]. The first approach provides complex analytical formulations which may require numerical methods for the computations. The second one gives approximate estimates of the statistical moments of maximum annual flood, useful for calculating the parameters of the distributions of interest; its practical applicability is paid in terms of the requirement of the existence of these moments. Alternatively, *Monte Carlo* methods can be run to estimate either flood quantiles or the moments (see, e.g., [274, 234]). This approach can be also used to assess the effects of potential changes in the drainage basin system (e.g. land-use, river regulation, and training) on extreme flood probabilities [245]. Further applications deal with the assessment of hydrologic sensitivity to global change, i.e. using downscaled precipitation scenarios from Global Circulation Models (GCM) to input deterministic models of basin hydrology [28, 29]. The key problem in simulation deals with stochastic modeling the input process (i.e. precipitation in space and time) and fine resolution data are needed [244]. The identification and estimation of reliable models of precipitation with fine resolution in space and time must face large uncertainties [299, 236, 116, 26].

The derived distribution approach is an attempt to provide a physically-based description of flood processes with an acceptable computational effort for practical applications. A simplified description of the physical processes is a necessity in order to obtain mathematically tractable models. Therefore, one must carefully consider the key factor controlling the flood generation and propagation processes. This method provides an attractive approach to ungauged basins.

Examples of analytical formulation of the *derived distribution* of *peak flood* and *maximum annual peak flood*, starting from a simplified description of rainfall and surface runoff processes, now follow.

ILLUSTRATION 1.38 (Analytical derivation of peak flood distribution). ►—
Let Δ be a given *time duration* (e.g., $\Delta = 1$ hour), and denote by P_Δ the *maximum rainfall depth* observed in a generic period of length Δ *within* the considered storm. We assume that P_Δ has the GP distribution as in Eq. (1.91). For simplicity, here, we consider the rainfall duration Δ as constant, and equal to the time of equilibrium of the basin t_c [284].

The *SCS-CN* method [294] is used to transform rainfall depth into rainfall excess. The total volume of *rainfall excess* P_e can be expressed in terms of the rainfall depth P as

$$P_e = P_e(P) = \begin{cases} \frac{(P-I_A)^2}{P+S-I_A}, & P > I_A \\ 0, & P \le I_A \end{cases} \tag{1.165}$$

where I_A is the rainfall lost as *initial abstraction*, and $S \ge 0$ is the *maximum soil potential retention*. Here S is expressed in *mm*, and is given by $S = 254\left(\frac{100}{CN} - 1\right)$, where *CN* is the *curve number*. Note that I_A is generally taken as $I_A \approx 0.2 \cdot S$. The curve number *CN* depends upon soil type, land-use and the *antecedent soil moisture conditions* (AMC). The U.S.D.A.-S.C.S. manual [294] provides tables to estimate the *CN* for given soil type, land-use, and the AMC type (dry, normal, and wet).

Note that if $P \le I_A$, then $P_e = 0$. Since P has a GP law, one obtains:

$$\mathbb{P}\{P_e = 0\} = \mathbb{P}\{P \le I_A\} = 1 - \left(1 + \frac{\gamma}{b}(I_A - a)\right)^{-\frac{1}{\gamma}}, \tag{1.166}$$

with a, b, and γ denoting the parameters of the GP distribution of P. The distribution of P_e has an *atom* (mass point) at zero. Using Eq. (1.165) we derive the conditional distribution of P_e given that $P > I_A$:

$$\begin{aligned} \mathbb{P}\left\{P_e \le x \middle| P > I_A\right\} &= \mathbb{P}\left\{P \le I_A + \frac{x + \sqrt{x^2 + 4xS}}{2} \middle| P > I_A\right\} \\ &= 1 - \frac{\left(1 + \frac{\gamma}{b}\left(I_A + \frac{x+\sqrt{x^2+4xS}}{2} - a\right)\right)^{-\frac{1}{\gamma}}}{\left(1 + \frac{\gamma}{b}(I_A - a)\right)^{-\frac{1}{\gamma}}}, \end{aligned} \tag{1.167}$$

for $x > 0$. Then, from Eq. (1.166) and Eq. (1.167) one obtains the derived distribution of rainfall excess:

$$F_{P_e}(x) = 1 - \left(1 + \frac{\gamma}{b}\left(I_A + \frac{x + \sqrt{x^2 + 4xS}}{2} - a\right)\right)^{-\frac{1}{\gamma}}, \qquad (1.168)$$

for $x \geq 0$, which is *right-continuous* at zero. Note that $\sqrt{x^2 + 4xS} \approx x$ for x large enough. Hence, for $x \gg 1$, the limit distribution of P_e is again a GP law with parameters $a_e = a - I_A$, $b_e = b$, and $\gamma_e = \gamma$. This result could also be derived recalling that the GP distribution is *stable* with respect to *excess-over-threshold* operations [37], and noting that Eq. (1.165) is asymptotically linear for $P \gg 1$.

Let Q denote the *peak flood* produced by a precipitation P according to the transformation

$$Q = Q(P) = \begin{cases} \phi \frac{(P-I_A)^2}{P+S-I_A}, & P > I_A \\ 0, & P \leq I_A \end{cases} \qquad (1.169)$$

with $\phi = A/t_c$, where A is the *area* of the basin and t_c is the *time of concentration* of the basin. The transform function is *non-linear* in P (but *linear* in P_e, since $Q = \phi P_e$), and invertible for $P > I_A$. Using Eq. (1.168), one computes the distribution of Q as

$$F_Q(q) = 1 - \left(1 + \frac{\gamma}{b}\left(I_A + \frac{q + \sqrt{q^2 + 4q\phi S}}{2\phi} - a\right)\right)^{-\frac{1}{\gamma}}, \qquad (1.170)$$

for $q \geq 0$. Note that $\sqrt{q^2 + 4q\phi S} \approx q$ for q large enough. Hence, for $q \gg 1$, the limit distribution of the peak flood Q is again a GP law with parameters $a_Q = \phi\,(a - I_A)$, $b_Q = \phi b$, and $\gamma_Q = \gamma$. Note that, the shape parameter of the flood distribution is the same as that of the rainfall distribution. ◀

ILLUSTRATION 1.39 (Analytical derivation of maximum annual peak flood distribution). ▶

Let us fix a *reference time period* $\tau > \Delta$: in the present case $\tau = 1$ year, and Δ is the time of equilibrium of the basin as in Illustration 1.38. Assuming that the sequence of rainfall storms has a Poisson chronology, the random number N_P of storms in τ is a Poisson r.v. with the parameter ν_P. Only a fraction of the N_P storms yields $P_e > 0$, able to generate a peak flood $Q > 0$. This corresponds to a random *Bernoulli selection* over the Poissonian chronology of the storms, and the random sequence of flood events has again a Poissonian chronology, with annual rate parameter ν_Q given by:

$$\nu_Q = \nu_{P_e} = \nu_P \mathbb{P}\{P > I_A\} = \nu_P \left(1 + \frac{\gamma}{b}\,(I_A - a)\right)^{-\frac{1}{\gamma}}, \qquad (1.171)$$

which, in turn, specifies the distribution of the random number N_Q of annual peak floods. If $I_A \leq a$ (i.e., if the minimum rainfall a is larger than initial abstraction I_A), then $\nu_Q = \nu_{P_e} = \nu_P$. Note that Eq. (1.171), if properly modified, provides the average number of peaks over a given threshold.

The initial abstraction I_A is a function of *soil properties*, *land-use*, and the AMC through the simple empirical relation $I_A = 0.2 \cdot S_{\text{AMC}}$; thus, the soil and land-use *do* influence the expected number of annual flood events. Rewriting Eq. (1.171), one gets

$$\frac{\nu_Q}{\nu_P} = \left(1 + \frac{\gamma}{b}\,(0.2 \cdot S - a)\right)^{-\frac{1}{\gamma}}. \tag{1.172}$$

Conditioning upon N_Q, the distribution of the *maximum annual peak flood* Q^* is found using Eq. (1.109):

$$G_{Q^*}(q) = e^{-\nu_Q\left(1-F_Q(q)\right)} \tag{1.173}$$

for all suitable values of q. If Q is asymptotically distributed as a GP r.v., then Q^* is (asymptotically) a GEV r.v., with the parameters $a'_Q = a_Q + \frac{b_Q}{\gamma_Q}\left(\nu_Q^{\gamma_Q} - 1\right)$, $b'_Q = b_Q\,\nu_Q^{\gamma_Q}$, and $\gamma'_Q = \gamma_Q$.

As in Illustration 1.38, the shape parameter of the flood distribution equals that of the rainfall distribution. Asymptotically, the curve of maximum annual flood quantiles is parallel to the curve of maximum annual rainfall quantiles. The *Gradex method* [123] shows a similar result. In fact, using a Gumbel distribution for the maximum annual rainfall depth, and assuming that, during the extreme flood event, the basin saturation is approached, the derived distribution of the specific flood volume is again a Gumbel law, with the location parameter depending on the initial conditions of the basin, and the scale parameter (*gradex*) equal to that of the rainfall distribution. ◄

The next Illustration shows the influence of the AMC on the distribution of flood peaks.

ILLUSTRATION 1.40 (The influence of AMC). ►

Wood [302] pointed out how the antecedent moisture condition AMC of the basin could be the most important factor influencing the flood frequency distribution. Here we consider the AMC classification given by *SCS-CN* method. It considers three AMC classes (I, II, and III) depending on the total 5-day antecedent rainfall and seasonality (*dormant* or *growing* season). Condition I describes a dry basin with a total 5-day antecedent rainfall less than 13 mm in the dormant season, and less than 36 mm in the growing season. Condition II deals with a total 5-day antecedent rainfall ranging from 13 to 28 mm in the dormant season, and from 36 to 53 mm in the growing season. Condition III occurs when the soil is almost saturated, with a total 5-day antecedent rainfall larger than 28 mm in the dormant season, and larger than 53 mm in the growing season.

We now consider the AMC as a random variable with a discrete probability distribution:

$$\begin{cases} \mathbb{P}\{\text{AMC}=\text{I}\} = \pi_{\text{I}} \geq 0 \\ \mathbb{P}\{\text{AMC}=\text{II}\} = \pi_{\text{II}} \geq 0 \\ \mathbb{P}\{\text{AMC}=\text{III}\} = \pi_{\text{III}} \geq 0 \\ \pi_{\text{I}} + \pi_{\text{II}} + \pi_{\text{III}} = 1 \end{cases} \tag{1.174}$$

Here $\{\pi_{\text{I}}, \pi_{\text{II}}, \pi_{\text{III}}\}$ are the probabilities of occurrence of the three AMC, that depend on climatic conditions. For example, Gray et al. [120] analyzed 17 stations in Kentucky and Tennessee to estimate these probabilities: AMC I was dominant (85%), whereas AMC II (7%), and AMC III (8%) were much less frequent. This analysis under different geography and climate obviously yields different results [271]. The distribution of the peak flood conditioned by the AMC distribution is derived by combining Eq. (1.170) and Eq. (1.174), yielding

$$F_Q(q) = \sum_{i=\text{I},\text{II},\text{III}} \pi_i \left(1 - \left(1 + \frac{\gamma}{b}\left(I_A + \frac{q+\sqrt{q^2+4q\phi S}}{2\phi} - a\right)\right)^{-\frac{1}{\gamma}}\right)_{\text{AMC}=i} \tag{1.175}$$

for $q \geq 0$. F_Q is the weighted sum of three terms, because of the dependence upon the AMC conditions. If AMC is constant, then all the probabilities π_i's but one are zero, and Eq. (1.175) gives Eq. (1.170) as in Illustration 1.38.

Using Eq. (1.175) in Eq. (1.109) yields the distribution F_{Q^*} of the maximum annual peak flood Q^* conditioned by AMC, resulting in

$$G_{Q^*}(q) = \exp\left(-\nu_Q\left(1 - \sum_{i=\text{I},\text{II},\text{III}} \pi_i \left(1 - \left(1 + \frac{\gamma}{b}\left(I_A + \frac{q+\sqrt{q^2+4q\phi S}}{2\phi} - a\right)\right)^{-\frac{1}{\gamma}}\right)_{\text{AMC}=i}\right)\right), \tag{1.176}$$

for all suitable values of q. Again, Eq. (1.176) gives the distribution of Illustration 1.39 for constant AMC. As an illustration, Figure 1.30 shows the function F_{Q^*} for five different AMC distributions.

The AMC distribution has a great influence on the flood frequency distribution: for example, passing from AMC I to AMC III, the 100-year flood for AMC III is more than three times larger than that for AMC I in the above illustration (398 vs. 119 m^3/s). Because the shape parameter of the flood distribution is that of the rainfall distribution, the variability of initial moisture condition only affects the scale and location parameters of the flood frequency curve, but not the shape parameter, and so it does not influence the asymptotic behaviour of the flood distribution. Various practical methods assume this conjecture as a reasonable working assumption, as,

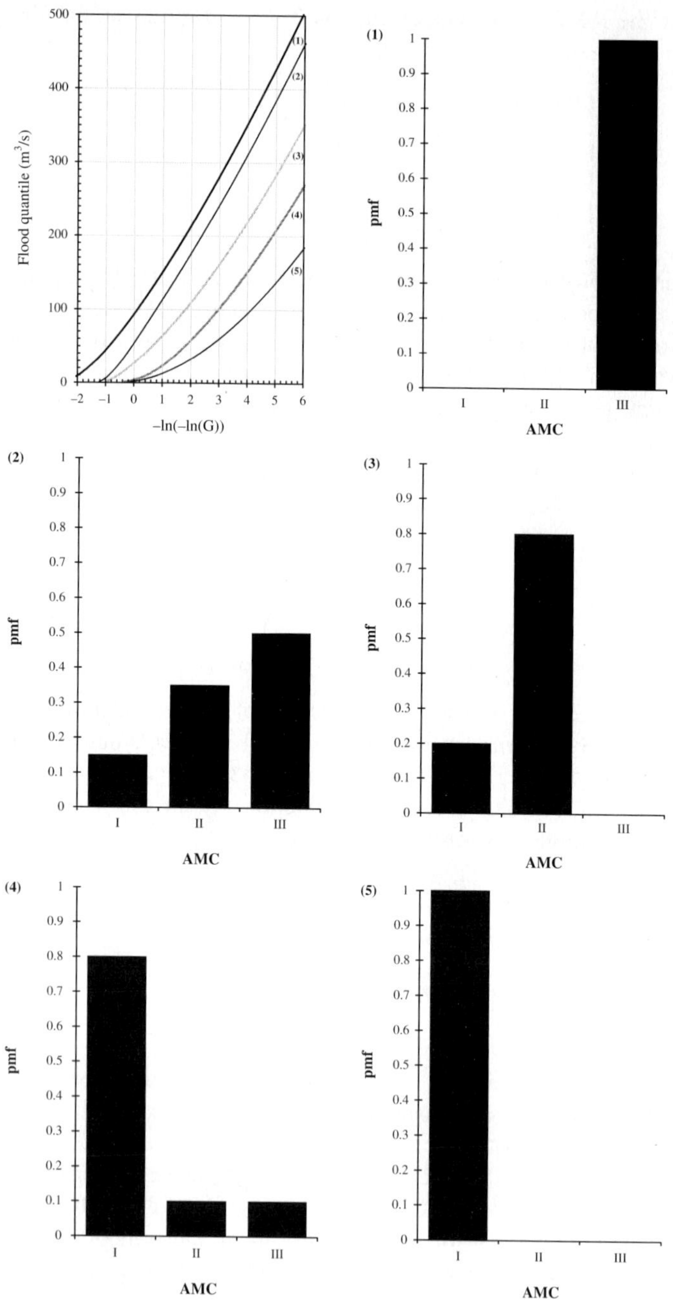

Figure 1.30. Probability distribution of maximum annual peak flood for five different AMC distributions. Here the parameters are: $\gamma = 0.031$, $b = 13.39$ mm, $a = 12.69$ mm, $\nu_Q = 9$ storm/year, $A = 34$ km^2, $t_c = 1.97$ hours, and $S_{\text{II}} = 76$ mm

e.g., in the *Gradex Method* [123], where a Gumbel probability law is used for the rainfall and the flood distribution, i.e. both laws have the shape parameter equal to zero. ◄

The derived distribution approach given in Illustrations 1.38–1.40 is applied below to three italian river basins [65].

ILLUSTRATION 1.41 (Derived flood distribution). ►

We consider here three river basins in Thyrrhenian Liguria, northwestern Italy, i.e. Bisagno at La Presa, Vara at Naseto, and Arroscia at Pogli. Table 1.12 gives some basin characteristics: area, relief, mean annual rainfall, soil type, land use, and information on rainfall and streamflow gauges, including the number of records.

Table 1.13 gives the parameters of the *derived distributions* of peak flood, Eq. (1.175), and maximum annual peak flood, Eq. (1.176).

According to Eq. (1.99), estimates of the rainfall parameters, a, b, and γ, for a duration $\Delta = t_c$, can be obtained from those of the parameters of the maximum

Table 1.12. Characteristics of the basins investigated

	Bisagno		Vara			Arroscia
Area	34.2 km^2		206 km^2			201 km^2
Relief	1063 m		1640 m			2141 m
Rainfall	167 cm/y		187 cm/y			116 cm/y
Soil	Limestone 62%		Sandy, Marly 22%			Sandy, Marly 36%
	Clay, Clayey 31%		Clay, Clayey 56%			Calcar.-Marly 58%
Land use	Trans. wood. shrub. 60%		Trans. wood. shrub. 59%			Trans. wood. shrub. 56%
	Agroforest 18%		Sown field in well-water 18%			Agroforest 11%
			Mixed forest 10%			Mixed forest 10%
Rain G.	**Scoffera**	**Viganego**	**Cento C.**	**Tavarone**	**Varese L.**	**Pieve di Teco**
# years	35	39	20	44	43	25
Flood G.	**La Presa**		**Naseto**			**Pogli**
# years	48		38			55

Table 1.13. Parameters of the derived distributions of peak flood, and maximum annual peak flood

Param.	**Unit**	**Bisagno**	**Vara**	**Arroscia**
a	[mm]	13.08	17.10	20.00
b	[mm]	13.80	14.40	28.53
γ	[–]	0.031	0.183	0.057
A	[km^2]	34.2	206	202
t_c	[h]	1.97	8.00	15.60
S_{II}	[mm]	80	89	99
π_I		0.25	0.26	0.27
π_{II}		0.12	0.16	0.08
π_{III}		0.63	0.58	0.65
ν_Q	[st./y]	9	13	9

annual rainfall distribution, a', b', and γ', for the same duration. However, since ν_P is not known, we have four parameters and three equations. De Michele and Salvadori [65] proposed the following equation in addition to Eq. (1.99):

$$\hat{b}_i \approx \frac{x^*_{(i+1)} - x^*_{(i)}}{\varphi^*_i - \varphi^*_{i+1}}, \tag{1.177}$$

where $x^*_{(i)}$ is the i-th order statistic from a sample of size n, the coefficient φ^*_i equals $\varphi^*_i = \ln(-\ln(i/(n+1)))$, and the index i is small. Eq. (1.177) can be combined with the last two formulas of Eq. (1.99) to obatin an estimate of ν_P. We note that the *average* of several estimates $\hat{b}_i$'s (for small indices i's) provides reliable estimates of the parameter b.

Estimates of a', b', and γ', for a duration $\Delta = t_c$, are obtained from the maximum annual rainfall depth data, for durations from 1 to 24 hours under the assumption of temporal scale invariance (see Illustration 1.18).

The maximum soil potential retention S_{II} (for AMC II) is obtained from thematic maps of soil type and land-use with resolution of 250 m. Cross-validation using observed contemporary rainfall and runoff data was also performed to assess the expected value of S_{II} (at the basin scale). This procedure is recommended by several authors (see, e.g., [120, 133, 271]). Then, S_{I} and S_{III} are computed from S_{II} as described in [220].

The distribution of the AMC for the three basins are estimated from the observed total 5-day antecedent rainfall for the available samples of precipitation and streamflow data. One notes that the AMC distributions are practically identical, thus indicating homogeneous climatic conditions for the three basins. In Figures 1.31–1.33 the flood quantiles for the derived distribution are compared with those obtained by fitting the GEV to the AFS.

For small return periods ($\leq$ 100 years), the derived distribution almost matches the GEV distribution fitted to the observed data, and both show a good agreement with the observations; in particular, the agreement is very good for Vara basin. From the analysis of Figures 1.31–1.33 it is evident that the statistical information extracted from the series of maximum annual peak floods (and represented by the distributions fitted on the data) is also available considering the rainfall data (for small return periods) using the derived distribution approach proposed here.

Overall, the derived distribution is able to represents, up to a practically significant degree of approximation, the flood quantile curves, and thus it performs fairly well when considering small return periods. ◀

1.4.10 Wildfires

A forest fire is an uncontrolled fire that occurs in trees more than 6 feet (1.8 meters) in height. Some fires are caused by combustion from surface and ground fires. On the other hand, a fire may spread through the upper branches of trees with

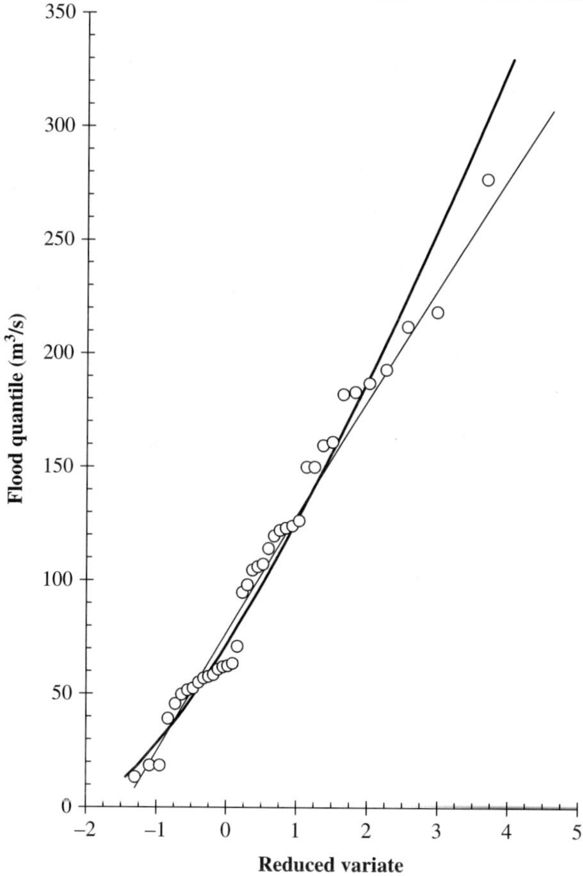

Figure 1.31. Flood frequency curves (in m^3/s) for the Bisagno river basin at La Presa: observations (*circles*), GEV fitted on data (*thin solid line*), derived distribution (*thick solid line*)

little effect on the ground or undergrowth. At high speeds, regardless of the level, such an event becomes a firestorm. An uncontrolled fire passing through any type of vegetation is termed a wildfire. In uninhabited lands the risk of fire depends on weather conditions. For example, drought, summer heat or drying winds can lead to ignition.

In recent years there have been major fires in Alaska, Australia, California, Greece, Ireland, Nicaragua and southern France. The summer bushfire season in Australia usually comes at the end of the year bringing death and destruction; in 1983, 2300 homes were destroyed in Victoria and South Australia. Fires sometimes originate from human negligence, arson and lightning. On the positive side, in Mediterranean and other temperate climates some species of plants depend on fires for their propagation and others survive in their presence.

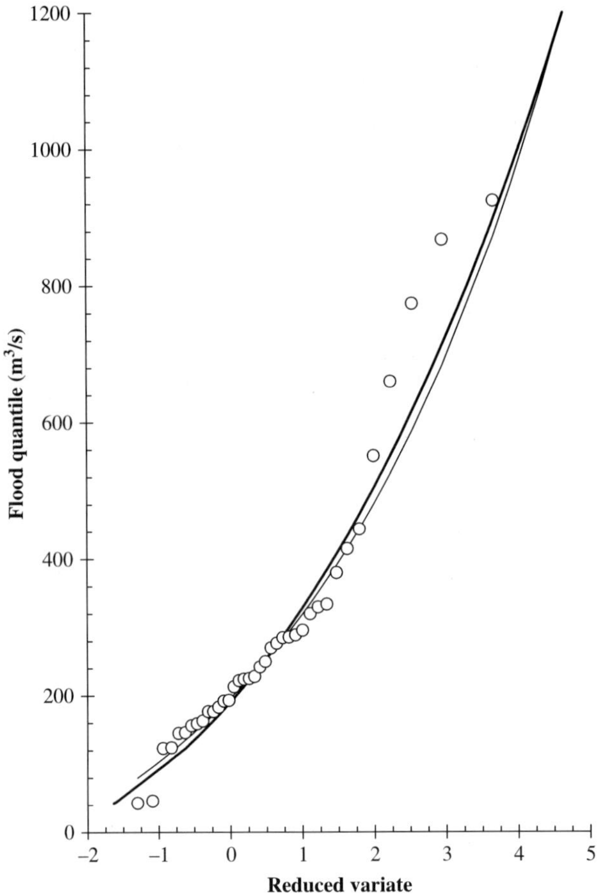

Figure 1.32. Flood frequency curves (in m^3/s) for the Vara river basin at Naseto: observations (*circles*), GEV fitted on data (*thin solid line*), derived distribution (*thick solid line*)

The hydrologic response to rainfall events changes consequent to a fire. For instance, Scott [266] reported on the consequences to fires in some South African mountain catchments. In the case of lands covered with scrubs there was no significant change in storm discharges but annual total flows increased by 16%, on average, in relation to reductions in transpiration and interception. In timber catchments, on the other hand, there were large increases in storm flows and soil losses. Total flows increased by around 12% in pine forests and decreased somewhat in eucalyptus catchments. Storm hydrographs were higher and steeper, following forest fires, with very little change in effective storm duration. This is attributed to changes in storm flow generation and increased delivery of surface flow. Similar conditions have occurred in Australian timber plantations and eucalyptus forests. Also, very high responses in stream flows have been observed in pine forests in Arizona, USA.

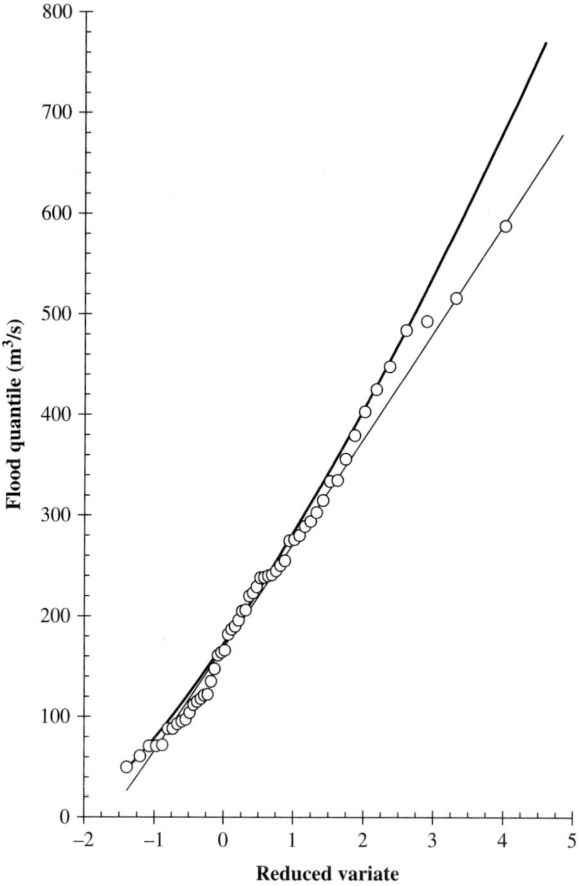

Figure 1.33. Flood frequency curves (in m^3/s) for the Arroscia river basin at Pogli: observations (*circles*), GEV fitted on data (*thin solid line*), derived distribution (*thick solid line*)

The likelihood of fires is higher under hot dry conditions when soil moisture levels are low, as already mentioned. In general, the hydrological effects of fire depend on several factors including the extent of soil heating and soil properties, in addition to the type of vegetation. The effects of the heating of the soil are two-fold. First, there is an increase in the water repellency in the soil. Second, the soil erodibility is higher. In several mountainous areas, there is an established sequence of uncontrolled fires and floods, resulting in erosion. The increased flooding and erosion are a consequence of the water repellency after the fire has had its effect.

Erosion rates increase by as much as 100 times soon after devastation by forest fires and their control costs millions of dollars. The flows are mainly of the debris type. Debris flows can be caused by low intensity rainfall. On the other hand, landslides occur on steep slopes after heavy rains of long duration following a fire.

The soil moisture is recharged and after saturation of the soil, the slopes become liable to slides. On average, gravity is considered to be a more important cause of erosion following a fire than water and wind. The angle of repose, that is the angle between the horizontal and the maximum slope assumed by a soil, can also be crucial with regard to causes of landslides after fires, considering that in the presence of deep rooted plants slopes can have steep angles of repose. Other factors leading to landslides are poor vegetation cover, weakness of the slope material, undermined slopes or unfavorable geology, sustained rainfall and seismic activity. Further details are found in Subsection 1.4.4.

Debris basin sedimentation data in southern California show wild fires to have a strong potential in increasing erosion in burned areas. Because of the transient character of this effect, it is difficult to assess its potential from data analysis, see Figure 1.34.

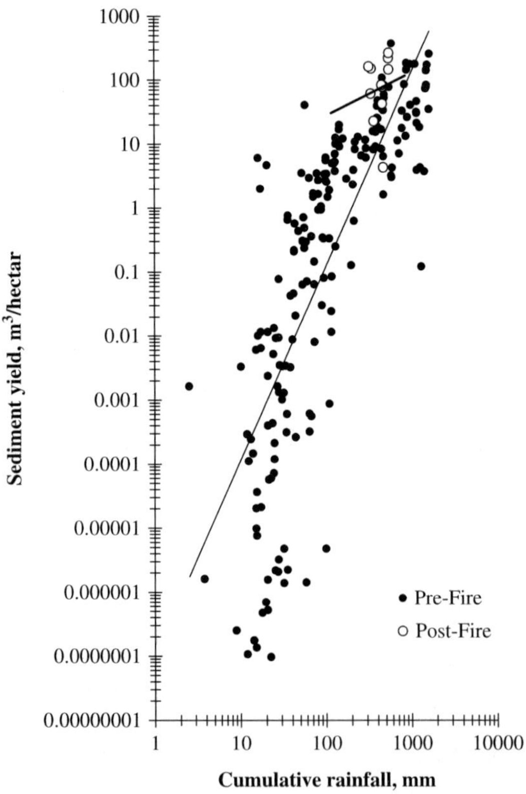

Figure 1.34. Empircal relationship between cumulative sediment yield per unit basin area and cumulative rainfall from nine basins in St. Gabriel mountains, California, USA (40 years of data). *Full dots* denote measurements in pre-fire condition, and *empty dots* denote the measurements in post-fire condition, after [248]

Rulli and Rosso [248] used a physically based model with fine spatial and temporal resolution to predict hydrologic and sediment fluxes for nine small basins in Saint Gabriel mountains under control (pre-fire), and altered (post-fire) conditions. Simulation runs show that the passage of fire significantly modifies the hydrologic response with a major effect on erosion. Long term smulations using observed houly precipitation data showed that the expected annual sediment increases from 7 to 35 times after a forest fire. This also occurs for low frequency quantiles, see Figure 1.35.

These results show the substantial increase of the probability that high erosional rates occur in burned areas, so triggering potential desertification and possibly enhancing the hazard associated with the occurrence of extreme floods and debris torrents. Further experiments in northern Italy [247] indicate that wildfires trigger much higher runoff rates than those expected from unburned soils. Also, the production of woody relics after a forest fire can highly increase the amount of

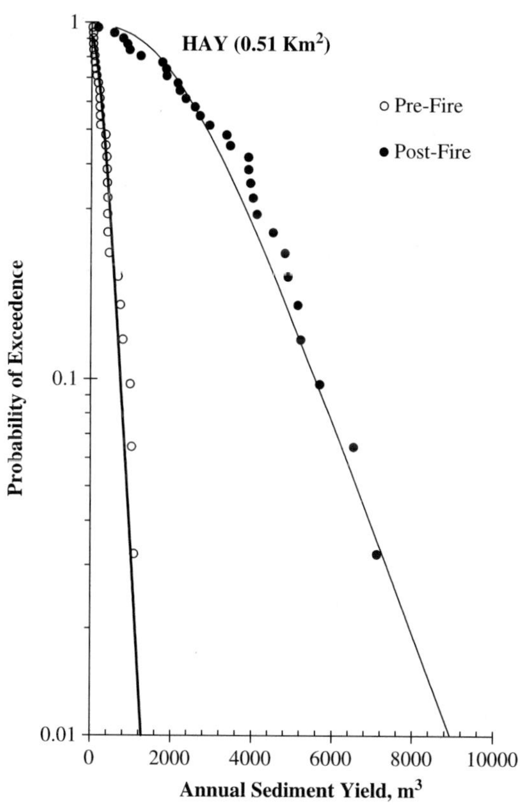

Figure 1.35. Probability of exceedence of annual annual sediment yield from Hay basin (southern California) under pre-fire and post-fire conditions. *Solid lines* denote the Gumbel probability distribution fitted to the simulated data, after [248]

woody debris during a flood [20]. One notes that woody debris can seriously enhance flood risks because of their interaction with obstacles and infrastructures throughout the river network and riparian areas.

Debris-flow initiation processes take a different form on hill slopes. Important considerations are the effects of erodibility, sediment availability on hillslopes and in flow channels, the extent of channel confinement, channel incision, and the contributing areas on the higher parts of the slopes. Cannon [34] found that 30 out of 86 recently burned basins in southern California reacted to the heavy winter rainfall of 1997–1998 with significant debris flows. Straight basins underlain by sedimentary rocks were most likely to produce debris flows dominated by large materials. On the other hand, sand- and gravel-dominated debris flows were generated primarily from decomposed granite terrains. It is known that for a fixed catchment area, there is a threshold value of slope above which debris flows are produced, and below which other flow processes dominate. Besides, it was found that the presence, or absence, of a water-repellent layer in the burned soil and an extensive burn mosaic hardly affected the generation of large grain size debris flows. The presence of water-repellant soils may have led to the generation of sand- and gravel-dominated debris flows.

How forest or wildfires arise, develop and spread depend, as already noted, on the type of prevailing vegetation and various meteorological factors acting in combination. A fire can cause damages that are highly costly to the forest, vegetation and the surrounding environment, and add to atmospheric pollution. It is of utmost importance to prevent the occurrence of such a fire and, if it arises, to control and contain the fire before it starts to spread.

The statistical aspects of forest or wildfires are not well known. This is because the data available are generally insufficient for an intensive study of risk assessment. Where data are available, risk assessments can be easily implemented following the methods given in this Chapter. There are some American and Canadian websites, and a French one: `www.promethee.com`

CHAPTER 2

MULTIVARIATE EXTREME VALUE THEORY

The mathematical theory of multivariate extremes is a relatively novel (but rapidly growing) field. Several areas are well-developed, and there are analogues of the Block and Threshold models discussed in Chapter 1. As for the univariate case, these models have only asymptotic justifications, and their suitability for any practical application must be checked with care.

The notion of *extremal event* in a multidimensional context is closely related to that of *failure region* in structural design, as defined in [48]. Practically, the multivariate observation $\boldsymbol{X} = (x_1, \ldots, x_d) \in \mathbb{R}^d$ is extreme if it falls into some failure region $\mathcal{F} \in \mathbb{R}^d$ having a small probability of being reached. For instance, $\mathcal{F}$ can be defined as $\left\{(x_1, \ldots, x_d) \in \mathbb{R}^d : \max_{1 \le i \le d} \{x_i\} > x^*\right\}$, for a given high threshold $x^* \gg 1$.

A first problem arising when working in a multidimensional context is the lack of a "natural" definition of extreme values: essentially, this is due to the fact that different concepts of ordering are possible. In addition, dimensionlity creates difficulties for both model validation and computation, and models are less fully prescribed by the general theory. In two and more dimensions there does not exist a simple distinction in three basic domains of attraction as in Chapter 1: for instance, there is no reason for the univariate marginals of a multivariate distribution to share the same type of limiting Extreme Value probability law — this will be modeled in later chapters by using copulas, where the marginals will no longer represent a problem.

Furthermore, in a multidimensional environment, the issue of dependence between different variables plays a fundamental role. Indeed, according to [46], quantifying dependence is a central theme in probabilistic and statistical methods for multivariate extreme values. Two limiting situations are possible: one where the extremes are dependent (is some sense), and the other where the extremes are independent (in the same sense). Actually, the strength of dependence may change by considering different "levels" of the process under investigation: for instance, it may become weaker for more extreme events, to the extent that the most extreme events are practically independent. We shall see in later chapters how to quantify the degree of dependence via the tail dependence coefficients, a copula feature (see

Section 3.4 and Section 5.3). A thorough discussion can be found in [46], where diagnostic measures for dependence are also developed.

This chapter is mostly a theoretical one, for we introduce some basic notions concerning multivariate extremes that will be re-interpreted later in a copula-based context. The approach we follow is the standard one, involving componentwise maxima: the corresponding definitions of multivariate extremes seem to be useful for several practical purposes, although some drawbacks will be emphasized. As fundamental bibliographical sources, both from a theoretical and a practical point of view, we indicate [187, 232, 36, 83, 155, 309, 169, 45], and references therein. A thorough review can be found in [201]; further references will be given throughout. Hereafter, for any integer $d > 1$, we use the vectorial notation in $\mathbb{R}^d$, i.e. $\boldsymbol{x} = (x_1, \ldots, x_d)$. Also, unless otherwise specified, multidimensional operations and inequalities are treated componentwise.

2.1. MULTIVARIATE EXTREME VALUE DISTRIBUTIONS

In multivariate Extreme Value analysis it is a standard practice to investigate vectors of *componentwise maxima* and *minima*, as defined below.

DEFINITION 2.1 (Vector of componentwise maxima and minima).
Let $\{(X_{i,1}, \ldots, X_{i,d})\}$, $i = 1, \ldots, n$, *be a sample of i.i.d. d-variate r.v.'s with joint distribution F. The corresponding vector of componentwise maxima* $\boldsymbol{M}_n$ *is defined as*

$$\boldsymbol{M}_n = (M_{n,1}, \ldots, M_{n,d}) = \left(\max_{1\le i\le n} \{X_{i,1}\}, \ldots, \max_{1\le i\le n} \{X_{i,d}\} \right). \tag{2.1}$$

Similarly, the vector of componentwise minima $\widetilde{\boldsymbol{M}}$ *is defined as*

$$\widetilde{\boldsymbol{M}}_n = (\widetilde{M}_{n,1}, \ldots, \widetilde{M}_{n,d}) = \left(\min_{1\le i\le n} \{X_{i,1}\}, \ldots, \min_{1\le i\le n} \{X_{i,d}\} \right). \tag{2.2}$$

As in the univariate case, the extremal behavior of multivariate maxima is based on the limiting performance of "block" maxima. However, it is important to realize that the maximum of each of the d different marginal sequences $X_{i,1}$'s,...,$X_{i,d}$'s may occur for different indices, say $i_1^*, \ldots, i_d^*$. As a consequence, $\boldsymbol{M}_n$ does not necessarily correspond to an observed sample value in the original series. For this reason the analysis of componentwise maxima may sometimes be of little use. Clearly, the same comment also holds for minima.

ILLUSTRATION 2.1 (Storm intensity–duration). ►

In [255] a seasonal analysis of the joint behavior of *storm intensity I* and *storm duration W* is used — see also Illustration 4.4 for more details. The data consist of seven years of hourly rainfall depth measurements collected at the Scoffera station, located in the Bisagno river basin (Thyrrhenian Liguria, northwestern Italy).

A non-rainy period lasting (at least) seven hours is used to separate two successive storms: a sequence of 691 storms were identified and extracted from the database. In Figure 2.1 we plot the available sample pairs.

In Figure 2.1 the seasonal analysis emphasizes different stochastic behaviors of the pair (I, W) in the four seasons. In particular, in winter the storms are characterized by small intensities ($I < 8$ mm/h) and long durations, while in summer they show short durations ($W < 40$ h) and large intensities. The behaviors in spring and fall are roughly "intermediate" between those in winter and summer. Also, largest values of one variable are always associated with smallest values of the other: see Illustrations B.1–B.2 for a thorough discussion. Most importantly, none of the vectors of componentwise maxima plotted (*empty circles*) correspond to any observed pair (I, W). ◄

As in the univariate case, a standard way to operate is to look for the existence of sequences of constants, $\{a_{n,i}\}$ and $\{b_{n,i} > 0\}$, $1 \le i \le d$, such that, for all $\boldsymbol{x} \in \mathbb{R}^d$, the function

$$\begin{aligned} G(x_1, \ldots, x_d) &= \lim_{n\to\infty} \mathbb{P}\left\{\frac{M_{n,1} - a_{n,1}}{b_{n,1}} \le x_1, \ldots, \frac{M_{n,d} - a_{n,d}}{b_{n,d}} \le x_d\right\} \\ &= \lim_{n\to\infty} F^n(a_{n,1} + b_{n,1}x_1, \ldots, a_{n,d} + b_{n,d}x_d), \end{aligned} \tag{2.3}$$

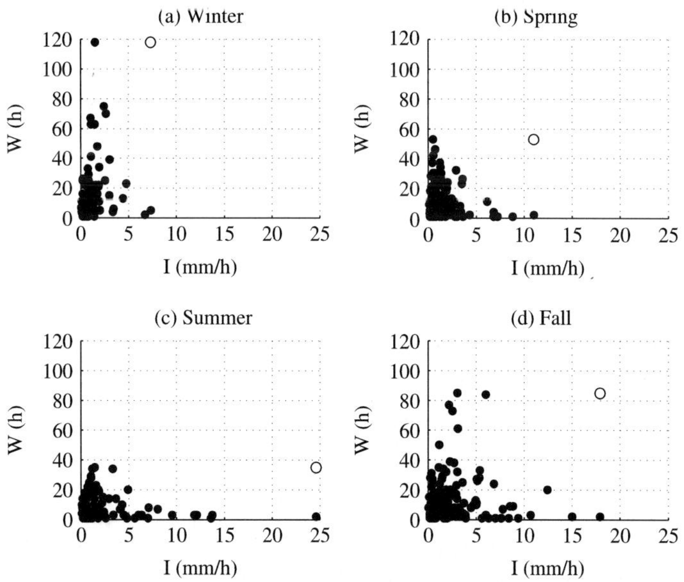

Figure 2.1. Observed pairs (I, W) (*full circles*) for the data analysed by [255] in each season. The *empty circles* represent the pairs $(\max_i \{I_i\}, \max_i \{W_i\})$

is a proper distribution with non-degenerate marginals. For minima, Eq. (2.3) is replaced by

$$\overline{G}(x_1,\dots,x_d) = \lim_{n\to\infty} \mathbb{P}\left\{\frac{\widetilde{M}_{n,1}-\tilde{a}_{n,1}}{\tilde{b}_{n,1}} \geq x_1,\dots,\frac{\widetilde{M}_{n,d}-\tilde{a}_{n,d}}{\tilde{b}_{n,d}} \geq x_d\right\} \tag{2.4}$$
$$= \lim_{n\to\infty} \overline{F}^n(\tilde{a}_{n,1}+\tilde{b}_{n,1}x_1,\dots,\tilde{a}_{n,d}+\tilde{b}_{n,d}x_d),$$

where the notation $\overline{G}$ and $\overline{F}$ indicates the *survival function* associated with G and F defined below.

DEFINITION 2.2 (Survival function). *The survival function $\overline{F}$ associated with the multivariate distribution F is given by*

$$\overline{F}(x_1,\dots,x_d) = \mathbb{P}\{X_1 > x_1,\dots,X_d > x_d\}. \tag{2.5}$$

NOTE 2.1. Evidently, in the univariate case $d = 1$,

$$\overline{F}(x) = 1 - F(x) \tag{2.6}$$

for all $x \in \mathbb{R}$. Unfortunately, in general this simple relationship fails in a multivariate context (see [36] for details and Illustration 5.6 for a practical example).

The notion of *Maximum Domain of Attraction* is fundamental in multivariate Extreme Value analysis.

DEFINITION 2.3 (Maximum Domain of Attraction (MDA)). *The distribution F is said to belong to the Maximum Domain of Attraction (MDA) of the Multivariate Extreme Value (MEV) distribution G if there exist sequences of constants, $\{a_{n,i}\}$ and $\{b_{n,i} > 0\}$, $1 \leq i \leq d$, such that Eq. (2.3) is satisfied. A similar definition can be given for minima.*

The following proposition provides an important result.

PROPOSITION 2.1. *The following equivalences hold for all $\boldsymbol{x} \in \mathbb{R}^d$ such that $G(\boldsymbol{x}) > 0$ and $\overline{G}(\boldsymbol{x}) > 0$.*

1. *Eq. (2.3) is equivalent to*

$$\lim_{n\to\infty} n\left[1 - F(a_{n,1}+b_{n,1}x_1,\dots,a_{n,d}+b_{n,d}x_d)\right] = -\ln G(\boldsymbol{x}). \tag{2.7}$$

2. *Eq. (2.4) is equivalent to*

$$\lim_{n\to\infty} n\left[1 - \overline{F}(\tilde{a}_{n,1}+\tilde{b}_{n,1}x_1,\dots,\tilde{a}_{n,d}+\tilde{b}_{n,d}x_d)\right] = -\ln \overline{G}(\boldsymbol{x}). \tag{2.8}$$

NOTE 2.2. For the same reasons pointed out in the univariate case (see Note 1.6), we need not study separately the behavior of the vector of *componentwise minima* $\widetilde{\boldsymbol{M}}_n$. Henceforth we concentrate on the analysis of maxima only.

It can be shown from *Pickands's representation* [187] that the condition stated by Eq. (2.3) is equivalent to convergence in distribution, and that MEV distributions are *continuous* (but not always absolutely continuous). A further fundamental point is as follows. Setting all the x_i's but one to $+\infty$ in Eq. (2.3) yields

$$\lim_{n\to\infty} F_i^n(a_{n,i}+b_{n,i}x_i) = \mathrm{G}_i(x_i), \quad i = 1, \ldots, d, \tag{2.9}$$

where F_i and G_i are, respectively, the i-th marginals of F and G. In turn, $F_i \in \mathrm{MDA}(\mathrm{G}_i)$, where G_i is a Type I, II, or III distribution, as prescribed by Theorem 1.7 or, equivalently, a member of the GEV family. As a consequence, the norming constants in Eq. (2.3) are precisely those calculated in Theorem 1.14 and the ensuing propositions.

As discussed in [169] (see also Section 5.1), in order to isolate the dependence features from the marginal distributional aspects, traditionally the components of both the distribution F and the corresponding MEV law G are transformed to standard marginals. It can be shown [232] that this does not pose difficulties. For technical convenience, it is customary to choose the standard Fréchet distribution (see Section 1.2) as marginals, i.e. the function $\phi : \mathbb{R} \to \mathbb{I}$ given by $\phi(x) = \exp(-1/x)$ for $x > 0$, and zero elsewhere. Clearly, its inverse is the function $\phi^{-1} : \mathbb{I} \to \mathbb{R}_+$ given by $\phi^{-1}(x) = -1/\ln x$.

Let $\boldsymbol{X}$ be a d-variate r.v. with distribution F and continuous marginals, and define

$$\boldsymbol{Y} = \phi^{-1}(F(\boldsymbol{X})) = -\frac{1}{\ln F(\boldsymbol{X})}, \tag{2.10}$$

i.e. $Y_i = \phi^{-1}(F_i(X_i)) = -1/\ln F_i(X_i)$, for all indices $1 \le i \le d$. Via the *Probability Integral Transform*, it follows that $\boldsymbol{Y}$ has standard Fréchet marginals. The following propositions shed light on the role of the above transformation.

PROPOSITION 2.2. *Let* G *be a multivariate distribution with continuous marginals* G_i*'s, and define*

$$\mathrm{G}^*(y_1, \ldots, y_d) = \mathrm{G}\left(\left(\frac{-1}{\ln \mathrm{G}_1}\right)^{(-1)}(y_1), \ldots, \left(\frac{-1}{\ln \mathrm{G}_d}\right)^{(-1)}(y_d)\right), \tag{2.11}$$

with $\mathbf{y} \ge 0$. *Then* G^* *has standard Fréchet marginals, and* G *is a MEV distribution if, and only if,* G^* *is also MEV distributed.*

Thus, the marginals of a MEV distribution can be standardized, yet preserving the extreme value properties.

PROPOSITION 2.3. *Let F be as in Eq. (2.10), and let F^* be the distribution of* $\mathbf{Y}$. *Also, let* G *and* G^* *be as in Proposition 2.2. If* $F \in MDA(\mathrm{G})$, *then* $F^* \in MDA(\mathrm{G}^*)$. *Conversely, if* $F^* \in MDA(\mathrm{G}^*)$, G^* *has non-degenerate marginals, and Eq. (2.9) holds, then* $F \in MDA(\mathrm{G})$.

Thus, the standardization is justified by showing that $F \in \mathrm{MDA}(\mathrm{G})$ if, and only if, $F^* \in \mathrm{MDA}(\mathrm{G}^*)$. In addition, as explained in [169], it helps in characterizing the MEV distribution G^*, as well as the domain of attraction conditions (see Section 2.2).

ILLUSTRATION 2.2 (Regionalization method (cont.)). ►

Flood frequency regionalization procedures available in the literature are based on the assumption of spatial independence between flood peaks at different sites (see Subsection 1.4.9 for an introduction). However, the meteo-climatic events generating floods may yield flood peaks at neighboring stations showing some degree of dependence, and this may affect the flood estimates.

Matalas and Benson [189] were the first to address the effect of intersite dependence in flood frequency analysis: they showed how the estimates of the coefficients of the regression method are not influenced by intersite dependence among sites, but their variance is affected. The effect of dependence was investigated in [277] for a *index-flood* procedure: as a result, the estimators of the distribution's statistics are unbiased under the presence of intersite dependence, but their variance may be seriously affected. Using a multivariate normal distribution, [140] studied the effect of spatial dependence on the estimates obtained, respectively, via the Generalized Extreme Value and the Log-Pearson type III distributions among others. The investigation of flood frequency regionalization using two multivariate Lognormal models was made in [221]: Maximum Likelihood estimation procedures were developed for the models considered, and the asymptotic variances of regional quantile estimators were derived analytically.

We outline here a MEV model generally used to study the problem of flood frequency regionalization via the *index-flood* method (see Illustration 1.36). Regional flood frequency analysis is the recommended procedure for estimating flood quantiles in small watersheds, for high return periods, by combining all the available information in a *homogeneous* area [49]. A prerequisite for this approach is the identification of a homogeneous flood region given that historical data in a number of river sites are available [128, 30, 127, 64].

An application of flood frequency regionalization in northwestern Italy is given here by considering some of the homogeneous regions identified in [64]. In particular:

- **Region A**, i.e. central Alps and Prealps that include Po sub-basins from Chiese to Sesia river basin (14 river gauged sites, with an average sample size of 23 years of annual data);
- **Region B**, i.e. western Alps and Prealps from Dora Baltea river to Rio Grana (14 river gauged sites, with an average sample size of 25 years of annual data);

- **Region C**, i.e. northwestern Apennines and Thyrrhenian basins that include Ligurian basins with outlet to the Thyrrhenian sea, and Po sub-basins from Scrivia to Taro river basins (27 river gauged sites, with an average sample size of 27 years of annual data).

The estimates of the parameters of the common regional univariate law may change if a specific multivariate dependence model is introduced to describe the data. Often the *logistic model* developed by [125] is adopted. The joint d-variate distribution G of the random vector $(X_1, \ldots, X_d)$ is given by

$$\mathrm{G}(x_1, \ldots, x_d) = e^{-\left[\sum_{i=1}^{d}(-\ln F_i(x_i))^{\theta}\right]^{1/\theta}}, \tag{2.12}$$

where F_i is the marginal distribution of X_i, and $\theta \in [1, \infty)$ is a *dependence parameter* — see Section C.2 for an interpretation in terms of copulas. In the present case, all the marginal F_i's are identical, and equal the common regional distribution, say F. The logistic model assumes that the dependence between each pair of sites is ruled only by θ, that can be viewed as a measure of the regional dependence [136]. A different approach will be presented in Illustration 5.6. The limit case $\theta \to 1$ corresponds to independence between the variables, whereas the limit case $\theta \to \infty$ corresponds to complete dependence (see Appendix B for a survey on the concepts of dependence and association). Note that the logistic model accounts for a positive association between the variables, the one of interest here. For a list of alternative models see, e.g., [144, 281, 47, 154, 166, 155].

The GEV distribution is chosen as the common law F for all the sites: thus, only four parameters need to be estimated (i.e., θ and the three GEV parameters a', b', γ'). In Table 2.1 we give the estimates of θ for all the regions considered. In the same Table we also compare the estimates of the GEV parameters both when fitted together with θ (i.e., using the four-dimensional MEV model given by Eq. (2.12) — "MEV" column), and when no dependence model is considered as in Table 1.10 ("IND" column). In the former case a Maximum Likelihood procedure is used.

As a general comment, the estimates of the parameters are practically the same for both the MEV and IND models. Only a slight difference is found considering the shape parameter in Region A. Overall, the regional dependence measured by θ is weak. However, it affects the variance of the quantile estimates for high return periods, as shown in Table 2.2.

Table 2.1. Estimates of the regional parameters — see text

	Region A		**Region B**		**Region C**	
Par.	**MEV**	**IND**	**MEV**	**IND**	**MEV**	**IND**
a'	0.760	0.745	0.656	0.635	0.646	0.643
b'	0.400	0.365	0.395	0.352	0.395	0.377
γ'	0.223	0.110	0.336	0.320	0.296	0.276
θ	1.412		1.350		1.193	

Table 2.2. Estimates of several τ-year normalized quantiles $\hat{x}_\tau$ and the corresponding variances of estimate using MEV and IND models — see text

Region	τ (years)	$\hat{\mathbf{x}}_\tau$ **MEV**	$\hat{\mathbf{x}}_\tau$ **IND**	$\mathbb{V}(\hat{x}_\tau)\cdot 10^3$ **MEV**	$\mathbb{V}(\hat{x}_\tau)\cdot 10^3$ **IND**
	5	1.472	1.340	9.712	2.020
	10	1.929	1.677	25.235	4.396
A	50	3.248	2.524	180.836	24.340
	100	3.970	2.931	381.119	50.178
	200	4.809	3.368	763.846	103.171
	5	1.426	1.313	12.699	3.539
	10	1.984	1.795	37.665	11.068
B	50	3.842	3.369	345.432	136.124
	100	4.995	4.329	810.030	393.261
	200	6.447	5.524	1813.87	1131.740
	5	1.392	1.343	5.296	1.613
	10	1.909	1.819	14.126	4.621
C	50	3.547	3.287	110.909	46.833
	100	4.519	4.139	247.593	124.676
	200	5.710	5.168	529.599	330.723

The results show that the ratio between MEV and IND quantiles is, on average, ≈ 1.7 in Region A, ≈ 1.1 in Region B, and ≈ 1.1 in Region C (and is an increasing function of the return period). Similarly, the ratio between MEV and IND variances is, on average, ≈ 6.6 in Region A, ≈ 2.6 in Region B, and ≈ 2.5 in Region C. This analysis indicates how the effect of spatial dependence on the variance of quantile estimates is non-negligible, also for small values of the dependence parameter θ. ◀

2.2. CHARACTERIZATION OF THE DOMAIN OF ATTRACTION

The concept of MDA for the multivariate case is less straightforward than in a univariate context, and its characterizations can be classified into those that are just necessary, or just sufficient, or both necessary and sufficient.

An interesting necessary characterization is due to de Haan [60], and involves *point processes*. Let $\{(X_{i,1},\dots,X_{i,d})\}$, $i\in\mathbb{N}$, be a sequence of i.i.d. d-variate r.v.'s with common distribution $F\in$ MDA(G) for some suitable G. Also, let

$$\mathcal{S}_d = \left\{ (w_1,\dots,w_{d-1}) : \sum_{i=1}^{d-1} w_i \le 1 \quad \text{and} \quad w_i \ge 0, 1\le i\le d-1 \right\} \tag{2.13}$$

be the $(d-1)$-dimensional unit *simplex*, and define $U_i(\cdot) = -1/\ln F_i(\cdot)$, $1\le i\le d$, and the transformation

$$T(y_1,\dots,y_d) = \left(\overline{y}, \frac{y_1}{\overline{y}},\dots,\frac{y_{d-1}}{\overline{y}}\right), \tag{2.14}$$

where $\overline{y} = \sum_{i=1}^d y_i$. The following result is fundamental.

THEOREM 2.1. *Let* $\mathcal{P}_n = \{(U_1(X_{i,1})/n, \ldots, U_d(X_{i,d})/n), i = 1, \ldots, n\}$ *be a point process. Then*

$$\mathcal{P}_n \xrightarrow{n\to\infty} \mathcal{P}, \tag{2.15}$$

where $\mathcal{P}$ *is a non-homogeneous Poisson process on* $\mathbb{R}^d_+$ *with intensity measure* λ^* *given by*

$$\lambda^* \circ T^{[-1]}(\mathrm{d}r, \mathrm{d}\boldsymbol{w}) = r^{-2}\,\mathrm{d}r\, H^*(\mathrm{d}\boldsymbol{w}), \tag{2.16}$$

where $r > 0$, $\boldsymbol{w} \in \mathcal{S}_d$, *and* H^* *is a non-negative measure on* $\mathcal{S}_d$ *satisfying*

$$H^*(\mathcal{S}_d) = d \quad \text{and} \quad \int_{\mathcal{S}_d} w_i\, H^*(\mathrm{d}\boldsymbol{w}) = 1 \tag{2.17}$$

for $i = 1, \ldots, d-1$.

As a consequence, the function G^* in Proposition 2.2 can be written as

$$G^*(y_1, \ldots, y_d) = \exp\{-V(y_1, \ldots, y_d)\}, \tag{2.18}$$

where

$$\begin{aligned} V(y_1, \ldots, y_d) &= \lambda^* \left\{([0, y_1] \times \cdots \times [0, y_d])^{\mathsf{C}}\right\} \\ &= \int_{\mathcal{S}_d} \max\left\{\frac{w_1}{y_1}, \ldots, \frac{1 - \sum_{i=1}^{d-1} w_i}{y_d}\right\} H^*(\mathrm{d}\boldsymbol{w}). \end{aligned} \tag{2.19}$$

In the literature V is usually referred to as the *exponent measure function* (or, sometimes, as the *dependence function* — see Definition 5.5 and the ensuing discussion).

NOTE 2.3. The constraints in Eq. (2.17) are the only ones on the non-negative measure H^*. Therefore, no finite parametrization exists for this measure.

According to [169], the point process approach illustrated above has an intuitive explanation in the two-dimensional case (see also [45] for a thorough discussion). In particular, the following facts are evident.

1. The transformations U_i's force a standardization of the $X_{i,j}$'s, $j = 1, \ldots, d$, to have unit Fréchet distributions.
2. The transformation T maps the vector $(U_1(X_{i,1}), \ldots, U_d(X_{i,d})) \in \mathbb{R}^d_+$ into pseudo-polar coordinates in $(0, \infty) \times \mathcal{S}_d$.
3. As $n \to \infty$, the components of $\mathcal{P}_n$ are made negligible by the scaling factor $1/n$, except those with unusually large values $X_{i,1}$, or $X_{i,2}$, or both (here $d = 2$). In the former cases, points are dragged down in $\mathbb{R}^2_+$ either to the horizontal or to the vertical axis. In the latter case, points with both exceptionally large components will survive in $\mathbb{R}^2_+$ away from the boundaries.

Thus, the limiting intensity measure λ^* describes the dependence structure between unusually large $X_{i,j}$'s, $j = 1, \ldots, d$, after standardization. Also, under the mapping induced by T, λ^* factorizes into a known function of the radial component r, and a non-negative measure H^* of the angular component $\boldsymbol{w}$. Practically, it is H^* that embodies the dependence structure of the extremes.

- If H^* concentrates its mass in the interior of $\mathcal{S}_d$, then it describes *strong* dependence. For instance, *total dependence* (see Section 2.3) corresponds to H^* having all its mass at $\{(1/d, \ldots, 1/d)\}$, i.e.

$$H^*(\{(1/d, \ldots, 1/d)\}) = d. \tag{2.20}$$

- If H^* concentrates its mass near the boundary of $\mathcal{S}_d$, then it describes *weak* dependence. For instance, *total independence* (see Section 2.3) corresponds to H^* having all its mass at the vertices, i.e.

$$H^*(\{(1/d, \ldots, 0)\}) = \cdots = H^*(\{(0, \ldots, 1/d)\}) = 1. \tag{2.21}$$

The issues of total dependence and independence will be made definitely clear in later chapters by introducing the copulas $\mathbf{M}_d$ and $\mathbf{\Pi}_d$.

As shown in [232, 169], the limiting intensity measure λ^* can be used to provide necessary and sufficient conditions for a multivariate distribution F^* to belong to the maximum domain of attraction of G^*; here the same notation as in Proposition 2.3 is used.

PROPOSITION 2.4. *The distribution $F^* \in MDA(G^*)$ if, and only if,*

$$t\,\mathbb{P}\left\{t^{-1}\boldsymbol{Y} \in B\right\} \xrightarrow{t\to\infty} \lambda^*(B) \tag{2.22}$$

for all relatively compact sets B for which the boundary of B has zero λ^*-measure.

A further necessary and sufficient condition [187, 169] involves the vector $\boldsymbol{Y}$ given by Eq. (2.10). Here G^* is expressed in terms of the conditional distribution of $\boldsymbol{Y}$ given that at least one of its components has exceeded the threshold t.

PROPOSITION 2.5. *The distribution $F^* \in MDA(G^*)$ if, and only if,*

$$\frac{-\ln F^*(ty_1, \ldots, ty_d)}{-\ln F^*(t, \ldots, t)} \xrightarrow{t\to\infty} \frac{-\ln G^*(y_1, \ldots, y_d)}{-\ln G^*(1, \ldots, 1)} \tag{2.23}$$

for each $y_i > 0$, $i = 1, \ldots, d$.

The theorems presented below [187] state necessary and sufficient conditions for a multivariate distribution F to belong to a given maximum domain of attraction.

THEOREM 2.2. *Let* G *be a d-variate MEV distribution such that its marginals are standard Gumbel probability laws. Then* $F \in MDA(\mathrm{G})$ *if, and only if,*

$$\lim_{t \to \omega_{F_1}} \frac{1 - F(\boldsymbol{a}(t)\boldsymbol{x} + \boldsymbol{b}(t))}{1 - F_1(t)} = -\ln \mathrm{G}(\boldsymbol{x}) \tag{2.24}$$

for all $\boldsymbol{x}$ *such that* $\mathrm{G}(\boldsymbol{x}) > 0$, *where* $a_i(t) = \overline{F}_i^{[-1]}\left[e^{-1}\overline{F}_i(t)\right] - \overline{F}_i^{[-1]}\left[\overline{F}_1(t)\right]$ *and* $b_i(t) = \overline{F}_i^{[-1]}\left[\overline{F}_1(t)\right]$, $i = 1, \ldots, d$.

THEOREM 2.3. *Let* G *be a d-variate MEV distribution such that its marginals are Fréchet probability laws with parameters* γ_i*'s,* $i = 1, \ldots, d$, *and let* $g_i(t) = \overline{F}_i^{[-1]}\left[\overline{F}_1(t)\right]$, *with* $t \in \mathbb{R}$ *and* $i = 2, \ldots, d$. *Then* $F \in MDA(\mathrm{G})$ *if, and only if,*

$$\lim_{t \to \infty} \frac{1 - F(tx_1, g_2(t)x_2, \ldots, g_d(t)x_d)}{1 - F_1(t)} = -\ln \mathrm{G}(\boldsymbol{x}) \tag{2.25}$$

for all $\boldsymbol{x}$ *such that* $\mathrm{G}(\boldsymbol{x}) > 0$.

THEOREM 2.4. *Let* G *be a d-variate MEV distribution such that its marginals are Weibull probability laws with parameters* γ_i*'s,* $i = 1, \ldots, d$. *Then* $F \in MDA(\mathrm{G})$ *if, and only if,*

1. *there exists* $\overline{\boldsymbol{x}} \in \mathbb{R}^d$ *such that* $F(\overline{\boldsymbol{x}}) = 1$ *and* $F(\boldsymbol{x}) < 1$ *if* $\boldsymbol{x} < \overline{\boldsymbol{x}}$, *and*
2. *given* $g_i(t) = \overline{x}_i - \overline{F}_i^{[-1]}\left[\overline{F}_1(\overline{x}_1 - t)\right]$, $i = 2, \ldots, d$,

$$\lim_{t \to 0^+} \frac{1 - F([tx_1, g_2(t)x_2, \ldots, g_d(t)x_d] + \overline{\boldsymbol{x}})}{1 - F_1(\overline{x}_1 - t)} = -\ln \mathrm{G}(\boldsymbol{x}) \tag{2.26}$$

for all $\boldsymbol{x}$ *such that* $\mathrm{G}(\boldsymbol{x}) > 0$.

Other conditions can be stated by assuming that the marginals of F are absolutely continuous [187]. Analogous results can be given for minima.

2.3. MULTIVARIATE DEPENDENCE

In the multivariate case the notions of dependence become numerous and complex, and their mutual relationships are difficult to understand. Below we introduce some basic concepts involving distribution functions. Further details, generalizations and extensions are given in Appendix B, where copulas are used to clarify the issue.

DEFINITION 2.4 (Positive orthant dependence). *Let* $d > 1$, *and let* $\boldsymbol{X} = (X_1, \ldots, X_d)$ *be a d-variate random vector.*

1. $\boldsymbol{X}$ *is positively lower orthant dependent (PLOD) if, for all* $\boldsymbol{x} \in \mathbb{R}^d$,

$$\mathbb{P}\{\boldsymbol{X} \le \boldsymbol{x}\} \ge \prod_{i=1}^{d} \mathbb{P}\{X_i \le x_i\}. \tag{2.27}$$

2. *$\boldsymbol{X}$ is positively upper orthant dependent (PUOD) if, for all $\boldsymbol{x} \in \mathbb{R}^d$,*

$$\mathbb{P}\{\boldsymbol{X} > \boldsymbol{x}\} \geq \prod_{i=1}^{d} \mathbb{P}\{X_i > x_i\}. \tag{2.28}$$

Overall, $\boldsymbol{X}$ is positively orthant dependent (POD) if, for all $\boldsymbol{x} \in \mathbb{R}^d$, both Eqs. (2.27)–(2.28) hold.

The definitions of *negative lower orthant dependence* (NLOD), *negative upper orthant dependence* (NUOD), and *negative orthant dependence* (NOD) can be introduced simply by reversing the sense of the corresponding inequalities in Definition 2.4.

NOTE 2.4. In the bivariate case the definitions of PLOD and PUOD are equivalent to that of *positive quadrant dependence* (PQD) given in Subsection B.1.1. In particular, PLOD and PUOD are the same for $d = 2$. However, this is not the case for $d \geq 3$.

If $\boldsymbol{X}$ has joint distribution F with marginals F_i, $i = 1, \ldots, d$, then Eq. (2.27) is equivalent to

$$F(x_1, \ldots, x_d) \geq F_1(x_1) \cdots F_d(x_d) \tag{2.29}$$

for all $\boldsymbol{x} \in \mathbb{R}^d$. Analogously, Eq. (2.28) is equivalent to

$$\overline{F}(x_1, \ldots, x_d) \geq \overline{F}_1(x_1) \cdots \overline{F}_d(x_d) \tag{2.30}$$

for all $\boldsymbol{x} \in \mathbb{R}^d$. A generalization in terms of copulas is shown in Subsection B.1.1. A further concept of dependence is that of *associated* variables [187].

DEFINITION 2.5 (Associated variables). *The r.v.'s $X_1, \ldots, X_d$ are said to be (positively) associated if, for every pair a, b of non-decreasing real-valued functions defined on $\mathbb{R}^d$,*

$$\mathbb{C}\left(a(X_1, \ldots, X_d), b(X_1, \ldots, X_d)\right) \geq 0 \tag{2.31}$$

whenever the relevant expectations exist.

This concept of (positive) dependence was introduced in [86], and is stronger than other forms of dependence (see Subsection B.1.1). The following theorem is fundamental, and emphasizes an important feature of MEV distributions.

THEOREM 2.5. *If the r.v.'s $X_1, \ldots, X_d$ have a MEV distribution, then they are associated.*

A further important feature of MEV distributions is as follows [285].

THEOREM 2.6. *A MEV distribution G satisfies the condition*

$$G(x_1, \ldots, x_d) \geq \prod_{i=1}^{d} G_i(x_i) \tag{2.32}$$

for all $\boldsymbol{x} \in \mathbb{R}^d$.

An analogous property for MEV copulas will be shown in Theorem 5.6. As pointed out in [169], the two extreme forms of limiting multivariate distributions correspond to the cases of (asymptotic) *total independence* and *total dependence* between the componentwise maxima. In the former case

$$G(x_1, \ldots, x_d) = \prod_{i=1}^{d} G_i(x_i), \tag{2.33}$$

while in the latter case

$$G(x_1, \ldots, x_d) = \min\left\{G_1(x_1), \ldots, G_d(x_d)\right\}, \tag{2.34}$$

for all $\boldsymbol{x} \in \mathbb{R}^d$. A generalization in terms of copulas will be given in Illustration 4.2. In particular, the study of independence in MEV distributions is greatly facilitated by the following result [16] — see also Proposition 5.5 for an interpretation in terms of copulas.

PROPOSITION 2.6. *Pairwise-independent r.v.'s having a joint MEV distribution are mutually independent.*

Thus, the study of asymptotic independence can be confined to the bivariate case — see Section 5.3, and also [187] for a thorough discussion. Some conditions for asymptotic independence are illustrated in [169], and are reported below.

PROPOSITION 2.7. *Asymptotic total independence arises if, and only if, Eq. (2.9) holds, and there exists $\boldsymbol{x} \in \mathbb{R}^d$ such that $0 < G_i(x_i) < 1$, $i = 1, \ldots, d$, and*

$$F^n(a_{n,1} + b_{n,1}x_1, \ldots, a_{n,d} + b_{n,d}x_d) \xrightarrow{n \to \infty} G_1(x_1) \cdots G_d(x_d). \tag{2.35}$$

Moreover, Eq. (2.33) holds for any $\boldsymbol{x} \in \mathbb{R}^d$ if, and only if,

1.

$$G(0, \ldots, 0) = G_1(0) \cdots G_d(0) = \exp(-d), \tag{2.36}$$

provided that the G_i's are standard Gumbel distributions; or

2.

$$G(1, \ldots, 1) = G_1(1) \cdots G_d(1) = \exp(-d), \tag{2.37}$$

provided that the G_i's are Fréchet distributions; or

3.

$$G(-1, \ldots, -1) = G_1(-1) \cdots G_d(-1) = \exp(-d), \tag{2.38}$$

provided that the G_i's are Weibull distributions.

Similar conditions hold for the case of asymptotic total dependence [169], and are reported below.

PROPOSITION 2.8. *Asymptotic total dependence arises if, and only if, Eq. (2.9) holds, and there exists $\boldsymbol{x} \in \mathbb{R}^d$ such that $0 < G_1(x_1) = \cdots = G_d(x_d) < 1$, and*

$$F^n(a_{n,1} + b_{n,1}x_1, \ldots, a_{n,d} + b_{n,d}x_d) \xrightarrow{n\to\infty} G_1(x_1). \tag{2.39}$$

For further conditions we refer to [187].

2.4. MULTIVARIATE RETURN PERIODS

The *return period* of a given event is usually defined as the average time elapsing between two successive realizations of the event. As discussed in Section 1.3, the return period of a prescribed event is generally adopted in applications as a common criterion for design purposes. Indeed, the return period provides a very simple and efficient means for risk analysis. Thus, one can use a single number to represent an important information.

In many situations, the analysis of the return period involves univariate cases — as, e.g., in Section 1.4. Unfortunately, this may lead to an overestimation or underestimation of the risk associated with a given event (see, e.g., the discussion and the example in Subsection 3.3.1). Natural events are often characterized by the joint behavior of several random variables, and these are usually non-independent. As a consequence, the relevant events should be defined in terms of two or more variables. Of course, this makes things complicated, since the family of pertinent events increases with the number of variables.

For the sake of clarity and simplicity, we shall now concentrate on the bivariate case. However, generalizations to higher dimensional settings will be evident. In bivariate frequency analysis an event with a given return period is not clearly defined, nor is it in multivariate analysis. Unfortunately, the literature on the subject is relatively sparse (for a review see [278, 308], and references therein), and in some studies the concepts are even misrepresented [307]. Clearly, incorrect interpretation of these concepts may lead to misinterpretation of frequency analysis results. For these reasons, we attempt to clarify the problem.

Let us consider a sequence $E_1, E_2, \ldots$ of *independent* events. We may assume that such events happen in the temporal domain at times $t_1 < t_2 < \cdots$ (i.e., we use a *temporal marked point process* as a model). Each event E_i is characterized by the joint behavior of a pair of r.v.'s $(X, Y) \sim F_{XY}$, where $X \sim F_X$ and $Y \sim F_Y$. The joint events of interest here can be expressed in terms of the following set of *marginal* events.

DEFINITION 2.6 (Marginal events). *The events* $E^{<}_{X,x}$, $E^{>}_{X,x}$, $E^{<}_{Y,y}$, *and* $E^{>}_{Y,y}$ *given by*

$$E^{<}_{X,x} = \{X \le x\}, \tag{2.40a}$$

$$E^{>}_{X,x} = \{X > x\}, \tag{2.40b}$$

$$E^{<}_{Y,y} = \{Y \le y\}, \tag{2.40c}$$

$$E^{>}_{Y,y} = \{Y > y\} \tag{2.40d}$$

are called marginal.

Using the (inclusive) OR operator "$\vee$" and the AND operator "$\wedge$", it is possible to combine these marginal events in several ways, as shown in Table 2.3. The eight cases reported define all the possible different types of *joint* events E_i of interest here.

Table 2.3. Possible joint events using the (inclusive) OR "$\vee$" and the AND "$\wedge$" operators

$\vee$	$\mathbf{E}^{<}_{\mathbf{X,x}}$	$\mathbf{E}^{>}_{\mathbf{X,x}}$
$\mathbf{E}^{<}_{\mathbf{Y,y}}$	$\{X \le x\} \vee \{Y \le y\}$	$\{X > x\} \vee \{Y \le y\}$
$\mathbf{E}^{>}_{\mathbf{Y,y}}$	$\{X \le x\} \vee \{Y > y\}$	$\{X > x\} \vee \{Y > y\}$
$\wedge$	$\mathbf{E}^{<}_{\mathbf{X,x}}$	$\mathbf{E}^{>}_{\mathbf{X,x}}$
$\mathbf{E}^{<}_{\mathbf{Y,y}}$	$\{X \le x\} \wedge \{Y \le y\}$	$\{X > x\} \wedge \{Y \le y\}$
$\mathbf{E}^{>}_{\mathbf{Y,y}}$	$\{X \le x\} \wedge \{Y > y\}$	$\{X > x\} \wedge \{Y > y\}$

Natural systems are affected by extreme events, which as already mentioned cause large damages to natural environments and lead to loss of many human lives. However, the severity of phenomena such as droughts and floods may be the consequence of a different behavior of the same cause, i.e. the precipitation: too small for a long period in the former case, too large in the latter case. In terms of the joint events reported in Table 2.3, if X denotes the (average) storm intensity and Y the storm duration, then $\{X \le x\} \wedge \{Y \le y\}$ could be the event of interest when investigating droughts, and $\{X > x\} \wedge \{Y > y\}$ when studying floods. Although the theoretical investigations outlined in the sequel could easily be generalized to all the eight types of joint events shown in Table 2.3, we concentrate on the *extremal* events defined below.

DEFINITION 2.7 (Extremal events). *The joint events* $E_{x,y}^{\vee}$ *and* $E_{x,y}^{\wedge}$ *given by*

$$E_{x,y}^{\vee} = \{X > x\} \vee \{Y > y\}, \tag{2.41a}$$

$$E_{x,y}^{\wedge} = \{X > x\} \wedge \{Y > y\} \tag{2.41b}$$

are called extremal.

In practice, an event is defined as "extreme" if (*i*) *either X or Y* exceed given thresholds (in which case $E_{x,y}^{\vee}$ is the relevant event), or (*ii*) *both X and Y* are larger than prescribed values (in which case $E_{x,y}^{\wedge}$ is the relevant event). For instance, in hydrology, a storm may be considered as extreme if *either* its (average) intensity is too large *or* its duration is too long. To make a further example, if flood-volume *and* flood-peak are both too high, the safety of a dam may be at risk.

Now, let T_i be the *interarrival time* between E_i and E_{i+1}, $i = 1, 2, \ldots$. As is natural, we assume that $T_i > 0$ (*almost surely*), and that $\mu_{T_i} = \mathbb{E}(T_i)$ exists and is finite; therefore $\mu_{T_i} > 0$. For instance, T_i may represent the time between the arrival of two successive rainfall storms in a given region. If we let $N_{x,y}^{\vee}$ and $N_{x,y}^{\wedge}$ denote, respectively, the number of events E_i between two successive realizations of $E_{x,y}^{\vee}$ and $E_{x,y}^{\wedge}$, and defining $T_{x,y}^{\vee}$ and $T_{x,y}^{\wedge}$ as, respectively, the interarrival time between two successive realizations of $E_{x,y}^{\vee}$ and $E_{x,y}^{\wedge}$, it turns out that

$$T_{x,y}^{\vee} = \sum_{i=1}^{N_{x,y}^{\vee}} T_i, \tag{2.42a}$$

$$T_{x,y}^{\wedge} = \sum_{i=1}^{N_{x,y}^{\wedge}} T_i. \tag{2.42b}$$

Assuming that the interarrival times T_i are i.i.d. (and independent of X and Y), via *Wald's Equation* [122, 241] it is easy to show that

$$\tau_{x,y}^{\vee} = \mathbb{E}\left(T_{x,y}^{\vee}\right) = \mathbb{E}\left(N_{x,y}^{\vee}\right)\mu_T, \tag{2.43a}$$

$$\tau_{x,y}^{\wedge} = \mathbb{E}\left(T_{x,y}^{\wedge}\right) = \mathbb{E}\left(N_{x,y}^{\wedge}\right)\mu_T, \tag{2.43b}$$

where T denotes any of the i.i.d. r.v.'s T_i. Clearly, $N_{x,y}^{\vee}$ and $N_{x,y}^{\wedge}$ have a Geometric distribution with parameters $p_{x,y}^{\vee}$ and $p_{x,y}^{\wedge}$ given by, respectively,

$$p_{x,y}^{\vee} = \mathbb{P}\left\{E_{x,y}^{\vee}\right\} = \mathbb{P}\{X > x \vee Y > y\}, \tag{2.44a}$$

$$p_{x,y}^{\wedge} = \mathbb{P}\left\{E_{x,y}^{\wedge}\right\} = \mathbb{P}\{X > x \wedge Y > y\}. \tag{2.44b}$$

Obviously, $p_{x,y}^{\vee} \geq p_{x,y}^{\wedge}$: in fact, it is sufficient that *either* $E_{X,x}^{>}$ *or* $E_{Y,y}^{>}$ happen for the realization of $E_{x,y}^{\vee}$, whereas it is necessary that *both* $E_{X,x}^{>}$ *and* $E_{Y,y}^{>}$ happen for the realization of $E_{x,y}^{\wedge}$. Then, using Eqs. (2.43), we obtain

$$\tau_{x,y}^{\vee} = \mu_T / p_{x,y}^{\vee}, \tag{2.45a}$$

$$\tau_{x,y}^{\wedge} = \mu_T / p_{x,y}^{\wedge}. \tag{2.45b}$$

Clearly, $\tau_{x,y}^{\vee} \leq \tau_{x,y}^{\wedge}$. The above results yield the following definition.

DEFINITION 2.8 (Return period). *The positive numbers* $\tau^{\vee}_{x,y}$ *and* $\tau^{\wedge}_{x,y}$ *given by Eqs. (2.45) define, respectively, the return periods of the events* $E^{\vee}_{x,y}$ *and* $E^{\wedge}_{x,y}$.

Note that $\tau^{\vee}_{x,y}$ and $\tau^{\wedge}_{x,y}$ are decreasing functions of the corresponding probabilities $p^{\vee}_{x,y}$ and $p^{\wedge}_{x,y}$: this is obvious, since the interarrival time gets longer for less probable events.

A practical problem, of importance in applications, is the identification of the events having an assigned return period. In general, its solution is not unique: in fact, it may happen that different realizations of X and Y yield the same return period. A general solution to the problem, based on copulas, will be outlined in Section 3.3.

CHAPTER 3

BIVARIATE ANALYSIS VIA COPULAS

In applications it is sometimes of interest to consider extreme events in a bivariate or multivariate context: indeed, the joint occurrence of high values is often of practical importance. One example is the relationship between flood peak and flow volume; another is that between the depths, durations and areal extents of rainfall. Many of the geophysical variables featured in Section 1.4 play a joint role in the dynamics of natural phenomena.

A possible way of investigating multivariate data consists of studying the dependence function and the marginals separately. Since copulas describe and model the dependence structure between random variables, independently of the marginal laws involved, in this chapter we shall introduce the mathematical theory of copulas. As we shall see, this approach will not only simplify the analysis of the phenomenon under investigation, but it will also give the possibility to introduce new parameters for the characterization of the extreme behavior of the system.

Contrary to traditional approaches in which extensions to the multivariate case are not often clear, the copula approach is easy to generalize to a d-dimensional framework, $d > 2$. Moreover, quite a few multivariate distributions found in the literature are direct extensions of well known univariate cases, and suffer from several limitations and constraints: for instance, the marginal distributions may belong to the same probability family, and the parameters of the marginals may also rule the dependence between the variables considered. On the contrary, the copula approach does not have such drawbacks.

A further advantage of using copulas is that complex marginal distributions, such as *finite mixtures* [289], which are receiving increasing attention in applications, can be applied easily to the model of interest (for practical examples see, e.g., [90, 68]). Incidentally, we observe that all the bivariate distributions currently used can be described in a straightforward manner in terms of suitable copulas. For an exhaustive list of models see, e.g., [145, 155, 207], and references therein.

3.1. 2-COPULAS

In this section we outline briefly the mathematics of 2-copulas needed in the sequel. All the theoretical justifications can be found in [264, 155, 207]; see also [267] for a survey. Let us commence with the definition of 2-copulas.

DEFINITION 3.1 (2-Copula). *Let* $\mathbb{I} = [0, 1]$. *A 2-copula is a bivariate function* $\mathbf{C}\colon \mathbb{I} \times \mathbb{I} \to \mathbb{I}$ *such that:*

1. *(uniform marginals) for all* $u, v \in \mathbb{I}$,

$$\mathbf{C}(u, 0) = 0, \quad \mathbf{C}(u, 1) = u, \quad \mathbf{C}(0, v) = 0, \quad \mathbf{C}(1, v) = v; \tag{3.1a}$$

2. *(2-increasing) for all* $u_1, u_2, v_1, v_2 \in \mathbb{I}$ *such that* $u_1 \le u_2$ *and* $v_1 \le v_2$,

$$\mathbf{C}(u_2, v_2) - \mathbf{C}(u_2, v_1) - \mathbf{C}(u_1, v_2) + \mathbf{C}(u_1, v_1) \ge 0. \tag{3.1b}$$

The following functions play an important role in characterizing copulas.

DEFINITION 3.2 (Horizontal and vertical sections). *Let* $\mathbf{C}$ *be a 2-copula, and let* $z \in \mathbb{I}$. *The function* $t \mapsto \mathbf{C}(t, z)$, $t \in \mathbb{I}$, *is called the horizontal section of* $\mathbf{C}$ *at* z. *Similarly, the function* $t \mapsto \mathbf{C}(z, t)$, $t \in \mathbb{I}$, *is called the vertical section of* $\mathbf{C}$ *at* z.

A 2-copula $\mathbf{C}$ is uniformly continuous on its domain, and its horizontal and vertical sections are all non-decreasing and uniformly continuous on $\mathbb{I}$. This latter property holds also for the *diagonal section* $\boldsymbol{\delta}_{\mathbf{C}}$ of $\mathbf{C}$ defined below (see also Subsection C.15.3).

DEFINITION 3.3 (Diagonal section). *The function* $\boldsymbol{\delta}_{\mathbf{C}}\colon \mathbb{I} \to \mathbb{I}$ *given by*

$$\boldsymbol{\delta}_{\mathbf{C}}(t) = \mathbf{C}(t, t) \tag{3.2}$$

is called the diagonal section of $\mathbf{C}$.

Note that, for any copula $\mathbf{C}$, $\boldsymbol{\delta}_{\mathbf{C}}(1) = 1$, and $\boldsymbol{\delta}_{\mathbf{C}}(t) \le t$ for all $t \in \mathbb{I}$. In addition, the following inequality holds:

$$0 \le \boldsymbol{\delta}_{\mathbf{C}}(t_2) - \boldsymbol{\delta}_{\mathbf{C}}(t_1) \le 2(t_2 - t_1) \tag{3.3}$$

for all $t_1, t_2 \in \mathbb{I}$, with $t_1 \le t_2$.

With the given definitions, it is not difficult to show that any finite *convex linear combination* of 2-copulas $\mathbf{C}_i$'s is itself a 2-copula. In fact, for $k \in \mathbb{N}$, let $\mathbf{C}$ be given by

$$\mathbf{C}(u, v) = \sum_{i=1}^{k} \lambda_i \mathbf{C}_i(u, v), \tag{3.4}$$

where $\lambda_i \ge 0$ for all indices, and $\sum_{i=1}^{k} \lambda_i = 1$. Then $\mathbf{C}$ is a proper 2-copula.

This can be made more general, e.g. by considering d-dimensional copulas — see later chapters, and extended to the case of a continuous mixing parameter as follows. Let $\{\mathbf{C}_\lambda\}$ be an infinite collection of copulas indexed by a continuous parameter $\lambda \in \mathbb{R}$. Now, suppose that λ is the observation of a continuous r.v. with distribution function L. Then, setting

$$\mathbf{C}(u, v) = \int_{\mathbb{R}} \mathbf{C}_\lambda (u, v) \, \mathrm{d}L(\lambda), \tag{3.5}$$

it is not difficult to show that $\mathbf{C}$ is a copula. Usually, L is referred to as the *mixing distribution* of the family $\{\mathbf{C}_\lambda\}$, and $\mathbf{C}$ is called the *convex sum* of $\{\mathbf{C}_\lambda\}$ with respect to L.

In applications, it is often quite useful to consider 2-copulas as restrictions to $\mathbb{I}^2$ of joint distribution functions whose marginals are Uniform laws on $\mathbb{I}$. The following definition is a natural consequence of this fact.

DEFINITION 3.4 (C-measure). *A 2-copula* $\mathbf{C}$ *induces a probability measure* $\mu_\mathbf{C}$ *on* $\mathbb{I}^2$*, called* $\mathbf{C}$*-measure, given by*

$$\mu_\mathbf{C}\{[0, u] \times [0, v]\} = \mathbf{C}(u, v). \tag{3.6}$$

Often $\mathbf{C}$*-measures are also called doubly stochastic measures.*

Intuitively, the $\mathbf{C}$-measure $\mu_\mathbf{C}$ of a (measurable) subset A of $\mathbb{I}^2$ is the probability that two r.v.'s (U, V) Uniform on $\mathbb{I}$, and having joint distribution function $\mathbf{C}$, take values in A. Note that, given a measurable subset A of $\mathbb{I}$,

$$\mu_\mathbf{C}\{A \times \mathbb{I}\} = \mu_\mathbf{C}\{\mathbb{I} \times A\} = \lambda\{A\}, \tag{3.7}$$

where λ denotes the ordinary Lebesgue measure on $\mathbb{I}$. As shown in [207], for any copula $\mathbf{C}$ let

$$\mathbf{C}(u, v) = A_\mathbf{C}(u, v) + S_\mathbf{C}(u, v), \tag{3.8}$$

where

$$A_\mathbf{C}(u, v) = \int_0^u \int_0^v \frac{\partial^2}{\partial s \partial t} \mathbf{C}(s, t) \, \mathrm{d}t \, \mathrm{d}s, \tag{3.9}$$

and

$$S_\mathbf{C}(u, v) = \mathbf{C}(u, v) - A_\mathbf{C}(u, v). \tag{3.10}$$

Note that $\partial^2 \mathbf{C}(s, t)/\partial s \partial t$ exists almost everywhere in $\mathbb{I}^2$. Concerning bivariate distributions in general, copulas have no "*atoms*" in $\mathbb{I}^2$, i.e. points whose $\mathbf{C}$-measure is positive. This yields the following definition.

DEFINITION 3.5 (Absolutely continuous and singular copulas). *If* $\mathbf{C} \equiv A_\mathbf{C}$ *on* $\mathbb{I}^2$*, i.e. if* $\mathbf{C}$ *has density* $\partial^2 \mathbf{C}(s, t)/\partial s \partial t$*, then* $\mathbf{C}$ *is absolutely continuous. If* $\mathbf{C} \equiv S_\mathbf{C}$ *on* $\mathbb{I}^2$*, i.e. if* $\partial^2 \mathbf{C}(s, t)/\partial s \partial t = 0$ *almost everywhere in* $\mathbb{I}^2$*, then* $\mathbf{C}$ *is singular. Otherwise,* $\mathbf{C}$ *has an absolutely continuous component* $A_\mathbf{C}$ *and a singular component* $S_\mathbf{C}$*.*

Note that in the latter case neither A_C nor S_C is a copula, since neither has Uniform marginals on $\mathbb{I}$. Clearly, the **C**-measure of the absolutely continuous component is $A_C(1,1)$, and that of the singular component is $S_C(1,1)$. It is interesting to note that suitable convex sums of singular copulas may nonetheless generate a family of absolutely continuous copulas [207].

ILLUSTRATION 3.1. ▶

The *independence* 2-copula $\mathbf{\Pi}_2$ given by $\mathbf{\Pi}_2(u,v) = uv$ (see Illustration 3.2) is absolutely continuous. In fact

$$A_{\mathbf{\Pi}_2}(u,v) = \int_0^u \int_0^v \frac{\partial^2}{\partial s \partial t} \mathbf{\Pi}_2(s,t)\, \mathrm{d}t\, \mathrm{d}s = \int_0^u \int_0^v 1\, \mathrm{d}t\, \mathrm{d}s = \mathbf{\Pi}_2(u,v)$$

for all $(u,v) \in \mathbb{I}^2$. Note that $\mathbf{\Pi}_2$ is the only 2-copula with *linear* horizontal and vertical sections (see Definition 3.2). ◀

The link between 2-copulas and bivariate distributions is provided by the following fundamental result [275]. Henceforth F_X, F_Y (respectively, F_U, F_V) will denote the marginal distribution functions of the r.v.'s X, Y (respectively, U, V), and Ran their Range.

THEOREM 3.1 (Sklar (2-dimensional case)). *Let F_{XY} be a joint distribution function with marginals F_X and F_Y. Then there exists a 2-copula* **C** *such that*

$$F_{XY}(x,y) = \mathbf{C}\left(F_X(x), F_Y(y)\right) \tag{3.11}$$

for all reals x, y. If F_X, F_Y are continuous, then **C** *is unique; otherwise,* **C** *is uniquely defined on* $\mathrm{Ran}(F_X) \times \mathrm{Ran}(F_Y)$.

Conversely, if **C** *is a 2-copula and F_X, F_Y are distribution functions, then the function F_{XY} given by Eq. (3.11) is a joint distribution with marginals F_X and F_Y.*

NOTE 3.1. Although no theoretical constraints exist on the choice of F_X, F_Y, for the sake of simplicity we shall limit our investigation to *continuous strictly increasing* marginals. This case is the one of utmost interest in applications. As a consequence, by virtue of Theorem 3.1, the copula representation will always be unique. However, only minor changes (involving the use of suitable *quasi-inverses* — see Eqs. (A.1)) are required in case F_X, F_Y do not satisfy such an assumption.

The *continuity assumption* discussed above is essential in order to derive the following result, that plays a fundamental role in practical applications.

COROLLARY 3.1 (Sklar inversion (2-dimensional case)). *Let* **C**, *F_{XY}, and F_X, F_Y be as in Theorem 3.1, and suppose that F_X, F_Y are continuous. If $F_X^{[-1]}$ and $F_Y^{[-1]}$ denote, respectively, the quasi-inverses of F_X and F_Y, then*

$$\mathbf{C}(u,v) = F_{XY}\left(F_X^{[-1]}(u), F_Y^{[-1]}(v)\right) \tag{3.12}$$

for any $(u,v) \in \mathbb{I}^2$.

The following example shows the properties of three fundamental 2-copulas.

ILLUSTRATION 3.2. ►

Three special 2-copulas deserve a particular attention.

1. The *Fréchet-Hoeffding lower bound* $\mathbf{W}_2$ given by

$$\mathbf{W}_2(u,v) = \max\{u+v-1, 0\}. \tag{3.13}$$

2. The *Fréchet-Hoeffding upper bound* $\mathbf{M}_2$ given by

$$\mathbf{M}_2(u,v) = \min\{u, v\}. \tag{3.14}$$

3. The *independence* (or *product*) 2-copula $\mathbf{\Pi}_2$ given by

$$\mathbf{\Pi}_2(u,v) = uv. \tag{3.15}$$

A family of 2-copulas which includes $\mathbf{W}_2$, $\mathbf{M}_2$ and $\mathbf{\Pi}_2$ is called *comprehensive*. The graphs of these three copulas are illustrated in Figure 3.1.

The copulas $\mathbf{W}_2$ and $\mathbf{M}_2$ provide general bounds, since for any 2-copula $\mathbf{C}$ and any pair $(u,v) \in \mathbb{I}^2$

$$\mathbf{W}_2(u,v) \leq \mathbf{C}(u,v) \leq \mathbf{M}_2(u,v). \tag{3.16}$$

In particular, when X and Y are continuous r.v.'s, the following characterization holds.

1. The variable Y is almost surely a strictly decreasing function of X if, and only if, $\mathbf{C}_{XY} = \mathbf{W}_2$. Alternatively, $\mathbf{W}_2$ is the distribution of the random vector $(U, 1-U)$, where U is Uniform on $\mathbb{I}$. Random variables with copula $\mathbf{W}_2$ are often called *counter-monotonic*.
2. The variable Y is almost surely a strictly increasing function of X if, and only if, $\mathbf{C}_{XY} = \mathbf{M}_2$. Alternatively, $\mathbf{M}_2$ is the distribution of the random vector (U, U), where U is Uniform on $\mathbb{I}$. Random variables with copula $\mathbf{M}_2$ are often called *co-monotonic*.
3. The 2-copula $\mathbf{\Pi}_2$ describes the absence of dependence between X and Y: in fact, if $F_{XY} = \mathbf{\Pi}_2(F_X, F_Y)$, then $F_{XY} = F_X F_Y$. Therefore, X and Y are independent if, and only if, their 2-copula is $\mathbf{\Pi}_2$.

In Chapter 4 the three functions $\mathbf{W}_2, \mathbf{M}_2, \mathbf{\Pi}_2$ will be extended to a multidimensional framework. ◄

Considering the special cases of counter-monotonic or co-monotonic r.v.'s, the following result holds [85].

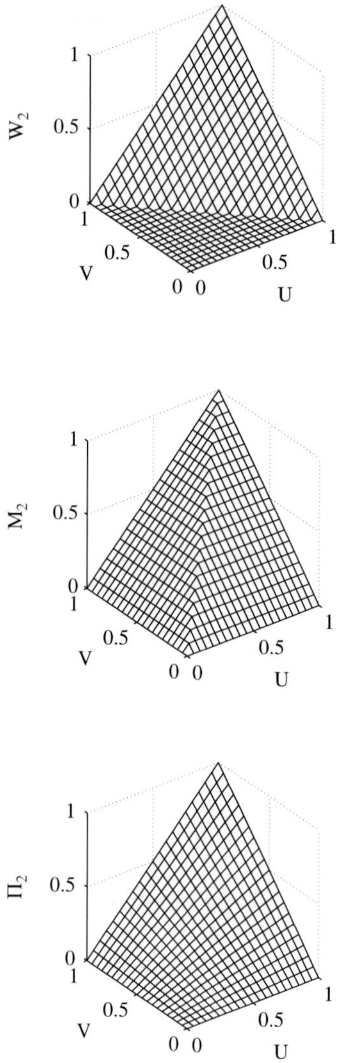

Figure 3.1. The 2-copulas $\mathbf{W}_2$, $\mathbf{M}_2$ and $\mathbf{\Pi}_2$, respectively from the top downwards

PROPOSITION 3.1. *Let (X, Y) have one of the copulas $\mathbf{W}_2$ or $\mathbf{M}_2$. Then there exist two monotonic functions $u, v : \mathbb{R} \to \mathbb{R}$, and a real r.v. Z, such that*

$$(X, Y) \sim (u(Z), v(Z)), \tag{3.17}$$

with u increasing and v decreasing in the former case, and both increasing in the latter. The converse of this result is also true.

The Fréchet-Hoeffding inequality given by Eq. (3.16) suggests the introduction of a *partial order* on the set of copulas. Clearly, the order is partial, since not every pair of copulas is comparable. More details can be found in Appendix B.

DEFINITION 3.6 (Partial order). *If* $\mathbf{C}_1, \mathbf{C}_2$ *are copulas, then* $\mathbf{C}_1$ *is smaller than* $\mathbf{C}_2$ *(or* $\mathbf{C}_2$ *is larger than* $\mathbf{C}_1$*), denoted by* $\mathbf{C}_1 \prec \mathbf{C}_2$ *(or* $\mathbf{C}_2 \succ \mathbf{C}_1$*), if* $\mathbf{C}_1(u, v) \leq \mathbf{C}_2(u, v)$ *for all* $u, v \in \mathbb{I}$.

An important feature of 2-copulas is that they are invariant under strictly increasing transformations, as stated by the following result.

PROPOSITION 3.2 (Invariance property). *Let* X *and* Y *be continuous r.v.'s with 2-copula* $\mathbf{C}_{XY}$. *If* α *and* β *are strictly increasing functions on, respectively,* $\mathrm{Ran}(X)$ *and* $\mathrm{Ran}(Y)$, *then*

$$\mathbf{C}_{\alpha(X)\beta(Y)} = \mathbf{C}_{XY}. \tag{3.18}$$

The following important properties are featured by 2-copulas.

PROPOSITION 3.3. *Let* X *and* Y *be continuous r.v.'s with 2-copula* $\mathbf{C}_{XY}$, *and let* α *and* β *be strictly monotone functions on, respectively,* $\mathrm{Ran}(X)$ *and* $\mathrm{Ran}(Y)$.

1. *If* α *is increasing and* β *decreasing then*

$$\mathbf{C}_{\alpha(X)\beta(Y)}(u, v) = u - \mathbf{C}_{XY}(u, 1 - v). \tag{3.19}$$

2. *If* α *is decreasing and* β *increasing then*

$$\mathbf{C}_{\alpha(X)\beta(Y)}(u, v) = v - \mathbf{C}_{XY}(1 - u, v). \tag{3.20}$$

3. *If* α *and* β *are both decreasing then*

$$\mathbf{C}_{\alpha(X)\beta(Y)}(u, v) = u + v - 1 + \mathbf{C}_{XY}(1 - u, 1 - v). \tag{3.21}$$

Given these invariance properties, and using the *Probability Integral Transform*, we may restrict our attention to the pair of r.v.'s (U, V) given by

$$\begin{cases} U = F_X(X) \\ V = F_Y(Y) \end{cases} \iff \begin{cases} X = F_X^{[-1]}(U) \\ Y = F_Y^{[-1]}(V) \end{cases}, \tag{3.22}$$

where $F_X^{[-1]}, F_Y^{[-1]}$ are the *quasi-inverses* of the corresponding distribution functions (see Eqs. (A.1)). Clearly, U and V are Uniform on $\mathbb{I}$, i.e. $U \sim \mathtt{U}(0, 1)$ and $V \sim \mathtt{U}(0, 1)$, and (U, V) has the same 2-copula as the pair (X, Y), i.e. $(U, V) \sim \mathbf{C}_{UV} = \mathbf{C}_{XY}$.

Hereafter we shall generally consider (X, Y) as the pair of r.v.'s effectively describing the phenomenon of interest. One of the relevant results of our approach will be to show how to make inferences on the behavior of (X, Y) by considering, instead, the pair (U, V). This may lead to simpler calculations. Indeed, the marginals of U and V are Uniform laws (and, hence, independent of any parameter). This may turn the original problem into a *non-parametric (distribution-free)* case, which is usually less difficult to solve.

ILLUSTRATION 3.3 (Order Statistics). ►

A simple application of 2-copulas concerns the *order statistics* (see Section 1.1). Let X, Y be continuous r.v.'s with copula $\mathbf{C}$ and marginals F_X, F_Y. Then $L = \min\{X, Y\}$ and $\mathrm{G} = \max\{X, Y\}$ represent the (extremal) order statistics for the pair (X, Y). The distributions of L and G are easy to calculate using copulas:

$$\mathbb{P}\{L \le t\} = F_X(t) + F_Y(t) - \mathbf{C}(F_X(t), F_Y(t)), \tag{3.23a}$$

$$\mathbb{P}\{\mathrm{G} \le t\} = \mathbf{C}(F_X(t), F_Y(t)). \tag{3.23b}$$

If $F_X = F_Y = F$, then

$$\mathbb{P}\{L \le t\} = 2F(t) - \boldsymbol{\delta}_{\mathbf{C}}(F(t)), \tag{3.24a}$$

$$\mathbb{P}\{\mathrm{G} \le t\} = \boldsymbol{\delta}_{\mathbf{C}}(F(t)), \tag{3.24b}$$

where $\boldsymbol{\delta}_{\mathbf{C}}$ is the diagonal section of $\mathbf{C}$ given by Eq. (3.2). In particular, if $X = U$ and $Y = V$, where U, V are Uniform r.v.'s on $\mathbb{I}$, then

$$\mathbb{P}\{L \le t\} = 2t - \boldsymbol{\delta}_{\mathbf{C}}(t), \tag{3.25a}$$

$$\mathbb{P}\{\mathrm{G} \le t\} = \boldsymbol{\delta}_{\mathbf{C}}(t), \tag{3.25b}$$

since $F(t) = t$ on $\mathbb{I}$. ◄

The study of *conditional distributions* is greatly facilitated by using copulas. In particular, we show in Subsection 3.3.4 how to calculate specific conditional return periods. For instance, the following conditional laws can easily be calculated using copulas [207]:

$$\mathbb{P}\{U \le u \mid V = v\} = \frac{\partial}{\partial v}\mathbf{C}(u, v) \in \mathbb{I}, \tag{3.26a}$$

$$\mathbb{P}\{U \le u \mid V \le v\} = \frac{\mathbf{C}(u, v)}{v} \in \mathbb{I}, \tag{3.26b}$$

$$\mathbb{P}\{U \leq u \mid V > v\} = \frac{u - \mathbf{C}(u, v)}{1 - v} \in \mathbb{I}, \tag{3.26c}$$

and similar expressions hold for the conditional distributions of V given U. Note that:

1. for any $u \in \mathbb{I}$, the partial derivative $\frac{\partial \mathbf{C}}{\partial v}$ exists for almost all $v \in \mathbb{I}$;
2. where $\frac{\partial \mathbf{C}}{\partial v}$ exists, its range is that of a distribution function, i.e.

$$0 \leq \frac{\partial}{\partial v}\mathbf{C}(u, v) \leq 1; \tag{3.27}$$

3. the function $u \mapsto \frac{\partial \mathbf{C}(u,v)}{\partial v}$ is well defined and non-decreasing almost everywhere in $\mathbb{I}$.

The above formulas represent the mathematical kernel for simulating copulas: see Appendix A, Appendix C, and also some of the exercises in [207] for special cases.

A further important point concerns the study of the *survival (excess) probabilities* in a multivariate context: again, copulas play an important role.

THEOREM 3.2 (Survival 2-Copula). *Let* $\mathbf{C}$ *be a 2-copula and let the bivariate function* $\overline{\mathbf{C}}\colon \mathbb{I}^2 \to \mathbb{I}$ *be given by*

$$\overline{\mathbf{C}}(u, v) = u + v - 1 + \mathbf{C}(1 - u, 1 - v). \tag{3.28}$$

Then $\overline{\mathbf{C}}$ *is a 2-copula called the survival 2-copula of* U *and* V.

It is easy to check that $\overline{\mathbf{C}}$ satisfies the conditions prescribed in Definition 3.1 for a 2-copula. In Figure 3.2 we illustrate the regions of the unit square $\mathbb{I}^2$ with the probability masses given by $\mathbf{C}(u, v)$ and $\overline{\mathbf{C}}(1 - u, 1 - v)$.

ILLUSTRATION 3.4 (Survival 2-Copula). ►

The survival 2-copula can be used to calculate the *joint survival* function $\overline{F}_{UV}$ of the pair (U, V) — see Definition 2.2 and the ensuing Note 2.1. Let us define the *marginal survival (excess)* functions

$$\begin{aligned} \overline{F}_U(u) &= \mathbb{P}\{U > u\} = 1 - u, \\ \overline{F}_V(v) &= \mathbb{P}\{V > v\} = 1 - v. \end{aligned} \tag{3.29}$$

Then, the *joint survival* function $\overline{F}_{UV}$ is given by

$$\overline{F}_{UV}(u, v) = \mathbb{P}\{U > u, V > v\} = \overline{\mathbf{C}}\left(\overline{F}_U(u), \overline{F}_V(v)\right). \tag{3.30}$$

Note that the arguments of $\overline{\mathbf{C}}$ must be survival functions. ◄

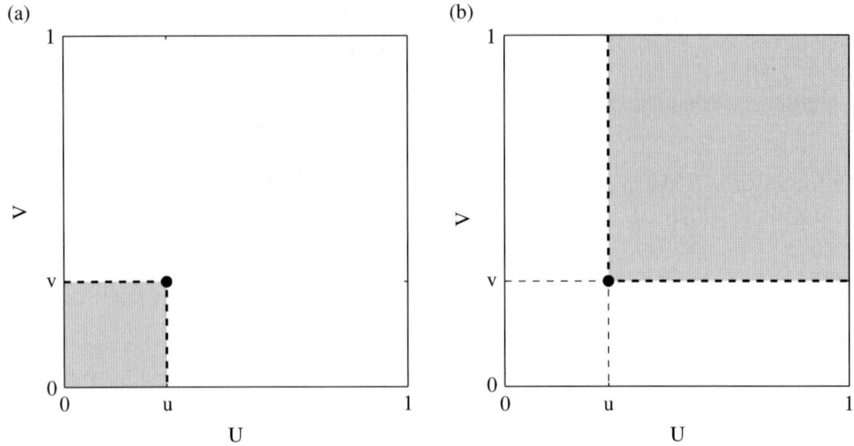

Figure 3.2. The regions of the unit square (*shaded areas*) having the probability masses given by (*a*) $\mathbf{C}(u, v) = \mathbb{P}\{U \le u, V \le v\}$ and (*b*) $\overline{\mathbf{C}}(1-u, 1-v) = \mathbb{P}\{U > u, V > v\}$

In the univariate case, the *empirical distribution function* represents a useful tool for investigating the behavior of the variables of interest. Similarly, the *empirical copulas* [69] defined as follows provide valuable information about the joint behavior of pairs of r.v.'s associated via a 2-copula $\mathbf{C}$.

DEFINITION 3.7 (Empirical copula). *Let $\{(R_k, S_k)\}$ be the ranks associated with the random sample $\{(X_k, Y_k)\}$, $k = 1, \ldots, n$. The corresponding empirical copula $\mathbf{C}_n$ is defined as*

$$\mathbf{C}_n(u, v) = \frac{1}{n} \sum_{k=1}^{n} \mathbf{1}\left(\frac{R_k}{n+1} \le u, \frac{S_k}{n+1} \le v \right), \tag{3.31}$$

where $u, v \in \mathbb{I}$ and $\mathbf{1}$ is an indicator function.

NOTE 3.2. An alternative definition of the empirical copula is given in [207] as follows:

$$\mathbf{C}_n\left(\frac{i}{n}, \frac{j}{n}\right) = \frac{N_{ij}}{n}, \tag{3.32}$$

where N_{ij}, $1 \le i, j \le n$, is the number of pairs (x, y) in the sample such that $x \le x_{(i)}$ and $y \le y_{(j)}$, with $x_{(i)}, y_{(j)}$ denoting the order statistics from the sample.

As in the univariate case, the empirical copula practically counts the number of pairs that satisfy given constraints, in order to provide an approximation to the copula linking the pair (X, Y). Furthermore, empirical copulas play an important role in several procedures for fitting copulas to experimental data (see below).

NOTE 3.3 (Goodness-of-fit procedures). Actually, fitting copulas to empirical observations is still an open problem in Statistics, and several *goodness-of-fit* procedures have recently been proposed to this end [109]. These can be divided into three broad classes:

1. those based on the *Probability Integral Transformation* of Rosenblatt [240] (see [24, 15]);
2. those involving *kernel smoothing* (see [94, 212, 257]);
3. those derived from continuous functionals of the *empirical copula process* (see [114, 113]).

It must be noted that procedures based on Rosenblatt's transformation involve conditioning on successive components of the random vector of interest, and depend upon the order in which this conditioning is done. In addition, none of the works based on this approach investigate how the asymptotic distribution of the test statistic is affected by (*a*) parameter estimation and (*b*) absence of knowledge of the marginal distributions.

Furthermore, although Scaillet [257] has recently streamlined the process, kernel-based goodness-of-fit testing procedures described by Fermanian [94] involve many arbitrary choices: kernel type, window length, weight function, and so on. Clearly, this makes their application quite cumbersome. Similar criticisms apply to the work of Panchenko [212].

Thus, apparently, at present the most feasible, practical, and realistic solution is represented by goodness-of-fit procedures based on the empirical copula process. However, the research is in progress, and further methods are expected in the near future.

ILLUSTRATION 3.5 (Sea storm characterization). ►

In [68] a characterization of the sea storm dynamics via an *equivalent triangular storm* approach [21] is provided. Four relevant variables are considered: (*1*) the significant wave height H (in metres) of the equivalent triangular storm, (*2*) the storm duration D (in hours), (*3*) the waiting time I between two successive "active" storm phases (in hours), and (*4*) the storm wave direction A (in deg), assuming it equals the direction of the significant wave height. The data are collected at the Alghero wave buoy (Sardinia, Italy), for a period of 12 years: 415 consecutive independent sea storm events are extracted from the available data base.

The triple (H, D, I) is of particular importance, for it provides the relevant information about the storm energetic content (via the pair (H, D)), and the timing of the storm sequence (via the pair (D, I)) — see Illustration 3.10. Two different families of 2-copulas (see Appendix C) were fitted to the pairs (H, D) and (D, I). In order to visually check the goodness of fit, in Figure 3.3 we compare the level curves of the theoretical and the empirical copulas for the two pairs.

Overall, the agreement is valuable: straight lines and jumps are simply due to the presence of "ties" (i.e., identical pairs of values), that affect the estimation

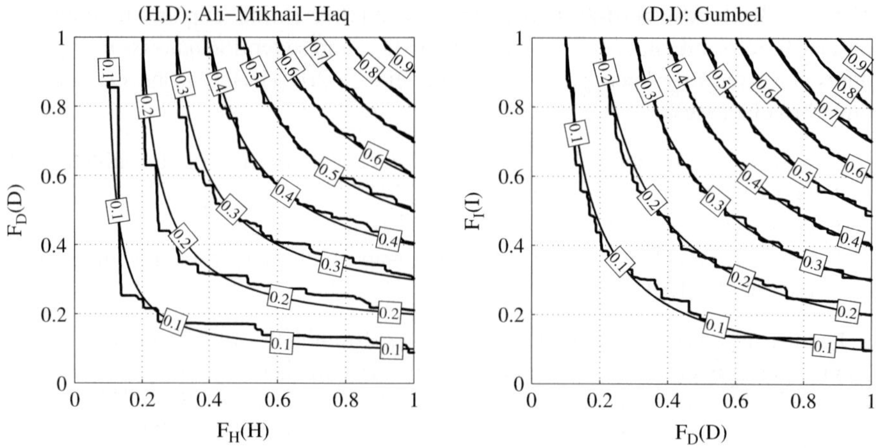

Figure 3.3. Comparison between the level curves of the theoretical copulas fitted to the available observations (*thin* lines), and those of the empirical copulas constructed using the same data (*thick* lines). The probability levels are as indicated (*labels*), as well as the pairs considered and the copulas used (*titles*)

of the probabilities of interest, and to the finite "discrete" nature of the sample values. However, such a problem only affects lower probability levels, characterized by the majority of the pairs featuring "ties". Note that what is actually compared is the distribution of the ranks (properly normalized into $\mathbb{I}$): this way of comparing bivariate distributions is non-parametric, in the sense that the marginal laws of the variables investigated are not used to construct the empirical copulas. For this reason the domain is the unit square $\mathbb{I}^2$, and no units are given on the axes. Lastly, none of the 2-copulas considered corresponds to the independence copula $\mathbf{\Pi}_2$. ◄

3.2. ARCHIMEDEAN COPULAS

A particular subclass of copulas, called *Archimedean*, that features many useful properties, will play an important role in our work. According to [111], Archimedean copulas provide a host of models that are versatile in terms of both the nature and strength of the association they induce between the variables. For instance, they have been used successfully in connection with the notion of "frailty" [210]. However, diagnostic procedures that can help delineate circumstances where Archimedean copulas are adequate have yet to be analyzed. Firstly we need the following definition.

DEFINITION 3.8 (Pseudo-inverse). *Let* $\gamma\colon \mathbb{I} \to [0, \infty]$ *such that* γ *is continuous and strictly decreasing, with* $\gamma(1) = 0$*; also let* γ^{-1} *denote the ordinary inverse* *function of γ. The pseudo-inverse of γ is the function $\gamma^{[-1]}\colon [0,\infty] \to \mathbb{I}$ given by*

$$\gamma^{[-1]}(t) = \begin{cases} \gamma^{-1}(t), & 0 \le t \le \gamma(0) \\ 0, & \gamma(0) \le t \le \infty \end{cases}. \tag{3.33}$$

Note that $\gamma^{[-1]}$ is continuous and non-increasing on $[0,\infty]$, and strictly decreasing on $[0,\gamma(0)]$. Also, $\gamma^{[-1]}(\gamma(t)) = t$ on $\mathbb{I}$, and

$$\gamma\left(\gamma^{[-1]}(t)\right) = \min\left\{t, \gamma(0)\right\}. \tag{3.34}$$

Clearly, if $\gamma(0) = \infty$, then $\gamma^{[-1]} = \gamma^{-1}$. The following result implicitly defines the Archimedean 2-copulas.

THEOREM 3.3. *Let $\mathbf{C}\colon \mathbb{I}\times\mathbb{I} \to \mathbb{I}$ be given by*

$$\mathbf{C}(u,v) = \gamma^{[-1]}\left(\gamma(u) + \gamma(v)\right), \tag{3.35}$$

where γ, $\gamma^{[-1]}$ are as in Definition 3.8. Then $\mathbf{C}$ is a 2-copula if, and only if, γ is convex.

Note that γ is convex if, and only if, $\gamma^{[-1]}$ is convex. Thus, we have the following definition.

DEFINITION 3.9 (Archimedean 2-Copulas). *Copulas given by Theorem 3.3 are called Archimedean, with (additive) generator γ. If $\gamma(0) = \infty$, then γ, as well as the corresponding 2-copula, are called strict.*

An Archimedean 2-copula $\mathbf{C}$ features the following properties.

1. $\mathbf{C}$ is *symmetric*, i.e., for all $u, v \in \mathbb{I}$,

$$\mathbf{C}(u,v) = \mathbf{C}(v,u). \tag{3.36}$$

   This means that (U, V) is a pair of *exchangeable* r.v.'s.
2. $\mathbf{C}$ is *associative*, i.e., for all $u, v, w \in \mathbb{I}$,

$$\mathbf{C}\left(\mathbf{C}(u,v), w\right) = \mathbf{C}\left(u, \mathbf{C}(v,w)\right). \tag{3.37}$$

3. If γ generates $\mathbf{C}$, then also $\gamma' = c\gamma$ is a generator of $\mathbf{C}$, where c is a positive constant.
4. The diagonal section $\boldsymbol{\delta}_{\mathbf{C}}$ of $\mathbf{C}$ satisfies $\boldsymbol{\delta}_{\mathbf{C}}(t) < t$ for all $t \in (0,1)$.

Given a generator γ, a simple way to construct new generators (and, consequently, families of Archimedean copulas) is to consider *interior* and *exterior power families*, as shown here.

PROPOSITION 3.4. *Let γ be a generator. Then*

1. ***(interior power)*** *$\gamma_\alpha(t) = \gamma(t^\alpha)$ is a generator for all $\alpha \in (0, 1]$;*
2. ***(exterior power)*** *$\gamma_\beta(t) = [\gamma(t)]^\beta$ is a generator for all $\beta \geq 1$.*

The families constructed via Proposition 3.4 are usually called the *alpha* or *beta family* associated with γ.

ILLUSTRATION 3.6. ▶

Let us investigate some details of the 2-copulas $\mathbf{\Pi}_2$, $\mathbf{W}_2$, and $\mathbf{M}_2$.

1. Let $\gamma = -\ln(t)$, $t \in \mathbb{I}$. Clearly, γ is a strict generator, and $\gamma^{-1} = e^{-t}$. Then, generating $\mathbf{C}$ via Eq. (3.35) yields
$$\mathbf{C}(u, v) = e^{-[(-\ln u)+(-\ln v)]} = uv = \mathbf{\Pi}_2(u, v).$$
Thus, $\mathbf{\Pi}_2$ is a strict Archimedean 2-copula.
2. Let $\gamma = 1 - t$, $t \in \mathbb{I}$. Clearly, γ is a non-strict generator, and $\gamma^{[-1]} = \max\{1 - t, 0\}$. Then, generating $\mathbf{C}$ via Eq. (3.35) yields
$$\mathbf{C}(u, v) = \max\{u + v - 1, 0\} = \mathbf{W}_2(u, v).$$
Thus, $\mathbf{W}_2$ is a non-strict Archimedean 2-copula.
3. The 2-copula $\mathbf{M}_2$ is not Archimedean: in fact, it does not satisfy property (4) above, since $\boldsymbol{\delta}_{\mathbf{M}_2}(t) = \min\{t, t\} = t \not< t$ for $t \in (0, 1)$.

◀

An important source of generators of Archimedean 2-copulas is represented by the inverses of *Laplace Transforms* of distribution functions [155, 207]. A generalization to the d-dimensional case will be shown in Section 4.2. Let F be the distribution of a positive r.v., and let ψ be the corresponding *Laplace Transform*:

$$\psi(s) = \int_0^\infty e^{-sx}\, \mathrm{d}F(x), \tag{3.38}$$

where $s \geq 0$. Note that the function $s \mapsto \psi(-s)$ is the *moment generating function* of F. Clearly, ψ is continuous and strictly decreasing, and $\psi(0) = 1$. Thus, its inverse ψ^{-1} is also strictly decreasing, and satisfies the boundary constraints $\psi^{-1}(0) = \infty$ and $\psi^{-1}(1) = 0$. Furthermore, ψ has continuous derivatives of all orders that alternate in sign, i.e. it is *completely monotonic* (see Definition 4.4 and the ensuing discussion). As shown in [188] (see also [155, 207]), the function $\mathbf{C}$ defined on $\mathbb{I}^2$ by

$$\mathbf{C}(u, v) = \psi^{-1}(\psi(u) + \psi(v)), \tag{3.39}$$

is a 2-copula. The result now shown will be generalized in Illustration 4.5.

ILLUSTRATION 3.7 (Clayton family). ▶

Let F be a Gamma distribution $\Gamma(1/\theta, 1)$, where $\theta > 0$. Then, the *Laplace Transform* of F is $\psi(s) = (1+s)^{-1/\theta}$, with $s \geq 0$. In turn, the function $\gamma(t) = \psi^{-1}(t) = t^{-\theta} - 1$, with $t \in \mathbb{I}$, generates (a subfamily of) the Clayton family shown in Section C.3. ◀

The Clayton family of Archimedean copulas has an interesting connection with the *sample range* [263, 207].

ILLUSTRATION 3.8 (Order Statistics (cont.)). ▶

Let $\{X_1, \ldots, X_n\}$ be a set of i.i.d. continuous r.v.'s with distribution F. Then $X_{(1)} = \min_{1\leq i\leq n}\{X_i\}$ and $X_{(n)} = \max_{1\leq i\leq n}\{X_i\}$ represent the (extremal) order statistics for the sample, whose distributions are given, respectively, in Subsections 1.1.1–1.1.2. The difference $R = X_{(n)} - X_{(1)}$ is called the sample range (see Illustration 1.5), and is often used in applications. Here we calculate the joint distribution H of $(X_{(1)}, X_{(n)})$, and the corresponding 2-copula $\mathbf{C}$.

For mathematical convenience, it is easier to calculate first the distribution H^- of $(-X_{(1)}, X_{(n)})$:

$$\begin{aligned} H^-(s,t) &= \mathbb{P}\left\{-X_{(1)} \leq s, X_{(n)} \leq t\right\} \\ &= \mathbb{P}\left\{-s \leq X_{(1)}, X_{(n)} \leq t\right\} \\ &= \begin{cases} (F(t) - F(-s))^n, & -s \leq t \\ 0, & -s > t \end{cases} \\ &= \left(\max\left\{F(t) - F(-s), 0\right\}\right)^n. \end{aligned} \tag{3.40}$$

Then, some algebraic manipulation yields the copula $\mathbf{C}^-$ of $(-X_{(1)}, X_{(n)})$:

$$\mathbf{C}^-(u,v) = \left(\max\left\{u^{1/n} + v^{1/n} - 1, 0\right\}\right)^n, \tag{3.41}$$

i.e. a member of the Clayton family, with parameter $\theta = -1/n$ (see Section C.3). From property (2) in Proposition 3.3, it follows that the vector $(X_{(1)}, X_{(n)})$ has copula $\mathbf{C}_{1n}$ given by

$$\begin{aligned} \mathbf{C}_{1n}(u,v) &= v - \mathbf{C}^-(1-u, v) \\ &= v - \left(\max\left\{(1-u)^{1/n} + v^{1/n} - 1, 0\right\}\right)^n. \end{aligned} \tag{3.42}$$

Clearly, $X_{(1)}$ and $X_{(n)}$ are not independent (that is, $\mathbf{C}_{1n} \neq \mathbf{\Pi}_2$). However, they are *asymptotically independent*, since

$$\lim_{n\to\infty} \mathbf{C}_{1n}(u,v) = v - \mathbf{\Pi}_2(1-u, v) = uv = \mathbf{\Pi}_2(u,v).$$

Indeed, the Clayton copula with parameter $\theta = 0$ corresponds to the independence 2-copula $\mathbf{\Pi}_2$. ◀

Although Archimedean copulas do not embrace the whole family of copulas, nevertheless they may suffice for modeling quite a few phenomena. In particular, they will play a fundamental role in Section 3.3, mainly because their *level curves (sets)* can be written explicitly.

DEFINITION 3.10 (Level curves). *Given* $0 < t \le 1$, *the curves*

$$L_t = \{(u, v) \in \mathbb{I}^2 : \mathbf{C}(u, v) = t\} \tag{3.43}$$

are called the level curves of **C**.

The line L_t connects the border points $(t, 1)$ and $(1, t)$, since $\mathbf{C}(t, 1) = \mathbf{C}(1, t) = t$ by virtue of Definition 3.1. Instead, for $t = 0$, $Z(\mathbf{C}) = L_0$ represents the *zero set* of **C**, i.e. the region of $\mathbb{I}^2$ where $\mathbf{C}(u, v) = 0$ or, equivalently, the region of the *impossible events*. Clearly, $Z(\mathbf{C})$ may either reduce to the two line segments $\{0\} \times \mathbb{I}$ and $\mathbb{I} \times \{0\}$, or have a positive area.

Now, let $0 < t \le 1$. In the Archimedean case, L_t is *convex* (but this does not hold for all copulas). Also, since **C** is constant on L_t by virtue of Eq. (3.43), it is clear from Eq. (3.35) that the points of L_t satisfy the relation

$$(u, v) \in L_t \iff \gamma(u) + \gamma(v) = \gamma(t). \tag{3.44}$$

Therefore, it is possible to express v as a function of u (and vice versa), for $t \le u \le 1$, simply calculating $v = L_t(u)$, where

$$L_t(u) = \gamma^{[-1]}(\gamma(t) - \gamma(u)) = \gamma^{-1}(\gamma(t) - \gamma(u)). \tag{3.45}$$

Note that replacing $\gamma^{[-1]}$ by γ^{-1} is justified, since $\gamma(t) - \gamma(u)$ is in the interval $[0, \gamma(0)]$. As an illustration, in Figure 3.4 we show the level curves of the 2-copulas $\mathbf{W}_2$, $\mathbf{M}_2$ and $\mathbf{\Pi}_2$.

A further important point is the following. Let

$$B_{\mathbf{C}}(t) = \{(u, v) \in \mathbb{I}^2 : \mathbf{C}(u, v) \le t\}, \quad 0 < t \le 1. \tag{3.46a}$$

Evidently, $B_{\mathbf{C}}(t)$ is the region in $\mathbb{I}^2$ lying on, or below and to the left of, the level curve L_t. Equivalently, in terms of the generator γ,

$$B_{\mathbf{C}}(t) = \{(u, v) \in \mathbb{I}^2 : \gamma(u) + \gamma(v) \ge \gamma(t)\}, \quad 0 < t \le 1. \tag{3.46b}$$

Since **C** induces in a natural way a probability measure on $\mathbb{I}^2$ (see Definition 3.4), it is possible to use it in order to introduce a suitable **C**-*measure* on the unit square. Clearly, the latter may provide a means to measure the sets $B_{\mathbf{C}}$. Its definition and expression are given as follows [207].

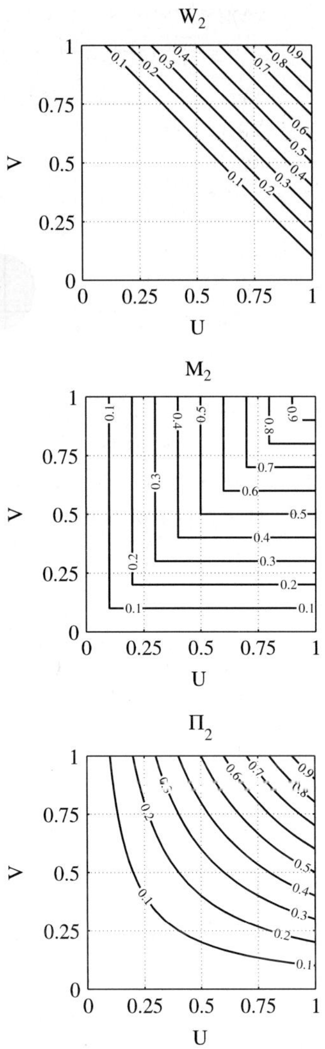

Figure 3.4. The level curves of the 2-copulas $\mathbf{W}_2$, $\mathbf{M}_2$ and $\mathbf{\Pi}_2$, respectively from the top downwards

THEOREM 3.4. *Let $0 < t \le 1$. The function $K_{\mathbf{C}}(t)$ denotes the* **C***-measure of the set $B_{\mathbf{C}}(t)$, and is given by*

$$K_{\mathbf{C}}(t) = t - \frac{\gamma(t)}{\gamma'(t^{+})}. \tag{3.47}$$

Note that γ' exists almost everywhere in $\mathbb{I}$. The next result gives an interpretation of the measure $K_{\mathbf{C}}$ [207].

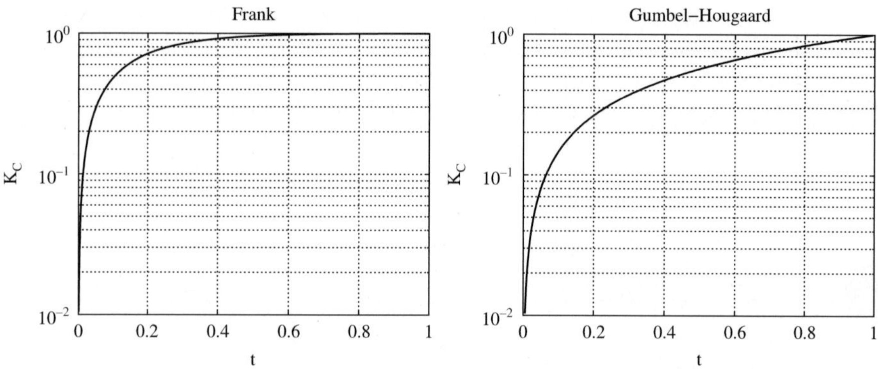

Figure 3.5. The function $K_{\mathbf{C}}$ for the Frank and the Gumbel-Hougaard 2-copulas

PROPOSITION 3.5. *Let $U \sim \mathrm{U}(0, 1)$ and $V \sim \mathrm{U}(0, 1)$ be r.v.'s whose joint distribution is the Archimedean copula $\mathbf{C}$ generated by γ. Then $K_{\mathbf{C}}$ is the distribution function of the r.v. $W = \mathbf{C}(U, V)$, i.e.*

$$F_W(w) = \mathbb{P}\{W \le w\} = K_{\mathbf{C}}(w), \quad 0 < w < 1, \tag{3.48}$$

with $F_W(w) = 0$ for $w \le 0$ and $F_W(w) = 1$ for $w \ge 1$.

The function $K_{\mathbf{C}}$ is sometimes called *Kendall's measure*. Note that the calculation of $K_{\mathbf{C}}$ can be extended to any copula $\mathbf{C}$ as follows:

$$K_{\mathbf{C}}(t) = t - \int_t^1 \frac{\partial}{\partial u} \mathbf{C}\left(u, v_{u,t}\right) \mathrm{d}u, \quad 0 < t \le 1, \tag{3.49}$$

where $v_{u,t} = \mathbf{C}_u^{-1}(t)$ and $\mathbf{C}_u(v) = \mathbf{C}(u, v)$. As an illustration, in Figure 3.5 we plot the function $K_{\mathbf{C}}$ for, respectively, the Frank 2-copula (see Section C.1) shown in Figure C.2, and the Gumbel-Hougaard 2-copula (see Section C.2) shown in Figure C.4.

As will be seen in Subsection 3.3.5, the function $K_{\mathbf{C}}$ represents a fundamental tool for calculating the return period of extreme events.

3.3. RETURN PERIODS VIA COPULAS

As already discussed in Section 1.3 (for the univariate case) and in Section 2.4 (for the multivariate case), the *return period* of a prescribed event is generally adopted in applications as a common criterion for design purposes, and provides a very simple and efficient means for risk analysis.

For the sake of clarity and simplicity, we shall now concentrate on the bivariate case. However, generalizations to higher dimensional settings will be evident. As a fundamental difference with standard approaches (see, e.g., Section 2.4), instead of considering a particular joint distribution F_{XY} with well specified marginals F_X and F_Y (as is usual in many applied works — which, in our opinion, denotes a limited view and prospect), in this Section we shall calculate shortly the return periods of the events of interest by using suitable copulas [251, 254]. This will solve the problem for a wider class of distributions than the function F_{XY} alone (i.e., *for all the joint distributions generated by the copula of interest*). We also stress that, in some cases, it may be even possible to derive the analytical expressions of the isolines of the return periods, both in the unconditional and in the conditional case: clearly, this represents an important (if not essential) piece of information in applications. In addition, we shall show how a new, well defined, probability distribution may be associated with the return period of specific events: this will lead to the definitions of *sub-*, *super-*, and *critical* events, as well as to those of *primary* and *secondary* return periods.

As done in Section 2.4, let us consider a sequence $E_1, E_2, \dots$ of *independent* events. We may assume that such events happen in the temporal domain at times $t_1 < t_2 < \cdots$ (i.e., we use a *temporal marked point process* as a model). As a difference with the approach outlined in Section 2.4, here each event E_i is characterized by the joint behavior of a pair of r.v.'s $(U, V) \sim \mathbf{C}_{UV}$, where $U \sim \mathtt{U}(0, 1)$ and $V \sim \mathtt{U}(0, 1)$. Note that there is no loss of generality by working on (U, V) instead of considering the pair (X, Y) which actually describes the stochastic dynamics of the phenomenon of interest. In fact, thanks to the invertible transformation (see Eqs. (3.22))

$$\begin{cases} u = F_X(x) \\ v = F_Y(y) \end{cases} \iff \begin{cases} x = F_X^{[-1]}(u) \\ y = F_Y^{[-1]}(v) \end{cases}, \tag{3.50}$$

where $(u, v) \in \mathbb{I}^2$ and $(x, y) \in \mathbb{R}^2$, and given the invariance property of copulas stated in Proposition 3.2, it is clear that $(X, Y) \sim \mathbf{C}_{UV}$. Thus all the copula-dependent results that follow will be the same for the pairs (U, V) and (X, Y). As a practical example, consider E_i to be a storm realization, characterized by an (average) intensity X and a duration Y, where X and Y are non-independent and statistically joined via the copula $\mathbf{C}_{UV}$ [66, 253].

Similarly to the approach adopted in Section 2.4, the joint events of interest here can be expressed in terms of a set of *marginal* events (see Definition 2.6, and change X (x) for U (u) and Y (y) for V (v)). Below, in Eqs. (3.51), we show how the statistics of the *extremal* event $\{E^{>}_{U,u} \vee E^{>}_{V,v}\}$ can be described via $\mathbf{C}$, and that of $\{E^{>}_{U,u} \wedge E^{>}_{V,v}\}$ via $\overline{\mathbf{C}}$. Using special functions of copulas — namely, the *dual-copula* and the *co-copula* [251, 207] — it is also possible to account for the statistics of other events (see, e.g., the discussion on droughts and floods in Section 2.4, and also [269]). However, this is not of immediate interest here.

Adopting the same notation as in Section 2.4, it is immediate to derive the following results concerning the probabilities of extremal events (see also Figure 3.2):

$$p^{\vee}_{u,v} = \mathbb{P}\{E^{\vee}_{u,v}\} = \mathbb{P}\{U > u \vee V > v\} = 1 - \mathbf{C}(u, v), \tag{3.51a}$$

$$p^{\wedge}_{u,v} = \mathbb{P}\left\{E^{\wedge}_{u,v}\right\} = \mathbb{P}\{U > u \wedge V > v\} = \overline{\mathbf{C}}\,(1-u, 1-v). \tag{3.51b}$$

Obviously, $p^{\vee}_{u,v} \geq p^{\wedge}_{u,v}$: in fact, it is sufficient that *either* $E^{>}_{U,u}$ *or* $E^{>}_{V,v}$ happen for the realization of $E^{\vee}_{u,v}$, whereas it is necessary that *both* $E^{>}_{U,u}$ *and* $E^{>}_{V,v}$ happen for the realization of $E^{\wedge}_{u,v}$. In turn,

$$\tau^{\vee}_{u,v} = \mu_T / p^{\vee}_{u,v}, \tag{3.52a}$$

$$\tau^{\wedge}_{u,v} = \mu_T / p^{\wedge}_{u,v}. \tag{3.52b}$$

Clearly, $\tau^{\vee}_{u,v} \leq \tau^{\wedge}_{u,v}$. The above results yield the following definition.

DEFINITION 3.11 (Return period). *The positive numbers $\tau^{\vee}_{u,v}$ and $\tau^{\wedge}_{u,v}$ given by Eqs. (3.52) define, respectively, the return periods of the events $E^{\vee}_{u,v}$ and $E^{\wedge}_{u,v}$.*

Note that $\tau^{\vee}_{u,v}$ and $\tau^{\wedge}_{u,v}$ are decreasing functions of the corresponding probabilities $p^{\vee}_{u,v}$ and $p^{\wedge}_{u,v}$: this is obvious, since the interarrival time gets longer for less probable events.

A practical problem, of importance in applications, is the identification of the events having an assigned return period. In general, its solution is not unique: in fact, it may happen that different choices of u and v (or, equivalently, different realizations of X and Y) yield the same return period. Thus, the problem is to identify in the unit square $\mathbb{I}^2$ (or, equivalently, in the plane $\mathbb{R}^2$ when considering the r.v.'s X and Y — see Eqs. (3.50)) the *isolines* of $\tau^{\vee}_{u,v}$ and $\tau^{\wedge}_{u,v}$, which correspond to curves of *isofrequency* for the events considered. However, it is important first to discuss and point out the differences between univariate and bivariate (and, more generally, multivariate) frequency analysis.

3.3.1 Univariate vs. Bivariate Frequency Analysis

Univariate frequency analysis is applicable when only one r.v. is significant in the design process. If a given event is multivariate, i.e. described by a set of non-independent r.v.'s, then univariate frequency analysis cannot provide complete assessment of the probability of occurrence. Actually, a better understanding of the statistical characteristics of such events requires the study of their joint distribution (for an illustration in a hydrological context see [308]). The (generalized) bivariate approach we propose may be useful not only in the design of structures, but also in many other practical situations concerning Natural Hazards.

Let us suppose that $(X, Y) \sim \mathbf{C}$ and, in particular, that X and Y are non-independent. Then, fix a probability level $q \in \mathbb{I}$ common to both variables (as is usual in many applications), and introduce the marginal q-level quantiles of X and Y given by $x_q = F_X^{[-1]}(q)$ and $y_q = F_Y^{[-1]}(q)$. Clearly, this is equivalent to fixing a common arbitrary return period τ for X and Y. Now, Eq. (3.2) states that the diagonal section $\boldsymbol{\delta}$ of a copula satisfies the inequality $\boldsymbol{\delta}(t) \leq t$ for all $t \in \mathbb{I}$. In turn, it is immediate to show that

$$\tau^{\vee}_{q,q} \leq \tau \leq \tau^{\wedge}_{q,q}. \tag{3.53}$$

As a consequence, $E^{\vee}_{x_q,y_q}$ and $E^{\wedge}_{x_q,y_q}$ might not be q-level bivariate events (with return period τ). More specifically: if $E^{\vee}_{x_q,y_q}$ were used as a "critical" design event, in order to have $\tau^{\vee}_{q,q} = \tau$ the marginal quantiles x_q, y_q (and hence q) might need to be increased. By the same token, if $E^{\wedge}_{x_q,y_q}$ were used as a "critical" design event, in order to have $\tau^{\wedge}_{q,q} = \tau$ the marginal quantiles x_q, y_q (and hence q) might need to be decreased. In other words: if x_q, y_q were used as design quantiles, then the resulting work could be either under-dimensioned (and would be at risk), or over-dimensioned (causing a waste of resources).

ILLUSTRATION 3.9 (Spillway design flood). ►

In order to illustrate the differences between a univariate and a bivariate (more generally, multivariate) approach, we analyse a well known hydrological problem: the estimation of the spillway design flood [67]. Its calculation is commonly based on the univariate quantiles x_q and y_q of, respectively, the maximum annual flood-volume X and the maximum annual flood-peak Y, for the same return period ϖ (or probability level q). Usually, $\varpi = 1000$ years (sometimes, $\varpi = 5000$ or 10000 years, depending upon the magnitude of damages that a failure could create, and the specific directives followed).

In the case study analysed by [67], the link between X and Y is modeled via a Gumbel-Hougaard's 2-copula (see Section C.2). Using the values of the parameters estimated in this work, in Table 3.1 we calculate the return periods $\tau^{\vee}_{q,q}$ (for $E^{\vee}_{x_q,y_q}$) and $\tau^{\wedge}_{q,q}$ (for $E^{\wedge}_{x_q,y_q}$), for several choices of ϖ.

Now, suppose that the desired return period is ϖ: evidently, as prescribed by the inequality in Eq. (3.53), the return period of the event $E^{\vee}_{x_q,y_q}$ is about 20% smaller than ϖ, while the return period of the event $E^{\wedge}_{x_q,y_q}$ is about 30% larger than ϖ. As already mentioned, in order to obtain joint events having a return period equal to ϖ, it is necessary to consider different quantiles: the last two columns in Table 3.1 show which quantiles (respectively, $q^{\vee}$ and $q^{\wedge}$) should be used for obtaining events $E^{\vee}_{q^{\vee},q^{\vee}}$ and $E^{\wedge}_{q^{\wedge},q^{\wedge}}$ having a return period equal to ϖ. Note that the correct values of the quantiles always satisfy the inequality

$$q^{\wedge} \le q \le q^{\vee}, \tag{3.54}$$

as in the discussion of Eq. (3.53).

In conclusion, the analysis of bivariate return periods is greatly facilitated by using copulas. For practical purposes, this may give a precise understanding of the

Table 3.1. Univariate and bivariate estimates for the case study shown in [67]

ϖ (years)	q	$\tau^{\vee}_{q,q}$ (years)	$\tau^{\wedge}_{q,q}$ (years)	$q^{\vee}$	$q^{\wedge}$
10	0.9	8.1	13.1	0.91946	0.86960
100	0.99	79.8	133.9	0.99202	0.98662
1000	0.999	797.1	1341.4	0.99920	0.99866
10000	0.9999	7970.2	13417.0	0.99992	0.99987

stochastic joint dependence of the variables of interest, which may be extremely difficult to achieve without adopting a copula approach. ◀

Furthermore, we show shortly how wrong assumptions about the independence of r.v.'s may yield poor estimates of derived functions. In addition, we show how copulas may help in solving problems where the hypothesis of independence is usually introduced to simplify the calculations, without any empirical justification.

ILLUSTRATION 3.10 (Sea storm characterization (cont.)). ▶

As discussed in Illustration 3.5, in [68] a characterization of the sea storm dynamics is provided by considering four relevant variables. In particular, the significant wave height H and the storm duration D deserve a special attention, for they rule the statistical behavior of the storm energetic content M, as shown here.

A Gumbel distribution closely fits the observations of H, while D is well described by a GEV distribution. Note that D shows a upper heavy tail, with a shape parameter less than $-1/2$: this means that the second order moment may not exist. In order to characterize the *storm energetic content*, the *sea storm magnitude* M is introduced:

$$\mathrm{M} = (H - \eta)D/2, \tag{3.55}$$

where $\eta = 2$ meters is a threshold used to identify and separate successive storms. The meaning of M is analogous to that of the storm depth (or volume) in rainfall modeling — see, e.g., [66, 253, 255]. From a practical point of view, M concentrates into a single number the relevant information concerning the energetic content of the sea storm. Here M is essentially the product of the duration D times the significant wave height H above the threshold η, according to the equivalent triangular storm model. Thus, M couples the dynamics of two fundamental storm variables, and provides a means to characterize the overall intensity of the phenomenon.

The analytical calculation of the distribution F_M of M is an involved problem, because H and D are non-independent: in the present case, the value of Kendall's τ_K (see Subsection B.2.2) is estimated as $\tau_K^{H,D} \approx 0.25$, and a 2-copula belonging to the Ali-Mikhail-Haq family (see Section C.4), with dependence parameter $\theta \approx 0.829$, is used to describe the statistical behavior of (H, D). As shown in [255], F_M can be numerically estimated by using a procedure exploiting copulas. Furthermore, the techniques outlined shortly are also of interest in their own right: they provide quite a general solution to the problem, and may have a broad application in many hydrological scenarios.

Let (R, S) be a random vector, where R and S have Uniform marginals on $\mathbb{I}$, and joint distribution $F_{RS} \sim \mathbf{C}_{HD}$. Via the *Probability Integral Transform*, the statistical behavior of the couple (H, D) — and hence of M — can be modeled as a function of (R, S). In fact, for fixed $v > 0$, we may write

$$\begin{aligned}\mathbb{P}\{M \le v\} &= \mathbb{P}\{H'D' \le v\} \\ &= \mathbb{P}\left\{F_{H'}^{-1}(R)F_{D'}^{-1}(S) \le v\right\} \\ &= \mathbb{P}\{R \le g_v(S)\}\end{aligned} \tag{3.56}$$

where $H' = H - \eta$ and $D' = D/2$. The function $s \mapsto g_v(s)$ given by

$$g_v(s) = F_{H'}\left(v/F_{D'}^{-1}(s)\right) \tag{3.57}$$

is monotonic and continuous on the compact subset $\mathbb{I}$. Clearly, g_v takes values on $\mathbb{I}$, and is bounded and integrable on the unit interval with respect to the variable s. Using Eq. (3.26a), and from the invariance property of copulas (see Proposition 3.2), we may then write

$$\psi_s(r) = \mathbb{P}\{R \le r \mid S = s\} = \frac{\partial}{\partial s}\mathbf{C}_{HD}(r, s), \tag{3.58}$$

which exists and is non-decreasing almost everywhere (in the Lebesgue sense) for $s \in \mathbb{I}$ (see the discussion ensuing Eqs. (3.26)). Note that ψ_s is a probability, and therefore belongs to $\mathbb{I}$. Thus, since the r.v. S is Uniform on $\mathbb{I}$, the last equality in Eq. (3.56) can be written as

$$\begin{aligned}F_M(v) &= \int_0^1 \mathbb{P}\{R \le g_v(S) \mid S = s\}\, \mathrm{d}s \\ &= \int_0^1 \psi_s(g_v(s))\, \mathrm{d}s.\end{aligned} \tag{3.59}$$

The last integrand is a function of s only, since v acts as a parameter. Therefore, the calculation of F_M reduces to a one-dimensional integration over $[0, 1]$. This is much easier (theoretically and computationally) than the two-dimensional integrations required by standard probabilistic techniques. In some cases the last integral can be solved explicitly, while in general a numerical approach is needed.

In Figure 3.6 we show the comparison between the empirical distribution (and survival) functions of M and those calculated using copulas, assuming H and D as dependent.

Evidently, the fit is valuable over the wide range considered: M takes on values in the interval 0–2580 mh. The disagreement in the right tail is more apparent than real. On the one hand, the plotting positions are adversely affected by the very low number of extreme events, and cannot estimate the correct probabilities. On the other hand, the vertical scale is logarithmic, and the disagreement is indeed negligible. For the sake of comparison, we also show the distribution of M in case H and D are considered as independent (i.e., with copula $\mathbf{\Pi}_2$). Evidently, the fit is not as good as in the previous case. Practically, the "independence" approximation

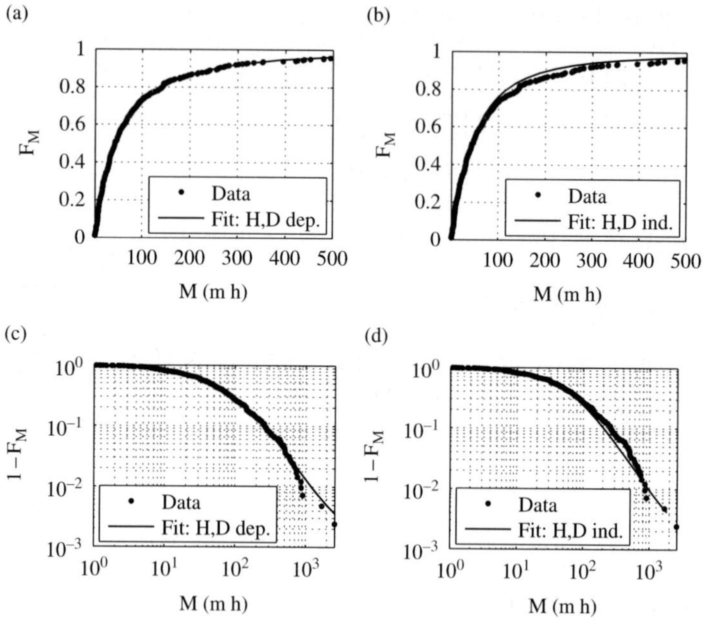

Figure 3.6. (*top*) Comparison between the empirical distribution function of M (*markers*) and the ones estimated using copulas, in case H and D are considered as dependent (*a*) or independent (*b*). (*bottom*) Comparison between the empirical survival function of M (*markers*) and the ones estimated using copulas, in case H and D are considered as dependent (*c*) or independent (*d*)

(generally used in applications) affects the calculation of the distribution of M, and yields wrong estimates of the relevant probabilities.

The above results are quite relevant, since they provide formulas for the calculation of F_M which can be used in applications. Most importantly, the above derivation is quite general, for it does not depend upon the copula and the marginals considered.

Let us now consider the *total storm duration* $S = D + I$. Using the estimates of the parameters reported in [68], it follows that $\mu_S = \mathbb{E}(S) \approx 259.6$ hours (the corresponding sample average is ≈ 258 hours, which is in good agreement with the theoretical result). In turn, the storm *return period* $\tau(p)$, associated with a fixed probability level $0 < p < 1$, can be calculated as

$$\tau(p) = \mu_S/(1-p). \tag{3.60}$$

In Table 3.2 we show the estimates of some return periods associated with extreme values of p. In addition, we compare the approximate estimations of several extreme quantiles of M by considering, respectively, H and D as independent or dependent. In all cases, the latter quantiles are about 24% larger than the ones calculated assuming H and D as independent. Such a difference may have relevant consequences when dealing with design evaluations for structures or operations at sea.

Table 3.2. Estimated values of the quantiles q_p of M for several extreme probability levels p: q'_p are calculated considering H and D as independent, whereas q''_p are calculated assuming H and D joined via a Ali-Mikhail-Haq 2-copula — see text. The return periods $\tau(p)$ associated with p are also indicated

p	**0.9**	**0.95**	**0.975**	**0.99**	**0.995**
q'_p (mh)	204	333	528	968	1592
q''_p (mh)	251	413	656	1206	1986
τ (years)	0.30	0.59	1.19	2.96	5.93

A further important point concerns the behavior of M for large values. If either H or D (or both) has a heavy upper tail — as in the present case considering D, then also M shows an asymptotic algebraic fall-off. Such a behavior is clearly present in Figure 3.6*cd*. This result was partially anticipated in [66, 253], by considering *ad hoc* "extremal" conditions, and is also discussed in [255]. Thus, a Pareto-like behavior of the variables characterizing the sea storm dynamics yields a heavy upper tail also for M. During the design phase, this may represent relevant information about the statistics of the phenomenon considered. ◀

3.3.2 The "OR" Case

In this Section we calculate the isofrequency curves of the "OR" return period $\tau^{\vee}_{u,v}$. Given Eqs. (3.52), it is clear that $\tau^{\vee}_{u,v}$ is constant in $\mathbb{I}^2$ whenever it applies to $p^{\vee}_{u,v}$; thus, it is sufficient to find the isolines of **C**. In general, the task is non-trivial from an analytical point of view, and may require the use of numerical routines. However, if we assume that **C** is Archimedean, then we obtain, for $0 < t \leq 1$,

$$p^{\vee}_{u,L_t(u)} = 1 - \mathbf{C}\,(u, L_t(u)) = 1 - t, \quad t \leq u \leq 1, \tag{3.61}$$

i.e. $p^{\vee}_{u,v}$ is constant on the (convex) level curve $v = L_t(u)$, which can be written explicitely using Eq. (3.45). As a consequence,

$$\tau^{\vee}_{u,L_t(u)} = \vartheta(t) = \frac{\mu_T}{1-t}, \tag{3.62}$$

where $\vartheta: \mathbb{I} \to (\mu_T, \infty]$ is one-to-one and increasing. Thus, the events $E^{\vee}_{u,L_t(u)}$ have a constant return period. In turn, the return period of the corresponding events for the r.v.'s (X, Y) is constant in $\mathbb{R}^2$ on the set

$$\begin{cases} x = F_X^{[-1]}(u) \\ y = F_Y^{[-1]}\left(L_t(F_X(x))\right) \end{cases}, \tag{3.63}$$

where $x \geq F_X^{[-1]}(t)$. Note that $p^{\vee}_{u,L_t(u)}$ reduces by increasing the (probability) level t, while the corresponding return period $\tau^{\vee}_{u,L_t(u)}$ increases: in fact, increasing t is equivalent to choosing more "extreme" events, which have longer interarrival times.

In Illustration 3.11 which follows we show the practical calculation of $\tau^{\vee}_{x,y}$ in the case of a flood-peak and flood-volume case study.

ILLUSTRATION 3.11 (Flood-peak and Flood-volume). ►

As discussed in Illustration 3.9, in [67] a Gumbel-Hougaard 2-copula (with upper-unbounded GEV marginals) is used to assess the spillway design flood of an existing Italian dam, by considering the maximum annual flood-volume and flood-peak as the hydrological variables of interest. The dependence parameter $\theta = 3.055$ is estimated using the relationship between θ and Kendall's τ_K given by Eq. (C.8). Note that the existence of Pearson's linear correlation coefficient is suspicious, since the estimates of the marginal parameters indicate that the second order moments may not exist.

As an illustration, in Figure 3.7 we plot both the isolines of $p^{\vee}_{u,v}$ and those of $\tau^{\vee}_{x,y}$, for four selected values (i.e., 10, 50, 100, 500 years); here X denotes the maximum annual flood-volume and Y the maximum annual flood-peak. Clearly, the use of annual values yields $\mu_T = 1$ year.

Figure 3.7 deserves several comments. The isolines plotted in *(a)* are the same for all the distributions F_{XY} generated by **C**, independently of the marginals used to model the behavior of X and Y. For instance, in [268] the Type B Bivariate Extreme Value Distribution with Gumbel marginals is adopted to characterize extreme floods in a basin in southern Taiwan. Such a bivariate distribution can simply be derived by using suitable Gumbel marginals in the Gumbel-Hougaard 2-copula. In addition, since **C** is Archimedean, and the expression of the generator is readily available, it is easy to derive the analytical equations of the level curves. These same formulas can then be used in case studies reported in [268, 67], as well as in many other situations.

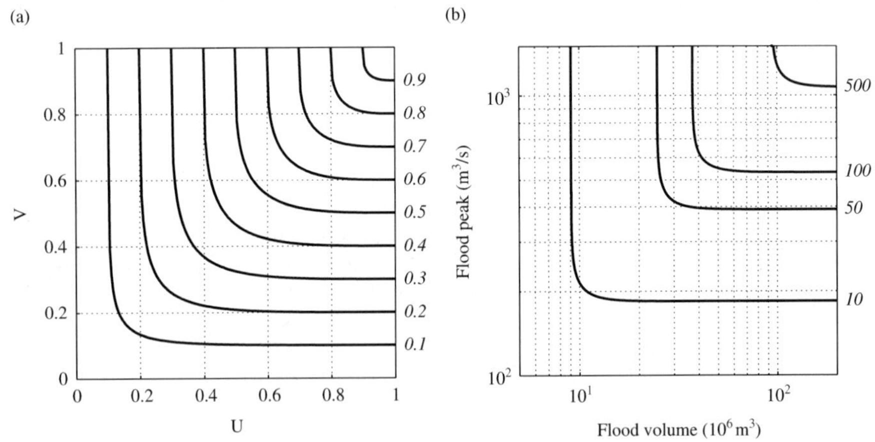

Figure 3.7. Level curves of (*a*) $p^{\vee}_{u,v}$ for the Gumbel-Hougaard 2-copula and (*b*) the "OR" return period $\tau^{\vee}_{x,y}$ for the data considered in [67]. The values of the probability levels t in (*a*) and of the return periods (in years) in (*b*) are as indicated (*italic*)

From a practical point of view, once the design return period has been decided, it is easy to calculate all the joint events $\{X > x\} \vee \{Y > y\}$ having the desired frequency. ◄

3.3.3 The "AND" Case

In this section we calculate the isofrequency curves of the "AND" return period $\tau^{\wedge}_{u,v}$. Essentially, finding the isolines of constant return period for the event $E^{\wedge}_{u,v}$ requires the same techniques used in the previous "OR" case, with the obvious changes. In general, the task is non-trivial from an analytical point of view, and may require the use of numerical routines. From Eqs. (3.28) and Eqs. (3.51) it is clear that the following relations are equivalent:

$$p^{\wedge}_{u,v} = 1 - t \Longleftrightarrow \overline{\mathbf{C}}\,(1-u, 1-v) = 1 - t \Longleftrightarrow u + v - \mathbf{C}\,(u, v) = t, \qquad (3.64)$$

where $0 < t \le 1$. As before, if $\overline{\mathbf{C}}$ were Archimedean, then we could easily find its level curves, as well as the corresponding isolines for $p^{\wedge}_{u,v}$ and $\tau^{\wedge}_{u,v}$.

Thus, let $\overline{\mathbf{C}}$ be Archimedean with generator γ, and consider Eq. (3.64). Then, the level curve $v = G_t(u)$ given by

$$v = G_t(u) = 1 - \gamma^{-1}\left(\gamma(1-t) - \gamma(1-u)\right), \quad 0 \le u \le t, \qquad (3.65)$$

yields the set in $\mathbb{I}^2$ where $p^{\wedge}_{u,v}$ (and, consequently, $\tau^{\wedge}_{u,v}$) is constant. In particular

$$\tau^{\wedge}_{u,G_t(u)} = \vartheta(t). \qquad (3.66)$$

Then, consider again Eq. (3.64) and suppose that $\overline{\mathbf{C}}$ is not Archimedean, but $\mathbf{C}$ is Archimedean, with generator γ. Introducing the new variables a, b given by

$$\begin{cases} a = \gamma(u) \\ b = \gamma(v) \end{cases} \Longleftrightarrow \begin{cases} u = \gamma^{[-1]}(a) \\ v = \gamma^{[-1]}(b) \end{cases}, \qquad (3.67)$$

the relationship in Eq. (3.64) can be written in a more tractable form involving only the univariate non-increasing function $\gamma^{[-1]}$, i.e.

$$\gamma^{[-1]}(a) + \gamma^{[-1]}(b) - \gamma^{[-1]}(a+b) = t. \qquad (3.68)$$

In general, the pairs (a, b) satisfying Eq. (3.68) cannot be determined analytically. However, given the beneficial properties of the function $\gamma^{[-1]}$, finding a numerical solution is not too difficult a task. Then, it is easy to find the isolines of constant return period in $\mathbb{R}^2$ of the event $E^{\wedge}_{x,y}$. Note that, given the symmetries in Eq. (3.68) — the roles of a and b are exchangeable — the search for a numerical solution, say, b as a function of a, may start at $a = \gamma(t)$, yielding $b = \gamma(0)$, including the case $b = +\infty$, and hence the solution $(u = t, v = 0)$, and then proceed by increasing a to $a = a'$, calculated via the equation

$$2\gamma^{[-1]}(a') - \gamma^{[-1]}(2a') = t.$$

Further solutions for b can then be found simply by interchanging a and b in the results previously obtained.

ILLUSTRATION 3.12 (Flood-peak and Flood-volume (cont.)). ►

In Illustration 3.11 it was shown how in [67] a Gumbel-Hougaard 2-copula (with upper-unbounded GEV marginals) is used to assess the spillway design flood of an existing Italian dam, by considering the maximum annual flood-volume and flood-peak as the hydrological variables of interest. Using the same values of the return periods (i.e., 10, 50, 100, 500 years), we plot in Figure 3.8 both the isolines of $p^{\wedge}_{u,v}$ and those of $\tau^{\wedge}_{x,y}$, where X denotes the maximum annual flood-volume and Y the maximum annual flood-peak.

The same comments as in Illustration 3.11 hold also in the present case. However, here $\overline{\mathbf{C}}$ is not Archimedean, and thus the level curves are calculated numerically. From a practical point of view, once the return period has been decided according to suitable design requirements, it is easy to calculate all the joint events $\{X > x\} \wedge \{Y > y\}$ having the desired frequency. The same approach can be used in both case studies reported in [268, 67], as well as in many other situations. ◄

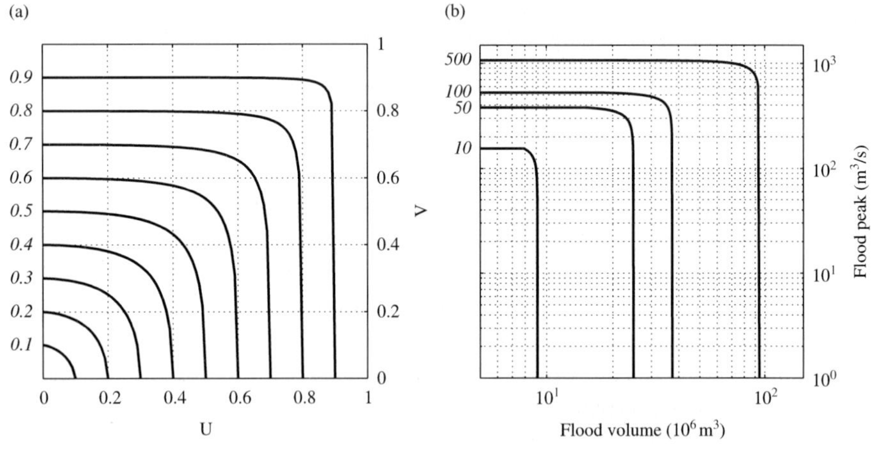

Figure 3.8. Level curves of (a) $p^{\wedge}_{u,v}$ for the Gumbel-Hougaard 2-copula and (b) the "AND" return period $\tau^{\wedge}_{x,y}$ for the data considered in [67]. The values of the probability levels t in (a) and of the return periods (in years) in (b) are as indicated (*italic*)

3.3.4 Conditional Return Periods

Using the same notation as above, it is possible to introduce the concept of *conditional return period*, which applies to particular conditional events that are of interest in applications. For instance, the evaluation of the hydrological risk concerning the safety of a dam needs to consider the event that the flood-peak Y is larger than a prescribed threshold y *given that* the flood-volume X equals or exceeds a design value x.

As already mentioned in Section 3.1, the study of conditional laws is facilitated by using copulas, from Eqs. (3.26). Herein we shall be concerned with a restricted class of conditional events: namely, we shall only consider the situation in which $V > v$ *given that* $U > u$. Note that the symmetric case obtained by swapping U and V, and exchanging their roles, will yield the same results, and thus it is not a cause for concern. However, conditional events other than the ones considered here could easily be investigated using the same techniques outlined as follows.

Adopting an operative approach, let us re-consider the temporal sequence of events $E_1, E_2, \ldots$ introduced at the beginning of Section 3.3. Conditioning with respect to the event $E^>_{U,u} = \{U > u\}$ means that we need first to restrict our attention only to those events E_i's where $U > u$, and hence extract from $\{E_i\}$ a well defined subsequence $\{E'_j\}$. Then, among the events of this latter sequence, we should only consider those in which $V > v$, and thus extract from $\{E'_j\}$ a further well defined subsequence $\{E''_k\}$. Our target is to calculate the average time elapsing between two successive realizations of the events in $\{E''_k\}$. Hence we introduce the following definition.

DEFINITION 3.12 (Conditional return period). *The conditional return period $\tau^>_{v|u}$ is the average interarrival time between two successive events in the sequence $\{E''_k\}$.*

An elementary reasoning shows that, in order to calculate $\tau^>_{v|u}$, it is necessary to operate a random Bernoulli sampling (selection) on the subsequence $\{E'_j\}$, with a probability of "success" equal to $\mathbb{P}\{E^>_{U,u} \wedge E^>_{V,v}\}$. As a consequence, since the return period of the events $E^>_{U,u}$'s is simply given by

$$\tau^>_{U,u} = \frac{\mu_T}{\overline{F}_U(u)} = \frac{\mu_T}{1-u} = \vartheta(u), \tag{3.69}$$

then, using Eqs. (3.51), $\tau^>_{v|u}$ can be calculated as

$$\tau^>_{v|u} = \frac{\tau^>_{U,u}}{p^\wedge_{u,v}} = \frac{\vartheta(u)}{\overline{\mathbf{C}}(1-u, 1-v)}. \tag{3.70}$$

Therefore, using the results of Subsection 3.3.3, it is possible to calculate the isolines of $\tau^>_{v|u}$ by solving the equation

$$(1-u)\,\overline{\mathbf{C}}(1-u, 1-v) = (1-u)\,(1-u-v+\mathbf{C}(u,v)) = 1-t, \tag{3.71}$$

with $0 < t \le 1$.

In general, finding the isolines of $\tau^{>}_{v|u}$ is a non-trivial task from an analytical point of view, and may require the use of numerical routines. However, if either $\mathbf{C}$ or $\overline{\mathbf{C}}$ is Archimedean, then the calculations may be simplified. In fact, let $\overline{\mathbf{C}}$ be Archimedean with generator γ. We may then rewrite the equation of interest as

$$(1-u)\,\gamma^{[-1]}\,(\gamma(1-u)+\gamma(1-v)) = 1-t, \tag{3.72}$$

which yields

$$\begin{aligned} v = H_t(u) &= 1 - L_{\frac{1-t}{1-u}}(1-u) \\ &= 1-\gamma^{-1}\left(\gamma\left(\frac{1-t}{1-u}\right)-\gamma(1-u)\right), \end{aligned} \tag{3.73}$$

where $0 \le u \le 1-\sqrt{1-t}$. Thus, the level curve $v = H_t(u)$ defines the set in $\mathbb{I}^2$ where $\tau^{>}_{v|u}$ is constant and takes on the value

$$\tau^{>}_{H_t(u)|u} = \vartheta(t). \tag{3.74}$$

Alternatively, suppose that $\mathbf{C}$ is Archimedean, with generator γ. Using again the variables a, b as given by Eqs. (3.67), we may rewrite the equation of interest as

$$\left(1-\gamma^{[-1]}(a)\right)\left(1-\gamma^{[-1]}(a)-\gamma^{[-1]}(b)+\gamma^{[-1]}(a+b)\right) = 1-t, \tag{3.75}$$

i.e. a more tractable form which only involves the univariate non-increasing function $\gamma^{[-1]}$. The same considerations expressed in the discussion ensuing Eq. (3.68) hold also in the present case.

ILLUSTRATION 3.13 (Flood-peak and Flood-volume (cont.)). ►

In Illustrations 3.11–3.12 we showed how in [67] a Gumbel-Hougaard 2-copula (with upper-unbounded GEV marginals) is used to assess the spillway design flood of an existing Italian dam, by considering the maximum annual flood-volume and flood-peak as the hydrological variables of interest. Using the same values of the return periods (i.e., 10, 50, 100, 500 years), we plot in Figure 3.9 both the isolines of $\tau^{>}_{v|u}$ and those of $\tau^{>}_{y|x}$, where X denotes the maximum annual flood-volume and Y the maximum annual flood-peak.

The same comments as in Illustration 3.11 hold also in the present case. However, here $\overline{\mathbf{C}}$ is not Archimedean, and thus the level curves are calculated numerically. From a practical point of view, once the conditional return period has been decided according to suitable design requirements, it is easy to calculate all the events $\{Y > y\}$, given that $\{X > x\}$, having the desired frequency. The same approach can be used in both case studies reported in [268, 67], as well as in many other situations. ◄

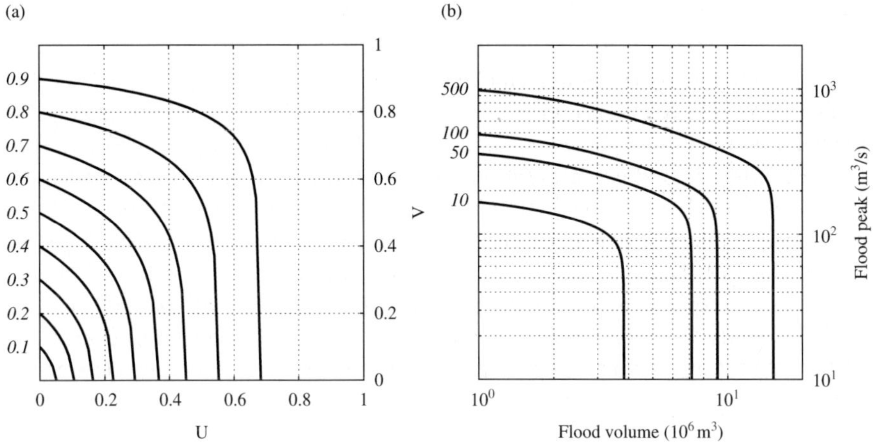

Figure 3.9. Level curves of (*a*) $\tau^{>}_{v|u}$ for the Gumbel-Hougaard 2-copula and (*b*) the conditional return period $\tau^{>}_{y|x}$ for the data considered in [67]. The values of the probability levels t in (*a*) and of the return periods (in years) in (*b*) are as indicated (*italic*)

3.3.5 Secondary Return Period

The use of copulas gives the possibility of introducing a new concept of relevance in applications: the *secondary return period.* Let us reconsider the results of Subsection 3.3.2 and the **C**-measure $K_{\mathbf{C}}$ of the region $B_{\mathbf{C}}$ introduced in Section 3.2. Then, the following equivalences hold:

$$(u, v) \in B_{\mathbf{C}}(t) \Longleftrightarrow \tau^{\vee}_{u,v} \leq \vartheta(t) \Longleftrightarrow w = \mathbf{C}(u, v) \leq t, \tag{3.76}$$

where $0 < t \leq 1$. In other words, $B_{\mathbf{C}}(t)$ is exactly the region in $\mathbb{I}^2$ where the return period $\tau^{\vee}_{u,v}$ is less than, or equal to, a well defined monotone function of t. Let $\mathcal{E}^{\vee} = \{\mathrm{E}^{\vee}_{u,v}\}$, $(u, v) \in \mathbb{I}^2$, denote the family of events of interest here. We can then introduce in $\mathcal{E}^{\vee}$ an *equivalence relation* as follows [251].

DEFINITION 3.13. *Let $0 < t \leq 1$ be fixed, and let $\mathrm{E}^{\vee}_1, \mathrm{E}^{\vee}_2 \in \mathcal{E}^{\vee}$. Then, $\mathrm{E}^{\vee}_1, \mathrm{E}^{\vee}_2$ are said to be t-equivalent ($\mathrm{E}^{\vee}_1 \overset{t}{\sim} \mathrm{E}^{\vee}_2$) if the corresponding return periods $\tau^{\vee}_1, \tau^{\vee}_2$ satisfy, respectively, the inequalities $\tau^{\vee}_1 \leq \vartheta(t)$ and $\tau^{\vee}_2 \leq \vartheta(t)$.*

Evidently, the relation "$\overset{t}{\sim}$" identifies, for each $0 < t \leq 1$, the equivalence class $\mathcal{E}^{\vee}_t$ given by all the *t-equivalent* events of $\mathcal{E}^{\vee}$, i.e. those events having a return period $\tau^{\vee}_{u,v} \leq \vartheta(t)$. In turn, the unit square $\mathbb{I}^2$ is partitioned into two disjoint classes, i.e. $\mathcal{E}^{\vee}_t$ and its complement $\overline{\mathcal{E}}^{\vee}_t$, collecting all the events having a return period $\tau^{\vee}_{u,v} > \vartheta(t)$. In addition, every threshold $\vartheta(t)$ can be uniquely associated with a probability $F_W(t)$ (see Proposition 3.5), which can be expressed in terms of the measure function $K_{\mathbf{C}}$ (see Theorem 3.4). Thus, the (survival) function

$$\overline{F}_W(t) = 1 - F_W(t) = \overline{K}_C(t) \tag{3.77}$$

yields the probability of finding an event belonging to $\overline{\mathcal{E}}_t^{\vee}$.

The above discussion has important consequences. If a "critical" threshold $\tau_*^{\vee}$ is fixed during the design phase (or, equivalently, if a critical level t_* is chosen), then the probability of finding an event with larger return period (and hence belonging to $\overline{\mathcal{E}}_{t_*}^{\vee}$) can be calculated explicitly via $\overline{K}_C$. At each realization of the process, two incompatible things may happen: the new event has either a return period $\tau_{u,v}^{\vee} \leq \tau_*^{\vee}$ (i.e., not larger than the "safety" threshold $\tau_*^{\vee}$), or $\tau_{u,v}^{\vee} > \tau_*^{\vee}$. Such a stochastic dynamics corresponds to a Bernoulli process, with probability of "failure" equal to $\overline{K}_C(t_*)$. Clearly, $\overline{\mathcal{E}}_{t_*}^{\vee}$ would represent the class of potentially dangerous events, the *outliers*, and it is possible to introduce an *ad hoc* return period for such destructive events.

DEFINITION 3.14 (Secondary return period). *Let* $0 < t_* \leq 1$ *be fixed. Then the quantity*

$$\varrho(t_*) = \frac{\mu_T}{1 - F_W(t_*)} = \vartheta(F_W(t_*)) = \vartheta(K_C(t_*)) \tag{3.78}$$

is called the secondary return period associated with the critical level t_*.

The label "secondary" is introduced to emphasize the difference with the "OR" return period $\tau_{u,v}^{\vee}$, which could be identified as "primary". In a way, ϱ is a *set* function: indeed, its domain could be chosen as (the family of) the classes $\overline{\mathcal{E}}_t^{\vee}$ in $\mathcal{E}^{\vee}$ identified by the relation $\overset{t}{\sim}$, with $0 < t \leq 1$. The range of ϱ is $(\mu_T, \infty]$, the same as ϑ. As an illustration, in Figure 3.10 we plot the secondary return period ϱ for, respectively, the Frank 2-copula (see Section C.1) shown in Figure C.2, and the Gumbel-Hougaard 2-copula (see Section C.2) shown in Figure C.4; here $\mu_T = 1$. Note how $\varrho(t) \to \infty$ as $t \to 1$.

The above results have practical significance. Using the primary return period as a criterion for design purposes only takes into account the fact that a prescribed critical event is expected to appear once in a given time interval. As opposed to this, on the one hand, $\overline{K}_C$ provides the *exact probability* that a potentially destructive event will happen at any given realization of the process under investigation, and, on the other hand, ϱ gives the expected time for such an outlier to occur. Evidently, these latter values may provide more useful information than the knowledge of the primary return period only. In addition, they may yield important hints for doing numerical simulations, as well as for a correct interpretation of the stochastic dynamics of the phenomenon. Perhaps, the use of $\overline{K}_C$ and the secondary return period would be a more appropriate approach to problems of (multivariate) risk-assessment.

The above results provide useful information for the design process. However, the new concepts introduced deserve further discussion. Suppose that a critical design threshold $\tau_*^{\vee} > \mu_T$ (or, equivalently, a critical level $t_* > 0$) is fixed: for instance, the

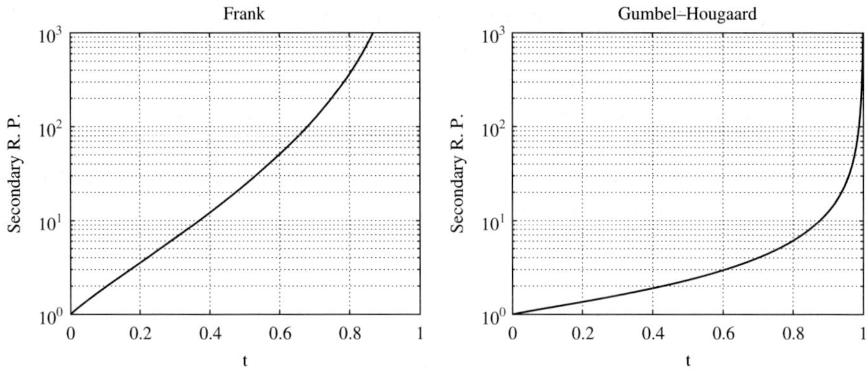

Figure 3.10. The function ϱ for the Frank and the Gumbel-Hougaard 2-copulas

primary return period $\tau_*^\vee$ could be the one chosen for sizing a given manufacture. Then, let us consider the unit square $\mathbb{I}^2$, as shown in Figure 3.11.

The *continuous line* connecting the points $(t_*, 1)$ and $(1, t_*)$ is the isoline of the events $E_{u,v}^\vee$ having a constant primary return period $\tau_{u,v}^\vee = \tau_*^\vee = \vartheta(t_*)$; one of such events is E_* (*circle*). Clearly, the points of this level curve satisfy the equation $v = L_{t_*}(u)$, which can be written explicitly in the Archimedean case. As a consequence, $B_C(t_*)$ is the region in $\mathbb{I}^2$ lying on, or below and to the left of, the level curve L_{t_*}. All the events in $B_C(t_*)$ have primary return period $\tau_{u,v}^\vee \leq \vartheta(t_*)$: in

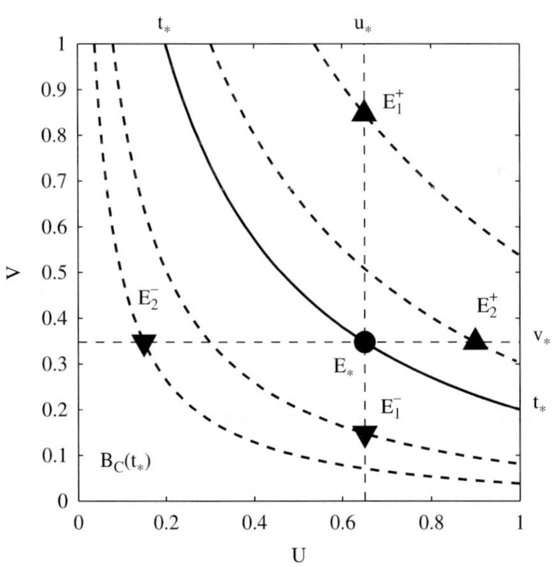

Figure 3.11. Illustration of *sub-*, *super-*, and *critical* events (see text)

particular, those events on L_{t_*} are *critical* (having primary return period $\tau^{\vee}_{u,v} = \tau^{\vee}_{*}$), while the others are *sub-critical*. In Figure 3.5 two different sub-critical events E^{-}_{1}, E^{-}_{2} are shown, plotted as *down-triangles*: note that the primary return period of E^{-}_{2} is smaller than the one of E^{-}_{1}. In addition, all the events on the level curve crossing E^{-}_{1} (*dashed line*) share the same primary return period as of E^{-}_{1}, while all the events on the level curve crossing E^{-}_{2} share the same primary return period as of E^{-}_{2}. Finally, all the events outside $B_C(t_*)$ have primary return period $\tau^{\vee}_{u,v} > \tau^{\vee}_{*}$, and represent the *super-critical* ones. In Figure 3.5 two different super-critical events E^{+}_{1}, E^{+}_{2} are shown, plotted as *up-triangles*: note that the primary return period of E^{+}_{2} is smaller than the one of E^{+}_{1}. In addition, all the events on the level curve crossing E^{+}_{1} share the same primary return period as of E^{+}_{1}, while all the events on the level curve crossing E^{+}_{2} share the same primary return period as of E^{+}_{2}.

A further fundamental consideration is as follows. The primary return period, the one usually adopted in applications, may only provide partial and vague information about the realization of the events of interest: in fact, it only predicts that a critical event is *expected* to appear once in a given time interval. However, it would be more important to be able to calculate (*1*) the probability that a super-critical (destructive) event will occur at any given realization of the process (e.g., at any storm), and (*2*) how long it takes, *on average*, for a super-critical event to appear. As a fundamental result, both questions can now be answered: the first by using $\overline{K_C}$, and the second by considering the function ϱ. Thus, the secondary return period provides precise indications for performing risk analysis in *Natural Hazards*, and may also yield useful hints for doing numerical simulations.

From a practical point of view, the introduction of the function $\overline{K_C}$ turns the difficult analysis of the bivariate dynamics of X and Y into a simpler one-dimensional problem. At any given realization of the process, only two mutually exclusive things may happen: either the event is super-critical, and hence potentially dangerous, or not. Actually, this is the main concern in practice. Thus, in applications, the investigation of the joint behavior of X and Y could be reduced to the analysis of the associated (univariate) Bernoulli process, in which the "failure" is simply the realization of a super-critical event. As explained above, given a critical design level t_*, all the super-critical events are enclosed in an "equivalence class" (practically, the region outside $B_C(t_*)$), whose probability can be calculated by using $\overline{K_C}$. Thus, whatever the values of X and Y occurring in the phenomenon of interest, it is now possible to discriminate between sub- and super-critical events, judge whether an event is super-critical or not by indicating which events are super-critical, estimate their probability of occurrence at any given realization of the process, and evaluate their average rate of recurrence via the secondary return period.

A final point concerns the study of the (marginal) quantiles of the events of interest. As before, assume that a critical design level t_* is fixed, and let L_{t_*} be the isoline of the critical events. Clearly, the coordinates (u, v) of any sub-critical event $E^{\vee}_{u,v}$ must satisfy the inequality $v < L_{t_*}(u)$, since the event must lie below and to the left of L_{t_*}: this is evident in Figure 3.5 considering the sub-critical event

E_1^-. Conversely, the coordinates of a super-critical event must satisfy the inequality $v > L_{t_*}(u)$, since the event must lie above and to the right of L_{t_*}: again, this is evident in Figure 3.5 considering the super-critical event E_1^+. Equivalently, in terms of the r.v.'s (X, Y) of practical interest, and considering a sub-critical event, for any arbitrary choice of the X's marginal quantile x_q,

$$x_q > x_* = F_X^{[-1]}(t_*),$$

the corresponding Y's marginal quantile y_q must satisfy the inequality

$$y_q < y_q^* = F_Y^{[-1]}(L_{t_*}(F_X(x_q))),$$

and the inequality $y_q > y_q^*$ when considering a super-critical event. Thus, a fundamental information is achieved: if $x_q > x_*$ is already given by design requirements, and a primary return period $\tau_*^{\vee}$ is desired, then all the events with $Y > y_q^*$ must be considered as *super-critical*, and hence as potentially dangerous. The same reasoning can be applied to X, simply by exchanging the roles of X and Y. In particular, note that the three events E_1^-, E_*, E_1^+ have the same X's marginal quantile $x_* = F_X^{[-1]}(u_*)$, but the corresponding Y's marginal quantiles satisfy the inequality $y_- < y_* = F_Y^{[-1]}(v_*) < y_+$: indeed, these events are, respectively, sub-critical, critical, and super-critical with respect to the threshold t_*. The same comments hold for the three events E_2^-, E_*, E_2^+ if we interchange the roles of X and Y.

ILLUSTRATION 3.14 (Flood-peak and Flood-volume (cont.)). ►

In Illustrations 3.11–3.13 we showed how in [67] a Gumbel-Hougaard 2-copula (with upper-unbounded GEV marginals) is used to assess the spillway design flood of an existing Italian dam, by considering the maximum annual flood-volume and flood-peak as the hydrological variables of interest.

In Figure 3.12 we plot the (survival) probability $\overline{K}_{\mathbf{C}}$ and the secondary return period ϱ as functions of $\vartheta(t)$; both are given on a log-log plane, in order to enhance the behavior of the functions plotted. Using $\vartheta(t)$ instead of t (obviously, with $t \in \mathbb{I}$) to parametrize the horizontal axis, yields a direct evaluation of the probability that an event $E_{u,v}^{\vee}$, having a primary return period $\tau_{u,v}^{\vee} > \vartheta(t)$, does occur at any given realization of the process, as well as the value of the corresponding secondary return period (i.e., its average recurrence time). The values of the probability level t are shown on the upper axis.

As a numerical illustration, in Table 3.3 we give some selected values of $\vartheta(t)$, $\overline{K}_{\mathbf{C}}$, and $\varrho(t)$. The last two columns provide, respectively, the probability of a super-critical event at any given realization of the process (with respect to either the level t or the primary return period $\vartheta(t)$), and the expected time for such an event to occur.

Note that the first two lines in Table 3.3 give the probabilities and the expected times of super-critical events belonging to the regions to the right of the corresponding level curves shown in Figure 3.7*b* (i.e., for 10 and 100 years). Furthermore, since **C** is Archimedean, all the calculations can be done analytically.

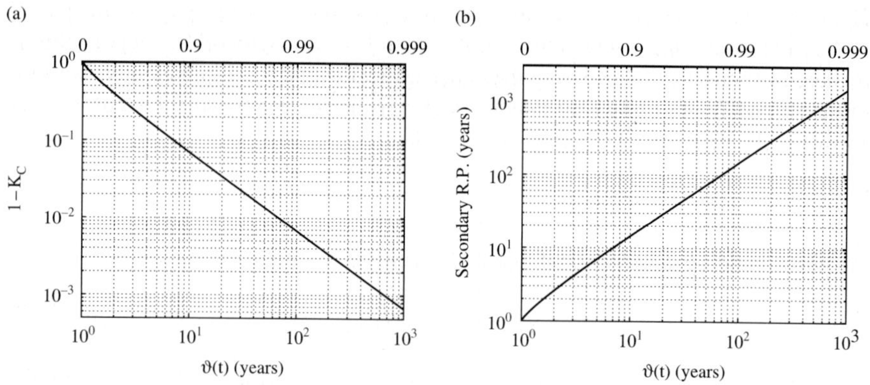

Figure 3.12. The (survival) probability $\overline{K}_C$ (*a*) and the secondary return period ϱ (*b*) as functions of $\vartheta(t)$, for the data considered in [67]

Evidently, without using copulas, the important information derived above will be difficult to achieve. The same approach can be used in both the case studies reported in [268, 67], as well as in many other situations. ◀

At first sight, the interpretation of ϱ may be somewhat counterintuitive, a fact that may obscure its practical usefulness. In order to clarify further, suppose that a critical probability level t_* is fixed by design requirements. Since γ' is negative (for the generator γ is a decreasing function), because of Eq. (3.47) and Eq. (3.77), it is clear that the probability of super-critical events, $\overline{K}_C(t_*)$, is smaller than $1-t_*$. Thus, events that are super-critical with respect to the probability level t_* occur with frequency smaller than $1-t_*$. Instead, one would intuitively expect that these events happen exactly with probability $1-t_*$. However, a thorough explanation involves measure-theoretic arguments, and goes beyond the scope of this book. In simple words, the primary return period is calculated using the critical probability level t_*, that corresponds to the constant height of **C** along the level curve L_{t_*}. On the contrary, the secondary return period involves the *measure* of the super-critical region, given by $\overline{K}_C(t_*)$. Clearly, the latter has no relationship with the height of **C**, for they assess the extent of different mathematical objects.

Table 3.3. Selected values of the (survival) probability $\overline{K}_C$ and the secondary return period ϱ for the data considered in [67]

t	$\boldsymbol{\vartheta(t)}$ **(years)**	$\overline{\mathbf{K}}_C\mathbf{(t)}$	$\boldsymbol{\varrho(t)}$ **(years)**
0.9	10	0.068961	14.5
0.99	100	0.006743	148.3
0.999	1000	0.000673	1486.3
0.9999	10000	0.000067	14865.8

ILLUSTRATION 3.15 (Sea storm characterization (cont.)). ►

As anticipated in Illustration 3.10, the significant wave height H and the storm duration D are used in [68] to define the energetic content M of a sea storm, as given by Eq. (3.55). In practical applications, like e.g. in structural fatigue or long-term coastal dynamics assessment, it is often of interest to consider as "critical" the bivariate event

$$E_{h,d}^{\vee} = \{H > h\} \vee \{D > d\},$$

where h, d are given thresholds. In Figure 3.13 we show the isofrequency curves of the events $E_{h,d}^{\vee}$, for four selected values of the primary return period $\tau_{h,d}^{\vee}$, i.e. one month, six months, one year, and two years. The corresponding values of the probability level t are shown as well. Also plotted are all the observed (H, D) pairs. As expected, the majority of the events have a short primary return period: indeed, the mean interarrival time between successive storms, μ_S, is only about 11 days. However, there exists one "extreme" storm with primary return period larger than one year. Now, suppose that the primary return period is used as a design variable, and that, according to specific risk assessment requirements, the threshold $\tau_*^{\vee} = 6$ months is considered as "critical". Then, one event is "super-critical" in Figure 3.13, and hence potentially dangerous. Actually, an investigation using simulated sequences of storms (see Illustration A.2), lasting several centuries, shows how similar (and even more) catastrophic events may occur.

In Figure 3.14 we plot the functions $\overline{K}_C$ and ϱ of interest here vs. $\tau^{\vee}$: both are drawn on a log-log plane, in order to enhance the behavior of the functions plotted. Using $\tau^{\vee}(t)$, instead of t (obviously, with $t \in (0, 1)$), to parameterize the horizontal

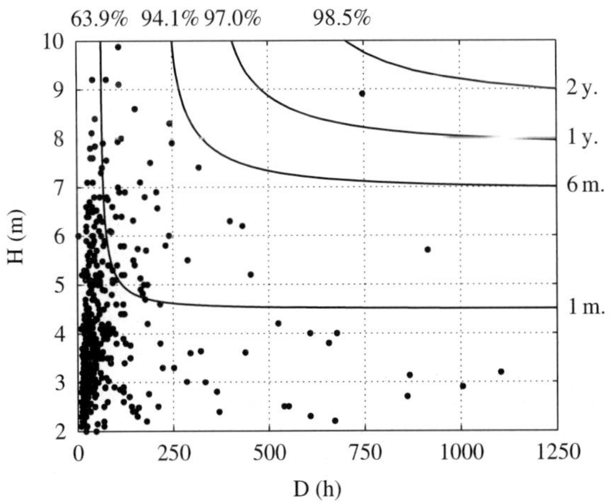

Figure 3.13. Isofrequency curves of the events $E_{h,d}^{\vee}$ (*lines*) for four selected primary return periods (*right* labels): one month, six months, one year, and two years. The *top* labels indicate the probability level associated with each curve. All the observed (H, D) pairs (*markers*) are also plotted

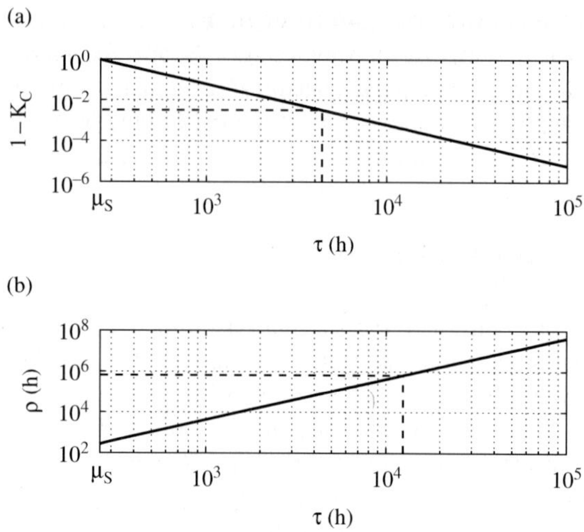

Figure 3.14. (*a*) The function $\overline{K}_C$ vs. $\tau^{\vee}$; the *dashed* line indicates the value $\overline{K}_C \approx 0.0032$, corresponding to $\tau^{\vee}_{*} = 6$ months. (*b*) The secondary return period ϱ vs. $\tau^{\vee}$; the *dashed* line indicates the value $\varrho \approx 74$ years, corresponding to the return period $\tau^{\vee} \approx 1.42$ years associated with the super-critical event shown in Figure 3.13. In both figures t ranges from 0.0005 to 0.9995

axis, yields a direct evaluation of the probability that a super-critical event, having primary return period larger than $\tau^{\vee}$, does occur at any given realization of the process, as well as the value of the corresponding secondary return period (i.e., its average recurrence time).

Now, consider the isofrequency curve corresponding to $\tau^{\vee}_{*} = 6$ months plotted in Figure 3.13: only one storm is super-critical with respect to such a threshold. As shown by Figure 3.14*a*, the actual probability of super-critical events is about 0.3%: this means that just one storm, out of 415, has to be expected. Moreover, the observed super-critical event looks rare (and, hence, potentially catastrophic): as shown by Figure 3.14*b*, storms with primary return period $\tau^{\vee}$ larger than ≈ 1.42 years (that of the super-critical event) occur, on average, only once in 74 years.

Working in two (or more) dimensions changes the traditional meaning and interpretation of the concept of return period. In the example considered here, all the (infinite) bivariate events on the level curve L_t have the same primary return period $\tau^{\vee}(t)$. However, from a practical point of view, these events may not be equivalent. In fact, a trivial analysis shows that, given $0 < t < 1$, their magnitude M may change along L_t, and has a minimum between the two extrema $u = t$ and $u = 1$, where one of the variables H or D diverges. Now, if $\tau(t)$ were used as a design variable, which of the events on L_t (equivalently, what values of H and D) should be chosen as "critical"? The answer is not obvious in a multivariate context, while in univariate frequency analysis just one variable suffices to define an event as critical.

Again copulas will help. As mentioned above, M has a minimum along L_t, say, $\epsilon(t)$. This corresponds to the *least energetic magnitude* of the events having common primary return period $\tau^{\vee}(t)$. In the present case, ϵ can be calculated analytically: it is sufficient to work on the derivative of M with respect to u, i.e. on the generator γ. Thus, a single variable having physical meaning, $\epsilon(t)$, can be associated to $\tau^{\vee}(t)$, for any probability level t. In Figure 3.15a we plot the function ϵ of interest here vs. $\tau^{\vee}$, on a log-log plane.

As expected, ϵ increases with $\tau^{\vee}$: asymptotically, the rate of growth follows a power-law (see also the discussion at the end of Illustration 3.10). Using $\tau^{\vee}(t)$, instead of t, to parameterize the horizontal axis yields a direct evaluation of the minimum energy associated with storms having primary return period $\tau^{\vee}(t)$. Evidently, ϵ may provide useful information for engineering design, as well as for risk and reliability assessment. For instance, Figure 3.15a shows that a magnitude $M \approx 932$ mh is the minimum energetic content to be expected when considering storms having a primary return period $\tau_*^{\vee} = 6$ months. Now, by combining the use of the secondary return period ϱ (or of the excess function $\overline{K}_C$) with ϵ, one may provide practical indications about the occurrence of dangerous events, as measured by M. Mathematically speaking, $\epsilon(t)$ is the "representative" element of all the events on L_t: this embeds the problem of identifying a critical bivariate event into a univariate setting. Then, ϱ and $\overline{K}_C$ can be used as standard tools for working out a "traditional" return period analysis. Note that this approach can be easily extended to any multivariate context. In Figure 3.15b we plot the function ϵ of interest vs. ϱ, on a log-log plane. All the comments given for Figure 3.15a hold also in the present case. In particular, a minimum energetic content $\epsilon \approx 2100$ mh has to be expected when considering storms having a return period $\tau \approx 1.42$ years,

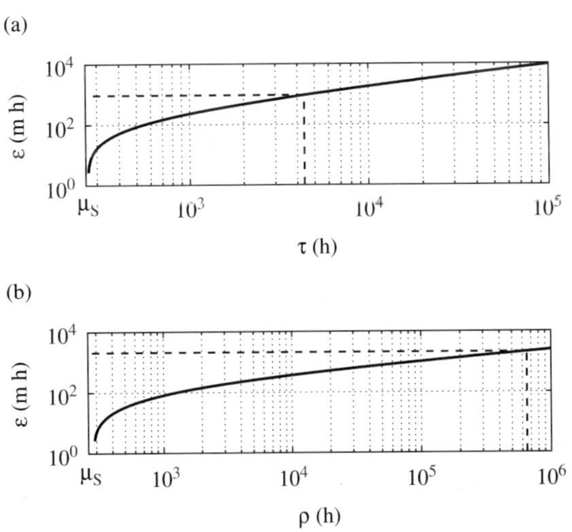

Figure 3.15. (a) The function ϵ vs. $\tau^{\vee}$. The *dashed* line indicates the value $\epsilon \approx 932$ mh, corresponding to $\tau_*^{\vee} = 6$ months. (b) The function ϵ vs. ϱ. The *dashed* line indicates the value $\epsilon \approx 2100$ mh, corresponding to $\varrho \approx 74$ years, the secondary return period associated with the super-critical event shown in Figure 3.13. In both figures t ranges from 0.0005 to 0.9995

the same as that of the super-critical event shown in Figure 3.13: this latter event has a magnitude M $\approx$ 2576 mh. ◀

3.4. TAIL DEPENDENCE

In the context of extrapolation in multivariate frequency analysis, it may be of great importance to be able to model the possible dependence of the extrema, i.e. the *tail dependence*. This quantity is a fundamental ingredient in order to estimate the risk adequately. As will be shown shortly, tail dependence is essentially a characteristic of the copula underlying a random vector. In turn, tail dependence, which relates to dependencies of extreme events, can be considered as a "scale-invariant" dependence measure, since the copula separates the dependence structure of multivariate distributions from its marginals. Clearly, the notion of tail dependence may provide useful indications for choosing a suitable family of copulas for modeling a given phenomenon [222].

The notion of tail dependence [153] for bivariate distributions relates to the amount of dependence in the upper-right-quadrant tail or lower-left-quadrant tail. Usually it is measured via the *tail dependence coefficients*, introduced by [270]. They reflect the limiting proportion of exceedance of one marginal over a quantile of a certain level, given that the other marginal has exceeded the same quantile.

We now present one of the possible definitions of tail dependence [155]. A thorough exposition can be found in [262]; for a survey on various estimators of tail dependence coefficients within a parametric, semiparametric, and non-parametric framework see [97]. A generalization to a d-dimensional context, $d > 2$, will be made in Section 5.3. A survey on other measures of tail dependence can be found in [42] — see also [96, 260].

DEFINITION 3.15 (Tail dependence (2-dimensional case)). *Let* $Z = (X, Y)$. *The random vector* Z *is upper tail dependent if*

$$\lambda_U = \lim_{t\to 1^-} \mathbb{P}\left\{X > F_X^{[-1]}(t) \mid Y > F_Y^{[-1]}(t)\right\} > 0, \tag{3.79}$$

provided that the limit exists. If $\lambda_U = 0$ *then* Z *is upper tail independent.* λ_U *is called the upper tail dependence coefficient.*

Similarly, the lower tail dependence coefficient λ_L *is defined as*

$$\lambda_L = \lim_{t\to 0^+} \mathbb{P}\left\{X \le F_X^{[-1]}(t) \mid Y \le F_Y^{[-1]}(t)\right\}, \tag{3.80}$$

provided that the limit exists. If $\lambda_L = 0$ *then* Z *is lower tail independent, and is lower tail dependent if* $\lambda_L > 0$.

The following result shows that tail dependence is a copula property.

PROPOSITION 3.6. *Let $Z = (X, Y)$ have copula* **C**. *Then*

$$\lambda_U = \lim_{t\to 1^-} \frac{1-2t+\mathbf{C}(t,t)}{1-t} = \lim_{t\to 1^-} \frac{1-2t+\boldsymbol{\delta}_{\mathbf{C}}(t)}{1-t} = 2-\boldsymbol{\delta}'_{\mathbf{C}}(1^-), \tag{3.81}$$

provided that the limit exists. Similarly,

$$\lambda_L = \lim_{t\to 0^+} \frac{\mathbf{C}(t,t)}{t} = \lim_{t\to 0^+} \frac{\boldsymbol{\delta}_{\mathbf{C}}(t)}{t} = \boldsymbol{\delta}'_{\mathbf{C}}(0^+), \tag{3.82}$$

provided that the limit exists.

NOTE 3.4. Using the survival copula $\overline{\mathbf{C}}$ of **C** (see Theorem 3.2), the numerator in Eq. (3.81) can be rewritten as $\overline{\mathbf{C}}(1-t, 1-t)$. In turn, λ_U can be expressed as

$$\lambda_U = \lim_{t\to 1^-} \frac{\overline{\mathbf{C}}(1-t,1-t)}{1-t} = \lim_{t\to 0^+} \frac{\overline{\mathbf{C}}(t,t)}{t} = \lim_{t\to 0^+} \frac{\boldsymbol{\delta}_{\overline{\mathbf{C}}}(t)}{t}, \tag{3.83}$$

provided that the limit exists. Clearly, upper tail dependence in the copula is equivalent to lower tail dependence in the survival copula, and vice versa.

Since the tail dependence coefficients can be expressed via copulas, many properties of copulas apply to these coefficients. For instance, they are invariant under strictly increasing transformations of the marginals. A further interesting point follows.

ILLUSTRATION 3.16. ▶

Let (U, V) be Uniform r.v.'s on $\mathbb{I}$, with the same copula **C** as (X, Y). As shown in Illustration 3.3, the function $t \mapsto 1-\overline{\mathbf{C}}(1-t, 1-t)$ is the distribution of the order statistics $L = \min\{U, V\}$, and the function $t \mapsto \boldsymbol{\delta}_{\mathbf{C}}(t)$ is the distribution of the order statistics $G = \max\{U, V\}$.

If the density of G exists, then λ_L represents its limiting value as $t \to 0^+$. Geometrically, λ_L is the slope of the one-sided tangent line at the origin to the graph $s = \boldsymbol{\delta}_{\mathbf{C}}(t)$. Similarly, if the density of L exists, then λ_U represents its limiting value as $t \to 1^-$. Geometrically, $2-\lambda_U$ is the slope of the one-sided tangent line to the graph $s = \boldsymbol{\delta}_{\mathbf{C}}(t)$ at 1.

As noted in [153], the condition $\lambda_U = 0$ is equivalent to the asymptotic independence of $\max\{X_i\}$ and $\max\{Y_i\}$, where $\{(X_i, Y_i)\}$ is a sample from a distribution with copula **C** — provided that the marginal limiting extreme distributions exist. ◀

As explained thoroughly in [205], Archimedean copulas play an important role in the study of tail dependence. For the sake of simplicity we only consider strict copulas (see Definition 3.9). If **C** is a strict Archimedean copula generated by γ, then $\mathbf{C}(t,t) = \gamma^{-1}(2\gamma(t))$; similarly, $\overline{\mathbf{C}}(1-t, 1-t) = 1-2t+\gamma^{-1}(2\gamma(t))$.

PROPOSITION 3.7. *Let* **C** *be a strict Archimedean copula generated by* γ. *Then*

$$\lambda_U = 2 - \lim_{t \to 1^-} \frac{1-\gamma^{-1}(2\gamma(t))}{1-t} = 2 - \lim_{t \to 0+} \frac{1-\gamma^{-1}(2t)}{1-\gamma^{-1}(t)}, \tag{3.84}$$

provided that the limit exists. Similarly,

$$\lambda_L = \lim_{t \to 0+} \frac{\gamma^{-1}(2\gamma(t))}{t} = \lim_{t \to \infty} \frac{\gamma^{-1}(2t)}{\gamma^{-1}(t)}, \tag{3.85}$$

provided that the limit exists.

Within the framework of tail dependence for Archimedean copulas, the following result is important [262]. Note that the one-sided derivatives of the generator at the domain boundaries exist, since γ is convex.

THEOREM 3.5. *Let* **C** *be an Archimedean copula generated by* γ. *Then*

1. *upper tail dependence implies* $\gamma'(1) = 0$, *and* $\lambda_U = 2 - (\gamma^{-1} \circ 2\gamma)'(1)$;
2. $\gamma'(1) < 0$ *implies upper tail independence;*
3. $\gamma'(0) > -\infty$ *or a non-strict generator implies lower tail independence;*
4. *lower tail dependence implies* $\gamma'(0) = -\infty$, *a strict generator, and* $\lambda_L = (\gamma^{-1} \circ 2\gamma)'(0)$.

ILLUSTRATION 3.17. ►

Consider the following families of copulas presented in Appendix C. Simple calculations show that:

- for the Frank family, $\lambda_L = \lambda_U = 0$;
- for the Gumbel-Hougaard family, $\lambda_L = 0$ and $\lambda_U = 2 - 2^{1/\theta}$;
- for the Ali-Mikhail-Haq family, $\lambda_L = \lambda_U = 0$.

◄

Using the *interior* and *exterior power families* (see Proposition 3.4), it is shown in [205] how to generate families of Archimedean copulas with arbitrary (positive) values of λ_L and λ_U.

PROPOSITION 3.8. *Let* **C** *be an Archimedean copula generated by* γ, *with lower and upper tail dependence parameters* λ_L *and* λ_U. *Then:*

1. *the lower and upper tail dependence parameters for the copula* $\mathbf{C}_\alpha$ *generated by* γ_α, *with* $\alpha \in (0, 1]$, *are, respectively,* $\lambda_L^{1/\alpha}$ *and* λ_U;
2. *the lower and upper tail dependence parameters for the copula* $\mathbf{C}_\beta$ *generated by* γ_β, *with* $\beta \geq 1$, *are, respectively,* $\lambda_L^{1/\beta}$ *and* $2 - (2 - \lambda_U)^{1/\beta}$.

As already mentioned (see Eq. (3.4)), a convex linear combination of copulas is itself a copula. As a consequence, due to the linearity of the limit operator, the tail dependence coefficients of the resulting copula are simply the convex linear combinations of those of the mixing copulas. This result will be generalized to multivariate copulas in Chapter 5 (see Section 5.3).

As for any parameter associated with the asymptotic behavior of a phenomenon, the estimation of the tail dependence coefficients is an involved task. Both parametric and non-parametric are available. In the former case, these can be estimated either by assuming a specific bivariate distribution [85], or a class of distributions [261], or a specific copula or a class of copulas [157]. For the non-parametric case see [222] (and references therein), where a case study is also presented: this latter approach is based on the empirical copula (see Definition 3.7), and is quite general, for no assumptions are made about the copula and the marginals. For a thorough review see [97].

We now illustrate the limit behavior of the probabilities involved in the estimation of the tail dependence coefficients by investigating a hydrological case study. For the sake of comparison, the same analysis is carried out by using simulated values.

ILLUSTRATION 3.18 (Tail dependence estimation). ►

In [252] a sample of 1194 storms, collected in the Bisagno drainage basin (Thyrrehenian Liguria, northwestern Italy), is analyzed. Measurements of *storm volume* $V = IW$ are derived by using the (average) *storm intensity* I and the (wet) *storm duration* W. The point of interest here is to check whether tail dependence is present in the available data. In Figure 3.16 we show the empirical (non-parametric) estimations of λ_L and λ_U for, respectively, $t \to 0^+$ and $t \to 1^-$: here Eqs. (3.81)–(3.82) are used, where $\mathbf{C}$ is approximated by means of the corresponding empirical copula.

Apparently, both λ_L and λ_U are null. However, λ_L is null for small enough values of t (say, $t < 0.1$) because no data are available in the corresponding region. Instead, the limit behavior of λ_U is unstable for decreasing values of $1 - t$, for it oscillates without converging: this is due to the limited number of available data in the region defined by $t \approx 1$ (say, $t > 0.97$), and illustrates a typical problem when trying to estimate the tail dependence coefficients. For the sake of comparison, we now consider a large sample of simulated data, and show how the estimation of λ_L and λ_U may change.

As discussed in Illustration 3.5, a characterization of the sea storm dynamics is provided in [68]. In particular, the significant wave height H and the storm duration D are used to describe the storm energetic content M. Clearly, the definition of M, and its meaning too, is similar to that of the storm volume $V = IW$ introduced above. Since H and D are non-independent, a 2-copula belonging to the Ali-Mikhail-Haq family (see Section C.4), with dependence parameter $\theta \approx 0.829$, is used to describe the statistical behavior of (H, D). Using the algorithm presented in Appendix A, a set of 50000 (H, D) pairs is simulated. Then, in Figure 3.17 we show the empirical (non-parametric) estimations of λ_L and λ_U. Since $\mathbf{C}$ is known, the theoretical exact

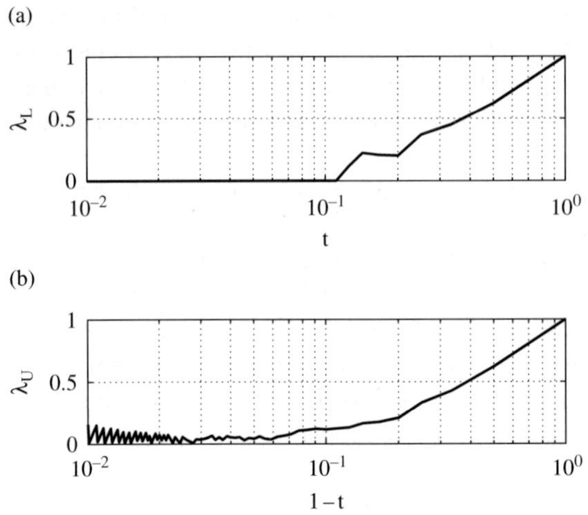

Figure 3.16. Non-parametric estimations of λ_L (*a*) and λ_U (*b*) for the data analysed in [252]

values of the functions involved in Eqs. (3.81)–(3.82) are also shown. Note that $\lambda_L = \lambda_U = 0$ for copulas belonging to the Ali-Mikhail-Haq family.

As expected, the theoretical estimations of λ_L and λ_U converge towards zero. Instead, the empirical ones show some instabilities as $t \to 0^+$ (respectively, $t \to 1^-$). Again, this is due to the limited number of available data in the regions of interest. However, as a difference with the case study previously investigated, problems for

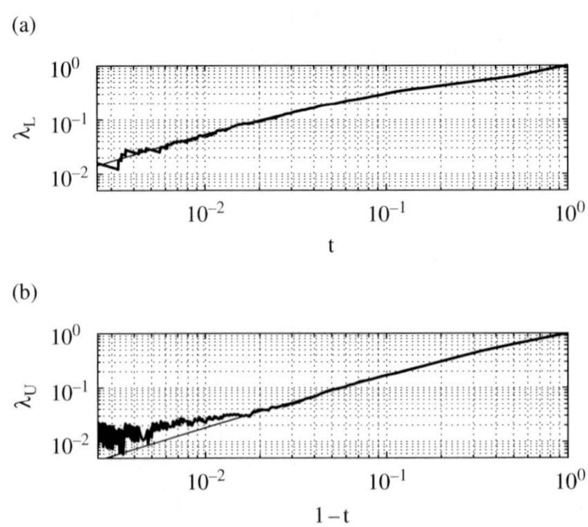

Figure 3.17. Non-parametric estimations of λ_L (*a*) and λ_U (*b*) for 50000 simulated pairs of the vector (H, D) analysed in [68]. The *thick* line is the empirical estimation, and the *thin* line is derived theoretically

λ_L show up for $t < 0.004$, while the estimation of λ_U is already a problem for $t > 0.98$. This shows how difficult it is to estimate the tail dependence coefficients from small samples, such as those available in hydrology. ◀

Another interesting point is as follows. As we shall see later in Section 5.3, the same tail dependence properties of a given 2-copula also hold for the Extreme Value limit of such a copula (as well as for the bivariate margins of a multivariate copula). This may provide useful indications for choosing a family of copulas as a model for a given phenomenon. Further details can be found in Appendix C.

CHAPTER 4

MULTIVARIATE ANALYSIS VIA COPULAS

Increasing the space dimension often makes things more complicated. Copulas are no exception. In fact, defining and constructing copulas in a d-dimensional framework, with $d > 2$, is much more involved than in the bivariate case, and still represents an open problem in Statistics. We now give a survey of the most important features and properties of multivariate copulas. A thorough discussion can be found in [155, 207]; further details are given in Appendix C.

As already mentioned, copulas are mainly considered in this book as a powerful tool for modeling natural phenomena involving non-independent variables. According to the "guidelines" reported in [155], some desirable properties of a (parametric) family of multivariate distributions should be:

1. *interpretability*;
2. *closure* under the taking of marginals (and, in particular, the bivariate marginals belonging to the same family);
3. flexibility and wide *range of dependence*, with the type of dependence structure suggested by the applications;
4. of *closed form* with expressions for the distribution and the density (or, at least, computationally feasible formulas).

At present, there are no known families of multivariate distributions sharing all the properties mentioned above. In general, it is not possible to satisfy all of these requirements. Thus, one may be asked in practice to partially abandon some of them, and decide on the relative importance of these properties in individual cases.

Given a 2-copula $\mathbf{C}$ and two univariate distribution functions F and G, according to Sklar's Theorem (see Theorem 3.1) we know that $\mathbf{C}(F(x), G(y))$ is always a bivariate distribution function. Thus, a simple approach to the construction of higher dimensional distributions is to extend such a procedure, and take F and G as multivariate distributions. Unfortunately, very little could be gained, as stated by the following "impossibility" theorem [112].

THEOREM 4.1. *Let $m, n \in \mathbb{N}$ such that $m + n \geq 3$, and suppose that $\mathbf{C}$ is a 2-copula such that $H(\mathbf{x}, \mathbf{y}) = \mathbf{C}(F(\mathbf{x}), G(\mathbf{y}))$ is a $(m+n)$-dimensional distribution function with marginals $H(\mathbf{x}, \infty) = F(\mathbf{x})$ and $H(\infty, \mathbf{y}) = G(\mathbf{y})$ for* *all m-dimensional distribution functions $F(\boldsymbol{x})$ and n-dimensional distribution functions $G(\boldsymbol{y})$. Then $\mathbf{C} = \mathbf{\Pi}_2$.*

In the same way as 2-copulas join univariate distributions, another simple approach to the construction of higher dimensional distributions could be to use 2-copulas to join other 2-copulas. However, here a "compatibility" problem appears [207], as shown below.

ILLUSTRATION 4.1 (Compatibility). ►

The following examples show some special cases of composition of 2-copulas.

1. Define the function

$$\mathbf{C}(u, v, w) = \mathbf{\Pi}_2(\mathbf{M}_2(u, v), w) = w \min\{u, v\}.$$

   Then $\mathbf{C}$ is a proper 3-copula.
2. Define the function

$$\mathbf{C}(u, v, w) = \mathbf{W}_2(\mathbf{M}_2(u, v), w) = \min\{u, v\} - \min\{u, v, 1-w\}.$$

   Then $\mathbf{C}$ is a proper 3-copula.
3. Define the function

$$\mathbf{C}(u, v, w) = \mathbf{W}_2(\mathbf{W}_2(u, v), w) = \max\{u+v+w-2, 0\}.$$

   Then $\mathbf{C} = \mathbf{W}_3$ is not a 3-copula.

Concerning the latter case, it is clearly impossible, in a set of three r.v.'s, for each variable to be almost surely a decreasing function of each of the remaining two (see Illustration 3.2). ◄

Thus, in general, the above approach also fails. If $\mathbf{C}_1$ and $\mathbf{C}_2$ are 2-copulas such that $\mathbf{C}_2(\mathbf{C}_1(u, v), w)$ is a proper 3-copula, then $\mathbf{C}_1$ is said to be *directly compatible* with $\mathbf{C}_2$ [226]. In the three-dimensional case, the compatibility problem is studied (and solved) in [77]. Before discussing d-dimensional copulas, we need to introduce the following concept.

DEFINITION 4.1 (Fréchet class). *The notation $\mathcal{F}(F_1, \ldots, F_d)$ indicates the family of all the d-dimensional distributions F with given univariate marginals $F_1, \ldots, F_d$, called the Fréchet class of $F_1, \ldots, F_d$.*

4.1. MULTIVARIATE COPULAS

In this Section we briefly outline the mathematics of multivariate copulas (i.e., when $d > 2$) needed in the sequel. All the theoretical justifications can be found in [155, 207]. Hereafter, we write $\boldsymbol{x} < \boldsymbol{y}$ when $x_i < y_i$ for all i. For $\boldsymbol{x} < \boldsymbol{y}$, the d-box $B = [x_1, y_1] \times \cdots \times [x_d, y_d]$ is denoted by $[\boldsymbol{x}, \boldsymbol{y}]$. The vertices of B are the

2^d points $\boldsymbol{c} = (c_1, \ldots, c_d)$, where each c_i is equal to either x_i or y_i. Also, a d-place real function F is a function whose domain, $\mathrm{Dom}\, F$, is a measurable subset of $\mathbb{R}^d$, and whose range, $\mathrm{Ran}\, F$, is a measurable subset of $\mathbb{R}$. The notion of *F-volume* is fundamental.

DEFINITION 4.2 ($\boldsymbol{F}$-volume). *Let $B = [\boldsymbol{x}, \boldsymbol{y}]$ be a d-box all of whose vertices are in* Dom *F. The F-volume of B is defined as*

$$V_F(B) = \sum \mathrm{Sgn}(\boldsymbol{c})F(\boldsymbol{c}), \tag{4.1}$$

where the sum is taken over all the 2^d vertices of B, and the Sign function Sgn *is given by*

$$\mathrm{Sgn}(\boldsymbol{c}) = \begin{cases} +1, & \text{if } c_i = x_i \text{ for an even number of } i\text{'s} \\ -1, & \text{if } c_i = x_i \text{ for an odd number of } i\text{'s} \end{cases}.$$

The function F is d-increasing if $V_F(B) \geq 0$ for all d-boxes B whose vertices lie in $\mathrm{Dom}\, F$. Let us now define copulas in a d-dimensional space.

DEFINITION 4.3 (Multivariate copula). *A d-copula is a function* $\mathbf{C}\colon \mathbb{I}^d \to \mathbb{I}$ *such that:*

1. *(uniform marginals) for all $\boldsymbol{u} \in \mathbb{I}^d$*

$$\mathbf{C}\,(\boldsymbol{u}) = 0 \tag{4.2}$$

if at least one coordinate of $\boldsymbol{u}$ is 0, *and*

$$\mathbf{C}\,(\boldsymbol{u}) = u_i \tag{4.3}$$

if all coordinates of $\boldsymbol{u}$ are 1 *except u_i;*

2. *(d-increasing) for all $\boldsymbol{x}, \boldsymbol{y} \in \mathbb{I}^d$ such that $\boldsymbol{x} \leq \boldsymbol{y}$*

$$V_{\mathbf{C}}([\boldsymbol{x}, \boldsymbol{y}]) \geq 0. \tag{4.4}$$

A d-copula $\mathbf{C}$ is uniformly continuous in its domain, and its one-dimensional horizontal and vertical sections (see Definition 3.2) are all non-decreasing and uniformly continuous on $\mathbb{I}$. If $\mathbf{C}$ is a d-copula, $d > 2$, its k-dimensional marginals (with $2 \leq k < d$) are obtained by fixing $d-k$ coordinates — the u_i's — equal to 1 in $\mathbf{C}\,(u_1, \ldots, u_d)$. The resulting functions are lower dimensional k-copulas defined over $\mathbb{I}^k$. Evidently, there are exactly $\binom{d}{k}$ k-dimensional marginal copulas of $\mathbf{C}$. Note that the "compatibility" problem reappears: in general, given an arbitrary set of $\binom{d}{k}$ k-copulas, seldom these are the k-marginals of a d-copula [207].

The link between d-copulas and multivariate distributions is provided by the following multidimensional version of Sklar's Theorem [275] (see also Theorem 3.1).

THEOREM 4.2 (Sklar (d-dimensional case)). *Let F be a joint distribution function with marginals $F_1, \dots, F_d$. Then there exists a d-copula* **C** *such that*

$$F(x_1, \dots, x_d) = \mathbf{C}\,(F_1(x_1), \dots, F_d(x_d)) \tag{4.5}$$

for all reals $\boldsymbol{x} \in \mathbb{R}^d$. If $F_1, \dots, F_d$ are all continuous, then **C** *is unique; otherwise,* **C** *is uniquely defined on* $\mathrm{Ran}(F_1) \times \cdots \times \mathrm{Ran}(F_d)$.

Conversely, if **C** *is a d-copula and $F_1, \dots, F_d$ are distribution functions, then the function F given by Eq. (4.5) is a joint distribution with marginals $F_1, \dots, F_d$.*

As in the two-dimensional case (see Corollary 3.1), the following result plays a fundamental role in practical applications.

COROLLARY 4.1 (Sklar inversion (d-dimensional case)). *Let* **C***, F, and $F_1, \dots, F_d$ be as in Theorem 4.2, and suppose that $F_1, \dots, F_d$ are continuous. If $F_1^{[-1]}, \dots, F_d^{[-1]}$ denote the quasi-inverses of $F_1, \dots, F_d$, then*

$$\mathbf{C}\,(u_1, \dots, u_d) = F\left(F_1^{[-1]}(u_1), \dots, F_d^{[-1]}(u_d)\right) \tag{4.6}$$

for any $\boldsymbol{u} \in \mathbb{I}^d$.

The following example shows that not all of the properties of 2-copulas automatically translate in higher dimensions (see Illustration 3.2).

ILLUSTRATION 4.2. ▶

Three special d-dimensional functions deserve particular attention.

1. The *Fréchet-Hoeffding lower bound* $\mathbf{W}_d$ is given by

$$\mathbf{W}_d(u_1, \dots, u_d) = \max\{u_1 + \cdots + u_d - (d-1), 0\}. \tag{4.7}$$

   $\mathbf{W}_d$ fails to be a d-copula for any $d > 2$ (see Illustration 4.1).
2. The *Fréchet-Hoeffding upper bound* $\mathbf{M}_d$ is given by

$$\mathbf{M}_d(u_1, \dots, u_d) = \min\{u_1, \dots, u_d\}. \tag{4.8}$$

   $\mathbf{M}_d$ is a d-copula for all $d \geq 2$.
3. The function $\mathbf{\Pi}_d$ is given by

$$\mathbf{\Pi}_d(u_1, \dots, u_d) = u_1 \cdots u_d. \tag{4.9}$$

   $\mathbf{\Pi}_d$ is a d-copula for all $d \geq 2$.

When the r.v.'s $X_1, \dots, X_d$ are continuous, it turns out that

1. X_i is almost surely a strictly increasing function of any of the others if, and only if, the d-copula of $(X_1, \dots, X_d)$ is $\mathbf{M}_d$;

2. $X_1, \ldots, X_d$ are independent if, and only if, their d-copula is $\mathbf{\Pi}_d$, and $\mathbf{\Pi}_d$ is called the *independence* copula.

The functions $\mathbf{W}_d$ and $\mathbf{M}_d$ provide general bounds for the Fréchet class $\mathcal{F}(F_1, \ldots, F_d)$. In fact, for any $F \in \mathcal{F}$ and all $\boldsymbol{x} \in \mathbb{R}^d$,

$$\mathbf{W}_d(F_1(x_1), \ldots, F_d(x_d)) \leq F(\boldsymbol{x}) \leq \mathbf{M}_d(F_1(x_1), \ldots, F_d(x_d)). \tag{4.10}$$

Note that the left-hand inequality in the above equation is the "best-possible": in fact, it can be shown [207] that, for any $d > 2$ and any $\boldsymbol{u} \in \mathbb{I}^d$, there exists a d-copula $\mathbf{C}$ such that $\mathbf{C}(\boldsymbol{u}) = \mathbf{W}_d(\boldsymbol{u})$. ◄

An interesting application of Fréchet-Hoeffding bounds is as follows (see also Illustration 3.3).

ILLUSTRATION 4.3 (Order Statistics). ►
Let $X_1, \ldots, X_d$ be continuous r.v.'s with copula $\mathbf{C}$ and marginals $F_1, \ldots, F_d$. Then consider the (extremal) order statistics $X_{(1)} = \min\{X_1, \ldots, X_d\}$ and $X_{(d)} = \max\{X_1, \ldots, X_d\}$ (see Section 1.1). For the distributions $F_{(1)}$ and $F_{(d)}$ of, respectively, $X_{(1)}$ and $X_{(d)}$, the following inequalities hold:

$$\max_{1 \leq i \leq d} \{F_i(t)\} \leq F_{(1)}(t) \leq \min\left\{\sum_{i=1}^{d} F_i(t), 1\right\}, \tag{4.11a}$$

$$\max\left\{\sum_{i=1}^{d} F_i(t) - (d-1), 0\right\} \leq F_{(d)}(t) \leq \min_{1 \leq i \leq d}\{F_i(t)\}, \tag{4.11b}$$

for all $t \in \mathbb{R}$. ◄

As a practical example, we now show a particular construction of a 3-copula used to model the fundamental variables characterizing rainfall storms. The multivariate copula considered is a special case of a general procedure, discussed later in Section 4.3, that exploits the links between copulas and conditional probabilities (see Eq. (3.26) and the ensuing discussion).

ILLUSTRATION 4.4 (Rainfall storm structure). ►
In [255] a statistical procedure to estimate the probability distributions of rainfall storm characteristics is presented. As is typical in event-based rainfall representations, it is shown how to distinguish between an "exterior" and an "interior" process. In particular, the emphasis is on the former one (a coarse representation of the rainfall process), characterizing the arrival, duration and average intensity of rainfall events at the synoptic scale.

The temporal dynamics of rainfall is modeled via a reward alternating renewal process, that describes wet and dry phases of storms. In particular, the wet phase is modeled as a rectangular pulse with dependent random duration and intensity, and both these variables are linked to the dry period following the rainy phase. All the marginal distributions are endowed with Generalized Pareto laws, and a seasonal analysis is made. Further details can be found in Illustration 2.1, and in a series of examples on storm intensity and duration presented in Appendix B.

For each storm four variables of interest are calculated: (*1*) the storm (average) intensity I (in mm/h); (*2*) the storm wet duration W (in hours); (*3*) the storm dry duration D (in hours) defining the non-rainy period between one storm and the next; (*4*) the storm volume $V = IW$ (in mm). An illustration of these variables is given in Figure 4.1.

A study of the pairwise relationships between I, W, D in terms of Kendall's τ_K and Spearman's ρ_S association measures (see Section B.2) shows that these three variables may not be independent. In turn, among many others tested, the 2-copulas belonging to the Frank family (see Section C.1) seem to provide a valuable fit to all the pairs (I, W), (I, D), and (W, D), for all the seasons. Now, the question is "how to join these three 2-copulas (a triple for each season) in a mathematically consistent way", given the compatibility problem mentioned earlier. In [255] the following 3-copula [38] is used, having a particularly simple and appealing structure:

$$\mathbf{C}_{IWD}(r, s, t) = t\,\mathbf{C}_{IW}\left(\frac{\mathbf{C}_{ID}(r, s)}{t}, \frac{\mathbf{C}_{WD}(s, t)}{t}\right), \tag{4.12}$$

where $r, s, t \in \mathbb{I}$. It is easy to verify that the trivariate function $\mathbf{C}_{IWD}$ is indeed a 3-copula, having three two-dimensional marginals given by the 2-copulas $\mathbf{C}_{IW}$, $\mathbf{C}_{ID}$, and $\mathbf{C}_{WD}$. The two arguments of $\mathbf{C}_{IW}$ in Eq. (4.12) involve 2-copulas: these control, respectively, the pairwise dynamics of (I, D) and (W, D). Furthermore, these arguments are conditional distributions (see Eq. (3.26)), and provide the link

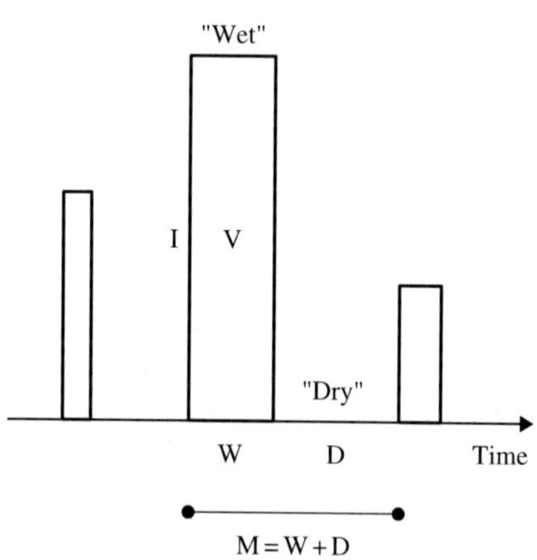

Figure 4.1. Illustration of the rainfall model adopted in [255]

between, respectively, I and W, given the behavior of D. In addition, the parameters of $\mathbf{C}_{IWD}$ are only those of its marginals, and no further estimations are required to fit $\mathbf{C}_{IWD}$. ◄

In the following Sections we illustrate briefly several approaches to the construction of multivariate copulas. Clearly, this only represents a partial account of all the results available in the literature. However, the techniques shown are either promising, or have already found interesting applications. Further approaches are described in Appendix C.

4.2. ARCHIMEDEAN COPULAS

Archimedean copulas were introduced in Section 3.2 in a two-dimensional context. We now investigate how to extend the idea to a d-dimensional framework, following the approach adopted in [207], where more mathematical details and references can be found. In general, while it is relatively simple to generate Archimedean d-copulas, they do have some limitations, that may reduce their practical usefulness. In fact, lower dimensional marginals of multivariate Archimedean copulas are generally identical. Also, there have usually only one or two parameters, which reduces the features and the generality of the dependence structure in these families.

As for 2-copulas, Archimedean d-copulas can be constructed via a suitable generator: however, additional properties are required. Let us commence by introducing the concept of *completely monotonic* function [300].

DEFINITION 4.4 (Completely monotonic function). *A function f is completely monotonic on an interval I if it is continuous there and has derivatives of all orders alternating in sign, i.e.*

$$(-1)^i \frac{\mathrm{d}^i}{\mathrm{d}x^i} f(x) \geq 0 \tag{4.13}$$

for all x in the interior of I and $i \in \mathbb{N}$.

The following result [162, 264] provides necessary and sufficient conditions for a strict generator γ to generate Archimedean d-copulas for all $d \geq 2$.

THEOREM 4.3 (Archimedean d-copula). *Let $\gamma\colon \mathbb{I} \to [0, \infty]$ such that γ is continuous and strictly decreasing, with $\gamma(0) = \infty$ and $\gamma(1) = 0$. If $\mathbf{C}\colon \mathbb{I}^d \to \mathbb{I}$ is given by*

$$\mathbf{C}(\boldsymbol{u}) = \gamma^{-1}\left(\gamma(u_1) + \cdots + \gamma(u_d)\right), \tag{4.14}$$

then $\mathbf{C}$ is a d-copula for all $d \geq 2$ if, and only if, γ^{-1} is completely monotonic on $[0, \infty)$.

The following illustration shows how the abovementioned theorem can be used to extend a family of 2-copulas to a multivariate framework.

ILLUSTRATION 4.5 (Clayton family). ▶

Consider the strict generator $\gamma_\theta(t) = t^{-\theta} - 1$, $\theta > 0$, with inverse $\gamma_\theta^{-1}(t) = (1+t)^{-1/\theta}$. As shown in Section C.3, γ generates a subfamily of the Clayton family. Since γ_θ^{-1} is completely monotonic on $[0, \infty)$, the subfamily can be extended to any dimension $d > 2$:

$$\mathbf{C}_\theta(\boldsymbol{u}) = \left(u_1^{-\theta} + \cdots + u_d^{-\theta} - (d-1)\right)^{-1/\theta}, \tag{4.15}$$

where $\theta > 0$. This result generalizes the outcome of Illustration 3.7. ◀

According to the notion of *partial order* discussed in Definition 3.6, the subfamily just introduced only contains copulas $\mathbf{C}$ larger than $\mathbf{\Pi}_d$, i.e. $\mathbf{C} \succ \mathbf{\Pi}$. Actually, this must occur whenever γ^{-1} is completely monotonic.

PROPOSITION 4.1. *If the inverse γ^{-1} of a strict generator γ of an Archimedean copula $\mathbf{C}$ is completely monotonic, then $\mathbf{C} \succ \mathbf{\Pi}$.*

Further useful results are given below [300, 92].

PROPOSITION 4.2. *Special rules to construct completely monotonic functions are as follows.*

1. *If f and g are completely monotonic, then so is their product fg.*
2. *If f is completely monotonic, and g is a positive function with a completely monotone derivative, then $f \circ g$ is completely monotonic.*
3. *If g is completely monotonic, and f is absolutely monotonic, that is*

$$\frac{\mathrm{d}^i}{\mathrm{d}x^i} f(x) \geq 0 \tag{4.16}$$

for $i \geq 0$, then $f \circ g$ is completely monotonic.

In particular, construction (2) ensures that e^{-g} is completely monotonic.

Evidently, Propositions 4.1–4.2 help in extending families of 2-copulas to higher dimensions. In particular, for all the values of the parameter for which the corresponding copulas are larger than $\mathbf{\Pi}$, some of the families given in Appendix C can be generalized for all $d > 2$.

ILLUSTRATION 4.6 (Gumbel-Hougaard family). ▶

Consider the generator $\gamma(t) = (-\ln t)^\theta$, $\theta \geq 1$, which generates the bivariate Gumbel-Hougaard family of copulas (see Section C.2). Using the results on completely monotonic functions given previously, it is easy to show that this family can be generalized in d-dimensions as

$$\mathbf{C}_\theta(\boldsymbol{u}) = e^{-\left[(-\ln u_1)^\theta + \cdots + (-\ln u_d)^\theta\right]^{1/\theta}}, \tag{4.17}$$

where $\theta \geq 1$. ◄

ILLUSTRATION 4.7 (Frank family). ►

Consider the generator $\gamma(t) = -\ln(\theta^t - 1) + \ln(\theta - 1)$, $\theta \geq 0$, which generates the bivariate Frank family of copulas (see Section C.1). All the generators of this family are strict. However, due to Proposition 4.1, in order to extend this family to a multivariate context, we must restrict the range of the parameter θ to the interval $(0, 1)$, where $\mathbf{C}_\theta \succ \mathbf{\Pi}$. Using the results on completely monotonic functions given previously, it is easy to show that the Frank family can be generalized in d-dimensions, for $\theta \in (0, 1)$, as

$$\mathbf{C}_\theta(\boldsymbol{u}) = \frac{1}{\ln\theta} \ln\left(1 + \frac{(\theta^{u_1} - 1)\cdots(\theta^{u_d} - 1)}{(\theta - 1)^{d-1}}\right). \tag{4.18}$$

Note that γ^{-1} fails to be completely monotonic when $\theta > 1$. ◄

The following result shows that the above procedures can be generalized to any *beta family* of generators (see Proposition 3.4) associated with a strict generator γ whose inverse is completely monotonic.

PROPOSITION 4.3. *Let γ be a strict generator whose inverse is completely monotonic on $[0, \infty)$, and let $\gamma_\beta(t) = [\gamma(t)]^\beta$ for all $\beta \geq 1$. Then γ_β^{-1} is completely monotonic on $[0, \infty)$.*

As anticipated in Section 3.2, an important source of generators of d-dimensional copulas is represented by the inverses of *Laplace Transforms* of distribution functions, as stated by the following result [92, 155, 207].

PROPOSITION 4.4. *A function ψ on $[0, \infty)$ is the Laplace Transform of a distribution function if, and only if, ψ is completely monotonic and $\psi(0) = 1$.*

ILLUSTRATION 4.8 (Clayton family (cont.)). ►

As shown in Illustration 3.7, if F is a Gamma distribution $\Gamma(1/\theta, 1)$, $\theta > 0$, then the *Laplace Transform* of F is $\psi(s) = (1 + s)^{-1/\theta}$, with $s \geq 0$. In turn, its inverse $\gamma(t) = \psi^{-1}(t) = t^{-\theta} - 1$, with $t \in \mathbb{I}$, generates a subfamily of the Clayton family (see Section C.3), as shown in Illustration 4.5. ◄

Complete monotonicity is indeed a strong requirement. However, using the weaker property of *m-monotonicity*, a partial version of Theorem 4.3 can be given as follows. First of all, a function f is m-monotonic on an interval I if Eq. (4.13) holds for $i = 0, \ldots, m$ in the interior of I. Then, the function $\mathbf{C}$ given by Eq. (4.14) is a d-copula for all $2 \leq d \leq m$ if γ^{-1} is m-monotonic on $[0, \infty)$.

4.3. CONDITIONAL MIXTURES

In general, it is possible to construct a family of d-variate distributions starting from two $(d-1)$-dimensional marginals having $d-2$ variables in common (see [155] for details). Here d can be any integer larger than two, and the procedures outlined below can be extended recursively to any dimension. The fundamental point is that these families of multivariate distributions can be made to interpolate from perfect conditional negative dependence to perfect conditional positive dependence, with conditional independence in between. Furthermore, the approach we investigate here yields distributions showing interesting properties concerning concordance and tail dependence.

We shall concentrate shortly on the 3- and 4-dimensional cases. The models we shall introduce represent a unifying method for constructing multivariate distributions with given 2-copulas for each bivariate marginal law. On the one hand, such an approach is rather simple from a theoretical point of view. On the other hand, it makes it easier to estimate the parameters characterizing the multivariate dependence structure. In addition, the construction of the overall multivariate distribution only requires suitable conditional probabilities. In turn, this simply reduces to the calculation of the partial derivatives of the 2-copulas involved (see Eq. (3.26)). Finally, note that the same construction can be used to generate multivariate families of survival functions. Clearly, the present approach gives the possibility to describe how each variable affects the behavior of the others. Physically based models can be constructed in this way.

For the sake of notational convenience, we now show how to construct multivariate distribution functions instead of copulas. However, because of Sklar's Theorem (see Theorem 4.2 and Corollary 4.1), the calculation of the corresponding copulas is feasible.

4.3.1 The 3-Dimensional Case

Let X, Y, Z be three r.v.'s. The trivariate family of interest here is given by

$$F_{XYZ}(x,y,z) = \int_{-\infty}^{y} \mathbf{C}_{XZ}\left(F_{X|Y}(x \mid t), F_{Z|Y}(z \mid t)\right) F_Y(\mathrm{d}t). \tag{4.19}$$

The arguments of the integrand are conditional distributions (namely, $F_{X|Y}$ and $F_{Z|Y}$), and can themselves be written in terms of copulas (see Eq. (3.26)):

$$\begin{aligned} F_{X|Y}(x \mid y) &= \mathbb{P}\{X \le x \mid Y = y\} \\ &= \mathrm{Q}(F_X(x), F_Y(y)), \end{aligned} \tag{4.20}$$

where

$$\mathrm{Q}(a,b) = \partial_b \mathbf{C}_{XY}(a,b) = \frac{\partial \mathbf{C}_{XY}(a,b)}{\partial b}. \tag{4.21}$$

A similar expression holds for $F_{Z|Y}$. Such a way of linking three variables is quite a natural one: in fact, $\mathbf{C}_{XZ}$ simply measures the amount of conditional dependence in X and Z, given the behavior of Y.

The "conditional" approach presented here offers even more advantages. As shown in Appendix A, the simulation of multivariate vectors, with given d-variate distributions, can be made simply by calculating suitable partial derivatives of the copulas of interest. In the present trivariate case, these copulas can be directly derived using the integral representation given by Eq. (4.19) and Sklar's Theorem (see Illustration 4.9). Most importantly, the integral operator (which may represent a problem for numerical computations) is removed when differentiating the copulas of interest: eventually, only the partial derivatives of 2-copulas are needed.

ILLUSTRATION 4.9 (Sea storm characterization (cont.)). ►

As shown in Illustration 3.5, in [68] the Authors provide a characterization of the sea storm dynamics involving four variables: the significant wave height H, the storm duration D, the waiting time I between two successive "active" storm phases, and the storm wave direction A.

The two triples (D, A, H) and (D, A, I) are of practical importance. The former provides the relevant information about the dependence of the storm energetic content M (defined via the pair (H, D) — see Eq. (3.55)) upon A, while the latter rules the dependence of the timing of storm (driven by the pair (D, I)) upon the wave direction. The interesting point is that, if F_{DAH} and F_{DAI} are constructed as above, then F_{DAH} and F_{DAI} are generated by working on the dependence of, respectively, the storm magnitude and the storm temporal occurences upon the wave direction. More generally, this approach gives the possibility of describing how each variable affects the behavior of the others. Thus, physically based models can be easily constructed.

As already mentioned, Eq. (4.19) and Sklar's Theorem can be used to calculate the 3-copulas of interest here, namely $\mathbf{C}_{DAH}$ and $\mathbf{C}_{DAI}$. In particular, $\mathbf{C}_{DAH}$ is simply given by

$$
\begin{aligned}
\mathbf{C}_{DAH}&(u_1, u_2, u_3)\\
&= F_{DAH}\left(F_D^{-1}(u_1), F_A^{-1}(u_2), F_H^{-1}(u_3)\right)\\
&= \int_{-\infty}^{F_A^{-1}(u_2)} \mathbf{C}_{DH}\left(F_{D|A}(F_D^{-1}(u_1) \mid x),\right.\\
&\qquad\qquad \left. F_{H|A}(F_H^{-1}(u_3) \mid x)\right) F_A(\mathrm{d}x)\\
&= \int_0^{u_2} \mathbf{C}_{DH}\left(\mathrm{D}_2\mathbf{C}_{DA}(u_1, x), \mathrm{D}_2\mathbf{C}_{HA}(u_3, x)\right) \mathrm{d}x,
\end{aligned} \tag{4.22}
$$

where D_2 is the partial derivative with respect to the second component, and $u_1, u_2, u_3 \in \mathbb{I}$. Clearly, several equivalent ways of constructing $\mathbf{C}_{DAH}$ are possible. Similar results are obtained by considering $\mathbf{C}_{DAI}$. As already pointed out, for simulation purposes only the partial derivatives of $\mathbf{C}_{DAH}$ and $\mathbf{C}_{DAI}$ are needed. In the present case, this simply reduces to the differentiation of the integral in Eq. (4.22), which then disappears, and need not be calculated (see Illustration A.2).

As an illustration, in Figure 4.2 we show a non-parametric comparison between the copulas constructed using Eq. (4.22), and the corresponding empirical entities. Here the variables H, D, and A are considered. The isolines of the rank distribution of (H, D) are plotted for two selected conditioning events concerning A. Clearly, this latter conditioning is necessary, for only "sections" of 3-copulas can be plotted in a 3-dimensional space. The 2-copula used for the pair (H, D) belongs to the Ali-Mikhail-Haq family (see Section C.4), while those used for (H, A) and (D, A) belong to the Frank family (see Section C.1). Overall, the agreement is valuable in all cases, except for some nuisance due to a limited sample size, and to "ties" at low probability levels. These plots should be compared to those in Figure A.5, where a larger data set of simulated sea storms is used.

As a further illustration, the same comparison as above is shown in Figure 4.3, where the variables D, I, and A are considered. The 2-copula used for the pairs (D, I) and (I, A) belong to the Gumbel-Hougaard family (see Section C.2), while that used for (D, A) belongs to the Frank family (see Section C.1). Overall, the fit is generally good, except for some disagreement due to a limited sample size at large probability levels. These plots should be compared to those in Figure A.6, where a larger data set of simulated sea storms is used.

The examples presented here clearly illustrate how difficult it is to work with multivariate distributions using too small sample sizes. Unfortunately, this is often the situation in practical applications. ◄

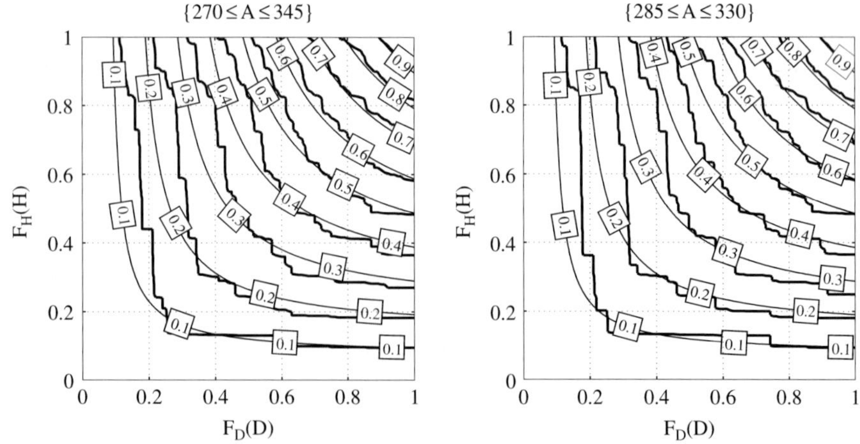

Figure 4.2. Comparison between the level curves of the theoretical copulas fitted to the available (D, A, H) observations (*thin* lines), and those of the empirical copulas constructed using the same data (*thick* lines). The probability levels are as indicated (*labels*), as well as the conditioning events $\{\bullet \leq A \leq \bullet\}$

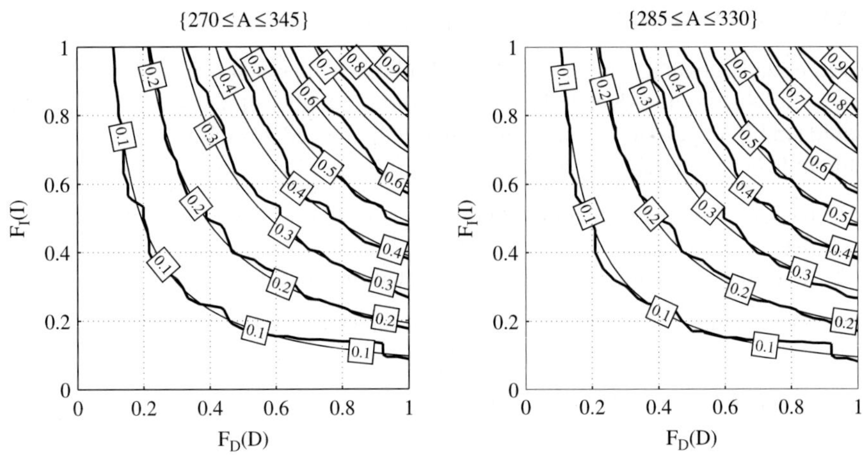

Figure 4.3. Comparison between the level curves of the theoretical copulas fitted to the available (D, A, I) observations (*thin* lines), and those of the empirical copulas constructed using the same data (*thick* lines). The probability levels are as indicated (*labels*), as well as the conditioning events $\{\bullet \le A \le \bullet\}$

4.3.2 The 4-Dimensional Case

For $d = 4$, the approach is similar to the trivariate case. Let X, Y, Z, W be four r.v.'s. The multivariate family of interest here is given by

$$F_{XYZW}(x, y, z, w) = \int_{-\infty}^{z} \int_{-\infty}^{y} \mathbf{C}_{XW}\left(F_{X|YZ}(x \mid \boldsymbol{r}), F_{W|YZ}(w \mid \boldsymbol{r})\right) F_{YZ}(\mathrm{d}s, \mathrm{d}t), \tag{4.23}$$

where $\boldsymbol{r} = (s, t)$. Again, the conditional distributions in Eq. (4.23) can be written in terms of copulas. For instance:

$$\begin{aligned} F_{X|YZ}(x \mid y, z) &= \mathbb{P}\{X \le x \mid Y = y, Z = z\} \\ &= \mathrm{R}(F_X(x), F_Y(y), F_Z(z)), \end{aligned} \tag{4.24}$$

where, using a simplified notation,

$$\mathrm{R}(a, b, c) = \frac{\partial_{b,c}\mathbf{C}_{XYZ}(a, b, c)}{\partial_{b,c}\mathbf{C}_{XYZ}(1, b, c)} = \frac{\partial_{b,c}\mathbf{C}_{XYZ}(a, b, c)}{\partial_{b,c}\mathbf{C}_{YZ}(b, c)}. \tag{4.25}$$

A similar expression holds for $F_{W|YZ}(w \mid y, z)$.

This theoretical modeling is physically appealing. Clearly, any variable of interest can be easily changed, and different combinations of practical interest are possible.

4.3.3 The General Case

An important advantage of the conditional mixtures approach is as follows: should a new variable be introduced, the construction of a higher dimensional distribution would simply require us to extend recursively the procedures outlined above. Thus, without loss of generality, and using a simplified but obvious notation, assume that

1. the $(d-1)$-dimensional distributions $F_{1\cdots d-1}$ and $F_{2\cdots d}$ have been defined, and that
2. they share a common $(d-2)$-dimensional distribution $F_{2\cdots d-1}$.

Then, the d-dimensional distribution $F_{1\cdots d}$ of interest here is given by

$$\begin{aligned} F_{1\cdots d}(\mathbf{y}) = \int_{-\infty}^{y_2} \cdots \int_{-\infty}^{y_{d-1}} & \mathbf{C}_{1d}\left(F_{1|2\cdots d-1}(y_1 \mid x_2, \ldots, x_{d-1}),\right. \\ & \left. F_{d|2\cdots d-1}(y_d \mid x_2, \ldots, x_{d-1})\right) F_{2\cdots d-1}(\mathrm{d}x_2, \ldots, \mathrm{d}x_{d-1}), \end{aligned} \tag{4.26}$$

where the conditional distributions $F_{1|2\cdots d-1}$ and $F_{d|2\cdots d-1}$ are derived as given using standard calculations from, respectively, $F_{1\cdots d-1}$ and $F_{2\cdots d}$.

A further interesting point is that, through the same construction, it is possible to generate families of multivariate distributions by means of survival functions. In fact, let $\overline{F}_i = 1 - F_i$, $i = 1, \ldots, d$, be (marginal) univariate survival functions. Then, bivariate survival distributions can be constructed as in Eq. (3.30) by using the survival 2-copula $\overline{\mathbf{C}}$ (see Theorem 3.2), i.e.

$$\overline{F}_{ij}(x_i, x_j) = \mathbb{P}\left\{X_i > x_i, X_j > x_j\right\} = \overline{\mathbf{C}}_{ij}\left(\overline{F}_i(x_i), \overline{F}_j(x_j)\right). \tag{4.27}$$

The general case requires the same construction as in Eq. (4.26), where now

1. all the distributions are replaced by the corresponding survival functions,
2. copulas are replaced by the corresponding survival copulas, and
3. the integrals have lower limits y_i's and upper limits ∞.

Thus,

$$\begin{aligned} \overline{F}_{1\cdots d}(\mathbf{y}) = \int_{y_2}^{\infty} \cdots \int_{y_{d-1}}^{\infty} & \overline{\mathbf{C}}_{1d}\left(\overline{F}_{1|2\cdots d-1}(y_1 \mid x_2, \ldots, x_{d-1}),\right. \\ & \left. \overline{F}_{d|2\cdots d-1}(y_d \mid x_2, \ldots, x_{d-1})\right) F_{2\cdots d-1}(\mathrm{d}x_2, \ldots, \mathrm{d}x_{d-1}). \end{aligned} \tag{4.28}$$

One can realize immediately that the above family is the same as that given by Eq. (4.26), with

$$\mathbf{C}_{ij}(u, v) = u + v - 1 + \overline{\mathbf{C}}_{ij}(1-u, 1-v).$$

CHAPTER 5

EXTREME VALUE ANALYSIS VIA COPULAS

Because copulas represent a fundamental tool to describe the structure of multivariate distributions (essentially via Sklar's Theorem — see Theorem 4.2), it is helpful to translate the results of classical Multivariate Extreme Value theory outlined in Chapter 2 in terms of copulas. The main references to this subject are [187, 105, 188, 155, 207], where further bibliography is indicated. Additional references are given throughout the text.

5.1. EXTREME VALUE COPULAS

In Chapter 2 it was shown that, if H is a multivariate EV distribution, then its marginals belong to the continuous GEV family (see Eq. (2.9) and the ensuing discussion). Here we focus our attention on the copula $\mathbf{C}$ associated with H. Note that, since the marginals H_i's are continuous, the copula representation is unique. We commence by showing an important result.

PROPOSITION 5.1. *Let* $H(\boldsymbol{x}) = \mathbf{C}\left(H_1(x_1), \ldots, H_d(x_d)\right)$ *be a MEV distribution. Then* $\mathbf{C}\left(\boldsymbol{u}^t\right) = \mathbf{C}^t\left(\boldsymbol{u}\right)$ *for all* $t > 0$. *In addition, if* $F \in MDA(H)$, *then* $F_i \in MDA(H_i)$ *for each marginal* F_i *of* F.

Thus, Proposition 5.1 provides a (partial) necessary characterization of the copulas underlying MEV distributions. Let us investigate the converse situation, i.e. suppose that $\mathbf{C}$, the d-copula associated with H, satisfies the above condition for all $t > 0$. In addition, let $F_i \in \text{MDA}(H_i)$, $i = 1, \ldots, d$, and define the following multivariate distributions:

$$H(\boldsymbol{x}) = \mathbf{C}\left(H_1(x_1), \ldots, H_d(x_d)\right),$$

$$F(\boldsymbol{x}) = \mathbf{C}\left(F_1(x_1), \ldots, F_d(x_d)\right).$$

Using standard calculations it is not difficult to show that $F \in \text{MDA}(H)$. In other words, Proposition 5.1 also provides a sufficient condition for $\mathbf{C}$ in order to be the copula of an EV distribution. This yields the following definition.

DEFINITION 5.1 (Extreme Value copula (I)). *A d-copula* **C** *satisfying the relationship*

$$\mathbf{C}\left(u_1^t, \ldots, u_d^t\right) = \mathbf{C}^t\left(u_1, \ldots, u_d\right) \tag{5.1}$$

for all $t > 0$ is called an Extreme Value copula (EVC). Also, if $F \in MDA(H)$, then the copula **C** *of H is called the limiting EVC of F.*

The following theorem summarizes the above results.

THEOREM 5.1 (EV distribution characterization). *Let H be a d-variate distribution with copula* **C**. *Then H is a MEV distribution if, and only if,*

1. *its marginals H_i's have a GEV distribution, and*
2. **C** *satisfies Eq. (5.1) for all $t > 0$.*

NOTE 5.1. As a consequence of Theorem 5.1, if **C** is an EVC, then the multivariate distribution H given by

$$H(\boldsymbol{x}) = \mathbf{C}\left(H_1(x_1), \ldots, H_d(x_d)\right),$$

where each marginal H_i's belongs to the GEV class, is an EV distribution. Clearly, it is necessary that the marginals have a GEV distribution.

We now give some examples of well known Extreme Value copulas.

ILLUSTRATION 5.1 (EV copulas). ►

There are quite a few families of copulas satisfying Eq. (5.1), i.e. suitable for representing an EVC.

1. **(Independence):** one sees that

$$\mathbf{\Pi}_d(\boldsymbol{u}^t) = \prod_{i=1}^{d} u_i^t = \mathbf{\Pi}_d^t(\boldsymbol{u}).$$

2. **(Gumbel-Hougaard):** given the expression of the Gumbel-Hougaard d-copula $\mathbf{C}_\theta$ (see Section C.2), $\theta \geq 1$, it follows that

$$\mathbf{C}_\theta\left(\boldsymbol{u}^t\right) = e^{-\left[(-t\ln u_1)^\theta + \cdots + (-t\ln u_d)^\theta\right]^{1/\theta}} = \mathbf{C}_\theta^t\left(\boldsymbol{u}\right).$$

It is worth noting [115] that the Gumbel-Hougaard family is the *only* Archimedean family of EV copulas. ◄

Note also that Eq. (5.1) emphasizes a fundamental feature of EV copulas. In the literature, this is known as *max-stability property*, and has its own definition (see also Definition 1.4).

DEFINITION 5.2 (Max-Stability). *A d-copula* **C** *is max-stable if it satisfies the relationship* $\mathbf{C}(u_1, \ldots, u_d) = \mathbf{C}^t\left(u_1^{1/t}, \ldots, u_d^{1/t}\right)$ *for all* $t > 0$ *and* $\mathbf{u} \in \mathbb{I}^d$.

The following illustration shows the relationship between max-stable copulas and the componentwise maxima of random vectors. For the sake of simplicity, only the two-dimensional case is presented [207].

ILLUSTRATION 5.2 (Order Statistics (cont.)). ►

Let (X_i, Y_i), $i = 1, \ldots, n$, be a sample of i.i.d. bivariate vectors with common joint distribution H, 2-copula $\mathbf{C}$, and marginals F (for the X_i's) and G (for the Y_i's). In applications, the joint distribution of the extremal order statistics $X_{(n)} = \max_{1\leq i\leq n}\{X_i\}$ and $Y_{(n)} = \max_{1\leq i\leq n}\{Y_i\}$ is often of interest. To this purpose, we calculate $H_{(n)}$, the law of $(X_{(n)}, Y_{(n)})$, and the corresponding 2-copula $\mathbf{C}_{(n)}$.

As is well known (see Section 1.1), the marginal laws of the component-wise maxima are $F_{(n)}(x) = F^n(x)$ and $G_{(n)}(y) = G^n(y)$. Thus

$$\begin{aligned} H_{(n)}(x, y) &= \mathbb{P}\left\{X_{(n)} \leq x, Y_{(n)} \leq y\right\} \\ &= H^n(x, y) \\ &= \mathbf{C}^n\left(F(x), G(y)\right) \\ &= \left[\mathbf{C}\left(\left[F_{(n)}(x)\right]^{1/n}, \left[G_{(n)}(y)\right]^{1/n}\right)\right]^n. \end{aligned} \tag{5.2}$$

In turn, given $u, v \in \mathbb{I}$,

$$\mathbf{C}_{(n)}(u, v) = \mathbf{C}^n\left(u^{1/n}, v^{1/n}\right). \tag{5.3}$$

The function $\mathbf{C}_{(n)}$ is a copula for all $n \in \mathbb{N}$, and represents the max-stable copula of $(X_{(n)}, Y_{(n)})$.

A further interesting point is as follows. Suppose that $\mathbf{C}$ is Archimedean with generator γ. Then Eq. (5.3) can be rewritten as

$$\mathbf{C}_{(n)}(u, v) = \left[\gamma^{[-1]}\left(\gamma(u^{1/n}) + \gamma(v^{1/n})\right)\right]^n. \tag{5.4}$$

Thus, $\gamma_{(n)}(t) = \gamma(t^{1/n})$ is the generator of $\mathbf{C}_{(n)}$, which, in turn, is a member of the interior power family generated by γ with parameter $\alpha = 1/n$ (see Proposition 3.4).

We now give some examples.

1. If $\mathbf{C} = \mathbf{\Pi}_2$, i.e. the vector (X_i, Y_i) has independent components, then $X_{(n)}$ and $Y_{(n)}$ must also be independent:

$$\mathbf{\Pi}_{(n)}(u, v) = \left(u^{1/n}v^{1/n}\right)^n = uv = \mathbf{\Pi}_2(u, v).$$

   This result is the same as that obtained in (1) in Illustration 5.1.

2. If X_i, Y_i are co-monotonic, then so are $X_{(n)}$ and $Y_{(n)}$:

$$\mathbf{M}_{(n)}(u, v) = \left(\min\left\{u^{1/n}, v^{1/n}\right\}\right)^n = \mathbf{M}_2(u, v).$$

3. If X_i, Y_i are counter-monotonic, then $X_{(n)}$ and $Y_{(n)}$ are not for any $n \geq 2$:

$$\mathbf{W}_{(n)}(u, v) = \left(\max\left\{u^{1/n} + v^{1/n} - 1, 0\right\}\right)^n,$$

which is different from $\mathbf{W}_2(u, v)$ — it is a member of the Clayton family shown in Section C.3.

4. Let $\mathbf{C}$ belong to the Marshall-Olkin family (see Section C.13), i.e.

$$\mathbf{C}_{\alpha,\beta}(u, v) = \min\left\{u^{1-\alpha}v, uv^{1-\beta}\right\},$$

with $0 < \alpha, \beta < 1$. Then, it follows that $\mathbf{C}_{(n)} = \mathbf{C}$ for any n.

◄

From the notion of max-stability, and the above illustration, it is possible to derive an alternative definition of an EV copula.

DEFINITION 5.3 (Extreme Value copula (II)). *A d-copula* $\mathbf{C}^*$ *is an Extreme Value copula if there exists a copula* $\mathbf{C}$ *such that*

$$\mathbf{C}^*(u_1, \ldots, u_d) = \lim_{n\to\infty} \mathbf{C}^n\left(u_1^{1/n}, \ldots, u_d^{1/n}\right) \tag{5.5}$$

for all $\boldsymbol{u} \in \mathbb{I}^d$.

NOTE 5.2. If the pointwise limit of a sequence of copulas exists at every point $\boldsymbol{u} \in \mathbb{I}^d$, then the limit must be a copula: indeed, the $\mathbf{C}$-volume of any d-rectangle in $\mathbb{I}^d$ will have a non-negative limit.

As already mentioned in Proposition 3.2, 2-copulas (and, more generally, d-copulas) are *invariant* under strictly increasing transformations of the marginals. More specifically: if $g_i : \mathbb{R} \to \mathbb{R}$, $i = 1, \ldots, d$, are all strictly increasing functions, then the vector $\boldsymbol{Y} = (g_1(X_1), \ldots, g_d(X_d))$ has the same copula as the vector $\boldsymbol{X} = (X_1, \ldots, X_d)$. The interesting point is that such an invariance property holds also for the limiting EVC of $\boldsymbol{Y}$.

PROPOSITION 5.2 (Invariance property). *Let* $\boldsymbol{X} = (X_1, \ldots, X_d)$, *and set* $\boldsymbol{Y} = (g_1(X_1), \ldots, g_d(X_d))$, *where the functions* $g_i : \mathbb{R} \to \mathbb{R}$, $i = 1, \ldots, d$, *are all strictly increasing. If* $\mathbf{C}_X^*$ *and* $\mathbf{C}_Y^*$ *denote, respectively, the EVC of* $\boldsymbol{X}$ *and* $\boldsymbol{Y}$, *then*

$$\mathbf{C}_Y^* = \mathbf{C}_X^*. \tag{5.6}$$

The invariance property mentioned above shows that the limiting EVC only depends upon the copula $\mathbf{C}$ of the multivariate distribution F, and is independent of the marginals of F. Thus, it makes sense to extend the concept of domain of attraction also to copulas.

DEFINITION 5.4 (Copula Domain of Attraction (CDA)). *A copula* $\mathbf{C}$ *is said to belong to the domain of attraction of an EVC* $\mathbf{C}^*$ *if*

$$\mathbf{C}\left(F_1, \ldots, F_d\right) \in MDA\left(\mathbf{C}^*\left(H_1, \ldots, H_d\right)\right), \tag{5.7}$$

where each marginal F_i*'s is continuous and belongs to the domain of attraction of a GEV law* H_i*. The copula* $\mathbf{C}^*$ *is also called the limiting copula of* $\mathbf{C}$*, and we write* $\mathbf{C} \in CDA(\mathbf{C}^*)$.

NOTE 5.3. Obviously, every EVC copula belongs to its own CDA. The marginals F_i's of F only play a role in fixing the marginals H_i's of H.

The theorem given below provides necessary and sufficient conditions to check whether or not a given copula is in the domain of attraction of some EVC. As already mentioned in Section 2.1 (see also [169]), in order to isolate the dependence features from the marginal distributional aspects, traditionally the components of both the distribution F and the corresponding MEV law H are transformed to standard marginals. It can be shown [232] that this does not pose difficulties. For technical convenience, it is customary to choose the standard Fréchet distribution ϕ as marginals (see Section 1.2 and the discussion preceding Eq. (2.10)).

THEOREM 5.2. *Let* $\mathbf{C}$ *and* $\mathbf{C}^*$ *be, respectively, the copulas given by*

$$\mathbf{C}\left(u_1, \ldots, u_d\right) = F\left(\phi^{-1}(u_1), \ldots, \phi^{-1}(u_d)\right), \tag{5.8a}$$

$$\mathbf{C}^*\left(u_1, \ldots, u_d\right) = H\left(\phi^{-1}(u_1), \ldots, \phi^{-1}(u_d)\right), \tag{5.8b}$$

where $\boldsymbol{u} \in \mathbb{I}^d$*. Then,* $\mathbf{C} \in CDA(\mathbf{C}^*)$ *if, and only if,*

1.

$$\lim_{t\to\infty} t\left(1 - \mathbf{C}\left(\boldsymbol{u}^{1/t}\right)\right) = -\ln \mathbf{C}^*\left(\boldsymbol{u}\right) \tag{5.9}$$

for all $\boldsymbol{u} \in \mathbb{I}^d$*, or, equivalently,*

2.

$$\lim_{t\to 1^-} \frac{1 - \mathbf{C}\left(\boldsymbol{u}^{1-t}\right)}{1-t} = -\ln \mathbf{C}^*\left(\boldsymbol{u}\right) \tag{5.10}$$

for all $\boldsymbol{u} \in \mathbb{I}^d$*, or, equivalently,*

3.

$$\lim_{t\to 0^+} \frac{1-\mathbf{C}(\boldsymbol{u}^t)}{1-\mathbf{C}^*(\boldsymbol{u}^t)} = 1 \tag{5.11}$$

for all $\boldsymbol{u} \in \mathbb{I}^d$, *or, equivalently,*

4.

$$d_{i_1,\ldots,i_k}(\boldsymbol{u}) = \lim_{n\to\infty} n\overline{F}\left(0,\ldots,n\phi^{-1}(u_{i_1}),\ldots,n\phi^{-1}(u_{i_k}),\ldots,0\right) < \infty \tag{5.12}$$

for all $\boldsymbol{u} \in \mathbb{I}^d$ *and all* $1 \le i_1 \le \cdots \le i_k \le d$.

If any of these statements is satisfied, then

$$\mathbf{C}^*(\boldsymbol{u}) = u_1 \cdots u_d \exp\left(\sum_{j=2}^{d}(-1)^j \sum_{1\le i_1\le\cdots\le i_k\le d} d_{i_1,\ldots,i_k}(\boldsymbol{u})\right), \tag{5.13}$$

where $\boldsymbol{u} \in \mathbb{I}^d$.

NOTE 5.4. Actually, Theorem 5.2 also provides a construction of $\mathbf{C}^*$.

An equivalent result is as follows.

PROPOSITION 5.3. *Let* $\mathbf{C}$ *be a copula and* $\mathbf{C}^*$ *be an EVC. Then,* $\mathbf{C} \in CDA(\mathbf{C}^*)$ *if, and only if,*

$$\lim_{t\to 0^+} \frac{1-\mathbf{C}(1-tu_1,\ldots,1-tu_d)}{t} = -\ln \mathbf{C}^*\left(e^{-u_1},\ldots,e^{-u_d}\right) \tag{5.14}$$

for all $\boldsymbol{u} \in \mathbb{I}^d$.

ILLUSTRATION 5.3. ►

The general expression for the Farlie-Gumbel-Morgenstern family of 2-copulas (see Section C.6) is

$$\mathbf{C}_\theta(u,v) = uv + \theta uv(1-u)(1-v),$$

where $u, v \in \mathbb{I}$ and $-1 \le \theta \le 1$. Using Proposition 5.3 it follows that

$$\begin{aligned}\lim_{t\to 0^+} \frac{1-\mathbf{C}_\theta(1-tu,1-tv)}{t} &= u+v \\ &= -\ln\left(e^{-u}e^{-v}\right) \\ &= -\ln \mathbf{\Pi}_2\left(e^{-u},e^{-v}\right).\end{aligned}$$

Thus, the CDA of the Farlie-Gumbel-Morgenstern 2-copula is that of the independence 2-copula $\mathbf{\Pi}_2$. ◄

The theorem below summarizes the results given before.

THEOREM 5.3. *Let F be a multivariate distribution with copula* $\mathbf{C}$*, and let H be an EV distribution with EV copula* $\mathbf{C}^*$*. Then,* $F \in MDA(H)$ *if, and only if,*

1. $F_i \in MDA(H_i)$ *for all i's, where* H_i *is a GEV distribution, and*
2. $\lim_{n\to\infty} \mathbf{C}^n\left(u_1^{1/n}, \ldots, u_d^{1/n}\right) = \mathbf{C}^*(u_1, \ldots, u_d)$ *for all* $\boldsymbol{u} \in \mathbb{I}^d$.

The above theorem has several consequences. On the one hand, suppose that $\mathbf{C} \in \mathrm{CDA}(\mathbf{C}^*)$, and let $F(\boldsymbol{x}) = \mathbf{C}(F_1(x_1), \ldots, F_d(x_d))$ and $H(\boldsymbol{x}) = \mathbf{C}^*(H_1(x_1), \ldots, H_d(x_d))$. By virtue of Eq. (5.7), it follows that $F \in \mathrm{MDA}(H)$, and thus, from (2) above, $\lim_{n\to\infty} \mathbf{C}^n\left(\boldsymbol{u}^{1/n}\right) = \mathbf{C}^*(\boldsymbol{u})$. On the other hand, suppose that this latter relationship holds for all $\boldsymbol{u} \in \mathbb{I}^d$. Then, taking $F_i \in \mathrm{MDA}(H_i)$, and defining F and H as above, Theorem 5.3 ensures that $F \in \mathrm{MDA}(H)$ and, in turn, that $\mathbf{C} \in \mathrm{CDA}(\mathbf{C}^*)$. Thus, a further important characterization follows (see also Definition 5.3).

THEOREM 5.4. *Let* $\mathbf{C}$ *be a copula, and let* $\mathbf{C}^*$ *be an EV copula. Then,* $\mathbf{C} \in CDA(\mathbf{C}^*)$ *if, and only if,*

$$\lim_{n\to\infty} \mathbf{C}^n\left(u_1^{1/n}, \ldots, u_d^{1/n}\right) = \mathbf{C}^*(u_1, \ldots, u_d) \tag{5.15}$$

for all $\boldsymbol{u} \in \mathbb{I}^d$.

NOTE 5.5. Oviously, using Eq. (5.1), every EVC belongs to its own CDA.

From Definition 5.2, the following corollary is a natural consequence.

COROLLARY 5.1. *A copula is max-stable if, and only if, it is an EVC.*

In Section C.15 we show how to generate two-parameters families of copulas. As an illustration, we now show how to exploit Theorem 5.4 in order to calculate the CDA of some of these families. Clearly, the same procedure can be extended to a broader context.

ILLUSTRATION 5.4. ►

Consider the two-parameter family of Archimedean copulas given by

$$\mathbf{C}_{\theta,\eta}(u, v) = \left(1 + \left[(u^{-\theta} - 1)^\eta + (v^{-\theta} - 1)^\eta\right]^{1/\eta}\right)^{-1/\theta},$$

where $u, v \in \mathbb{I}$, and $\theta > 0, \eta \geq 1$. As prescribed by Eq. (5.15), the limit $\lim_{n\to\infty} \mathbf{C}^n\left(u^{1/n}, u^{1/n}\right)$ must be calculated. Clearly, the role played by θ is not essential, and thus n can be changed to $n\theta$. By using elementary expansions, it is easy to show that the above limit is equivalent to

$$\lim_{n\to\infty} -n \ln\left(1 + \frac{\left[(n(u^{-1/n} - 1))^\eta + (n(v^{-1/n} - 1))^\eta\right]^{1/\eta}}{n}\right),$$

which reduces to $-[(-\ln u)^\eta + (-\ln v)^\eta]^{1/\eta}$. In turn,

$$\mathbf{C}^*(u, v) = \exp\left(-[(-\ln u)^\eta + (-\ln v)^\eta]^{1/\eta}\right),$$

and thus $\mathbf{C}_{\theta,\eta}$ belongs to the Gumbel-Hougaard CDA.

As a further illustration, consider the two-parameter family of copulas given by

$$\mathbf{C}_{\theta,\eta}(u, v) = \left(u^{-\theta} + v^{-\theta} - 1 - \left[(u^{-\theta} - 1)^{-\eta} + (v^{-\theta} - 1)^{-\eta}\right]^{-1/\eta}\right)^{-1/\theta},$$

where $u, v \in \mathbb{I}$, $\theta \geq 0$, and $\eta > 0$ (this is the family BB4 in [155] — see Section C.15). Proceeding as before,

$$\mathbf{C}^*(u, v) = uv \exp\left((-\ln u)^{-\eta} + (-\ln v)^{-\eta}\right),$$

and thus $\mathbf{C}_{\theta,\eta}$ belongs to the Galambos CDA. ◀

As already mentioned in Chapter 3 (see Eq. (3.4)), any convex linear combination of copulas is itself a copula. It is then of interest to calculate the limiting copula of such a combination. Let us consider a d-dimensional copula $\mathbf{C}$ given by a convex linear combination as in Eq. (3.4). Then, the EVC $\mathbf{C}^*$ of $\mathbf{C}$ is given by

$$\mathbf{C}^*(\boldsymbol{u}) = \prod_{i=1}^{k} \left(\mathbf{C}_i^*(\boldsymbol{u})\right)^{\lambda_i}, \tag{5.16}$$

where $\mathbf{C}_i^*$ denotes the EVC associated with the copula $\mathbf{C}_i$. It must be pointed out that, in general, even if all the the mixing copulas $\mathbf{C}_i$'s are EVC, this does not imply that their convex linear combination $\mathbf{C}$ is also an EVC, as shown by the following counter-example.

ILLUSTRATION 5.5. ▶

Let $\mathbf{C}_1 = \mathbf{\Pi}_d$ and $\mathbf{C}_2 = \mathbf{M}_d$: both these multivariate copulas are EVC (see, e.g., Illustration 5.2). Set $\mathbf{C} = \lambda\mathbf{C}_1 + (1-\lambda)\mathbf{C}_2$, where $\lambda \in (0, 1)$. Then, using Eq. (5.16), the EVC $\mathbf{C}^*$ of $\mathbf{C}$ is given by

$$\begin{aligned}\mathbf{C}^*(u) &= (\mathbf{\Pi}_d(\boldsymbol{u}))^\lambda (\mathbf{M}_d(\boldsymbol{u}))^{1-\lambda} \\ &= (u_1 \cdots u_d)^\lambda \min\left\{u_1^{1-\lambda}, \ldots, u_d^{1-\lambda}\right\}.\end{aligned} \tag{5.17}$$

Thus, $\mathbf{C}^*$ is a member of the Cuadras-Augé family, a subfamily of the Marshall-Olkin family of copulas (see Illustration 5.2, and also Section C.13 and Illustration C.4). Evidently, $\mathbf{C}^* \neq \mathbf{C}$, and $\mathbf{C}$ is not an EVC. ◀

In Definition 2.5 we introduced the notion of (positively) associated r.v.'s, and in the ensuing Theorems 2.5–2.6 we mentioned several important features of MEV distributions. Similar results can be stated for EVC's [155], recalling that copulas are themselves multivariate distributions (see also Subsection B.1.1).

THEOREM 5.5. *If* $\mathbf{C}^*$ *is an EVC, then* $\mathbf{C}^*$ *is associated.*

As a consequence, any MEV distribution is associated. Furthermore, the following inequality holds (see also Eqs. (B.3)–(B.4)).

THEOREM 5.6. *If* $\mathbf{C}^*$ *is an EVC, then*

$$\mathbf{C}^*(u) \geq \mathbf{\Pi}_d(\boldsymbol{u}) \tag{5.18}$$

for all $u \in \mathbb{I}^d$.

ILLUSTRATION 5.6 (Regionalization method (cont.)). ►

The issue of flood frequency regionalization was discussed in Illustration 2.2, where a simple MEV model (involving a single dependence parameter) was applied to several homogeneous regions in northwestern Italy. One of the limitations of the model is that the dependence between each pair of sites is ruled by the same parameter, and hence is constant over all the region. Here we show how to introduce further parameters, in order to improve the description of the dependence between the gauge stations.

For the sake of simplicity, only a set of three sites is considered, all belonging to Region C outlined in Illustration 2.2: (*1*) Airole - Piena, (*2*) Merelli - Centrale Argentina, and (*3*) Ponte Poggi - Ellera. These three stations provide a sample of 34 years of common annual maxima. The estimated values of Kendall's τ_K for the three pairs are: $\tau_K^{1,2} \approx 0.36$, $\tau_K^{1,3} \approx 0.20$, and $\tau_K^{2,3} \approx 0.22$. The distances d between the sites are: $d_{1,2} \approx 26.2$ km, $d_{1,3} \approx 88.4$ km, and $d_{2,3} \approx 72.8$ km. Apparently, the strength of the association (as measured by τ_K) is a decreasing function of the distance d. In turn, the assumption of a common degree of dependence over all the region could be questionable.

An interesting way to generate EV copulas is presented in Subsection C.15.2. By virtue of Proposition C.4, if $\mathbf{A}$ and $\mathbf{B}$ are d-dimensional EV copulas, then

$$\mathbf{C}_{\alpha_1,\ldots,\alpha_d}(\mathrm{u}) = \mathbf{A}(u_1^{\alpha_1},\ldots,u_d^{\alpha_d})\,\mathbf{B}(u_1^{1-\alpha_1},\ldots,u_d^{1-\alpha_d})$$

is also an EV d-copula with parameters $\alpha_1,\ldots,\alpha_d \in \mathbb{I}$. As shown in Illustration C.4, choosing $\mathbf{A}_\theta$ as a 3-dimensional Cuadras-Augé copula, with $\theta \in \mathbb{I}$, and $\mathbf{B} = \mathbf{\Pi}_3$, generates the following family of EV copulas:

$$\mathbf{C}^*_{\boldsymbol{\beta}}(u) = \prod_{i=1}^{3} u_i^{1-\beta_i} \min\left\{u_1^{\beta_1}, u_2^{\beta_2}, u_3^{\beta_3}\right\}, \tag{5.19}$$

where $\beta_i = \theta\alpha_i \in \mathbb{I}$. The point is that the 2-dimensional marginals of $\mathbf{C}^*$ are Marshall-Olkin bivariate copulas (see Section C.13) with parameters (β_i, β_j), where $i, j = 1, 2, 3$ and $i \neq j$. As a consequence (see Eq. (C.51a))

$$\tau_K^{i,j} = \frac{1}{1/\beta_i + 1/\beta_j - 1}. \tag{5.20}$$

Evidently, $0 \leq \tau_K^{i,j} \leq 1$, as it must be since $\mathbf{C}^*$ is an EVC, and given the results of Theorems 5.5–5.6. The parameters $\beta_1, \beta_2, \beta_3$ can then be calculated by using the estimated values of τ_K:

$$\frac{1}{\beta_i} = \frac{1}{2}\left(1 + \frac{1}{\tau_K^{i,j}} + \frac{1}{\tau_K^{i,k}} - \frac{1}{\tau_K^{j,k}}\right), \tag{5.21}$$

where (i, j, k) is a permutation of $(1, 2, 3)$. Note that not all of the triples $(\tau_K^{1,2}, \tau_K^{1,3}, \tau_K^{2,3})$ are admissible, given the constraints $\beta_i \in \mathbb{I}$, $i = 1, 2, 3$. In the present case the resulting values are: $\beta_1 \approx 0.4735$, $\beta_2 \approx 0.6045$, and $\beta_3 \approx 0.2604$.

The overall statistical behavior of the phenomenon within the region under investigation can be studied in several ways. For instance, an event of practical interest is given by

$$E_q^> = \{X_1 > x_q, X_2 > x_q, X_3 > x_q\}$$

where X_i denotes the observation at the i-th site, and x_q is the q-order quantile, with $q \in (0, 1)$. Extreme events correspond to $q \approx 1$. As thoroughly discussed in Section 3.3, the return period $\tau_q^>$ of $E_q^>$ is not simply given by $1/(1-q)$, since $E_q^>$ is a trivariate event. Instead, the calculation of the probability

$$p_q^> = \mathbb{P}\{E_q^>\} \tag{5.22}$$

is required, since $\tau_q^> = 1/p_q^>$. Note that dealing with annual maxima implies that $\mu_T = 1$ year, i.e. only one event per year is considered (see Note 1.14). The traditional regionalization approach assumes that the stations are independent and share the same probability law. Since the X_i's are i.i.d., then $p_q^> = (1-q)^3$. However, the $\tau_K^{i,j}$'s are significantly different from zero, i.e. the variables are not independent. In particular, let us assume that they are joined via the EV copula $\mathbf{C}^*$, and therefore calculate $p_q^>$ as

$$\begin{aligned} p_q^> = 1 - u_1 - u_2 - u_3 - \mathbf{C}_\beta^*(u_1, u_2, u_3) + \\ \mathbf{C}_\beta^*(u_1, u_2, 1) + \mathbf{C}_\beta^*(u_1, 1, u_3) + \mathbf{C}_\beta^*(1, u_2, u_3), \end{aligned} \tag{5.23}$$

with $u_1 = u_2 = u_3 = q$. It is then of interest to compare the values of $p_q^>$ calculated either under the standard independence assumption or using the copula approach, as well as the corresponding return periods $\tau_q^>$. The results are presented in Table 5.1.

As expected, the event $E_q^>$ is much more probable when the variables are not independent and joined via $\mathbf{C}^*$. Roughly speaking, since the X_i's are positively associated (see Appendix B), the fact that one variable is large affects the statistical behavior of the others, which are also expected to be large. As a consequence, the return periods are much smaller under the copula assumption, i.e. the extreme events occur more frequently. Thus, if the independence hypothesis is used when

Table 5.1. Values of $p_q^>$ calculated either under the independence assumption ("IND") or using the copula approach ("$\mathbf{C}^*$"), for different choices of the quantile of order q. The corresponding return periods $\tau_q^>$ (in years) are also shown

	$\mathbf{p}_q^>$		$\tau_q^>$	
q	**IND**	$\mathbf{C}^*$	**IND**	$\mathbf{C}^*$
0.9	0.001	0.02889	1000	34.614
0.95	0.000125	0.013693	8000	73.03
0.975	$1.5625 \cdot 10^{-5}$	0.0066734	64000	149.85
0.99	10^{-6}	0.0026297	10^6	380.27
0.995	$1.25 \cdot 10^{-7}$	0.0013084	$8 \cdot 10^6$	764.3
0.999	10^{-9}	0.00026065	10^9	3836.5

it is false (as in the present case), there might be a considerable underestimation of the occurrence of catastrophic events.

A similar rationale can be used to reanalyze the results shown in Illustration 2.2. In fact, the d-variate MEV distribution G introduced in Eq. (5.12) is simply given by a Gumbel-Hougaard d-copula (see Section C.2), with identically distributed GEV marginals F_i's. ◄

5.2. DEPENDENCE FUNCTION

An alternative way to describe the dependence structures of multivariate limits can be achieved via the introduction of the so-called *dependence functions* (see Theorem 2.1 and the ensuing discussion). As an additional result, by the same construction it is also possible to derive alternative representations of Extreme Value copulas, as well as alternative formulations of (some of) the domain of attraction conditions. We now illustrate the bivariate case, with the simplest mathematical representation. The literature on this subject is extensive: see, e.g., [286, 287, 217, 70, 288, 151, 153, 155, 2, 207].

Let $\mathbf{C}^*$ be an EV copula, and suppose that $\mathbf{C}^*$ represents the survival copula (see Theorem 3.2) of the r.v.'s X and Y, both endowed with a standard Exponential distribution. Then,

$$\overline{F}_X(x) = e^{-x}\,\mathbf{1}(x > 0) \quad \text{and} \quad \overline{F}_Y(y) = e^{-y}\,\mathbf{1}(y > 0),$$

respectively, and the joint survival function $\overline{H}$ is

$$\overline{H}(x, y) = \mathbb{P}\{X > x, Y > y\} = \mathbf{C}^*\left(e^{-x}, e^{-y}\right). \tag{5.24}$$

Using the fact that $\mathbf{C}^*$ is max-stable, it follows that, for any $s > 0$,

$$\overline{H}(sx, sy) = \left(\mathbf{C}^*\left(e^{-x}, e^{-y}\right)\right)^s = \left(\overline{H}(x, y)\right)^s. \tag{5.25}$$

Let us introduce the concept of a *dependence function* (see also Eq. (5.18) and the ensuing discussion).

DEFINITION 5.5 (Dependence function). *Let* $\mathbf{C}^*$ *be an EVC. The function* $A : \mathbb{I} \to [1/2, 1]$ *given by*

$$A(t) = -\ln \mathbf{C}^*\left(e^{-(1-t)}, e^{-t}\right) \tag{5.26}$$

is called the dependence function of the EVC $\mathbf{C}^*$.

Consider now the following change of variables:

$$\begin{cases} x = s(1-t) \\ y = st \end{cases} \iff \begin{cases} s = x + y \\ t = \frac{y}{x+y} \end{cases}, \tag{5.27}$$

where $s > 0$ and $t \in (0, 1)$. Then, Eq. (5.24) can be rewritten as

$$\begin{aligned} \overline{H}(x, y) &= \overline{H}(s(1-t), st) \\ &= \left(\overline{H}(1-t, t)\right)^s \\ &= \left(\mathbf{C}^*\left(e^{-(1-t)}, e^{-t}\right)\right)^s \\ &= \exp\left(-sA(t)\right) \\ &= \exp\left(-(x+y)A\left(\frac{y}{x+y}\right)\right). \end{aligned} \tag{5.28}$$

Since $\mathbf{C}^*(u, v) = \overline{H}(-\ln u, -\ln v)$, the following fundamental result holds (see [155] for details).

THEOREM 5.7 (Dependence function representation). *Let* $\mathbf{C}^*$ *be an EVC. Then*

$$\mathbf{C}^*(u, v) = \exp\left(\ln(uv)\, A\left(\frac{\ln v}{\ln(uv)}\right)\right), \tag{5.29}$$

for an appropriate choice of the dependence function A. In particular, the following constraints must be satisfied:

1. $\max\{t, 1-t\} \le A(t) \le 1$;
2. *A is convex.*

Incidentally, note that the above constraints yield $A(0) = 1$ and $A(1) = 1$. Thus, any function satisfying constraint (2) above, whose graph lies in the shaded region of Figure 5.1, may represent a dependence function.

Evidently, such a family of functions is infinite dimensional, giving a large freedom in the construction of the EVC $\mathbf{C}^*$. At the same time, this leads to difficulties, for no finite parametrization exists for such a family. In practical applications, only parametric subfamilies are used. However, by a careful choice, it is possible to ensure that a wide enough subclass of the entire limit family is approximated. We now present some possible choices for A.

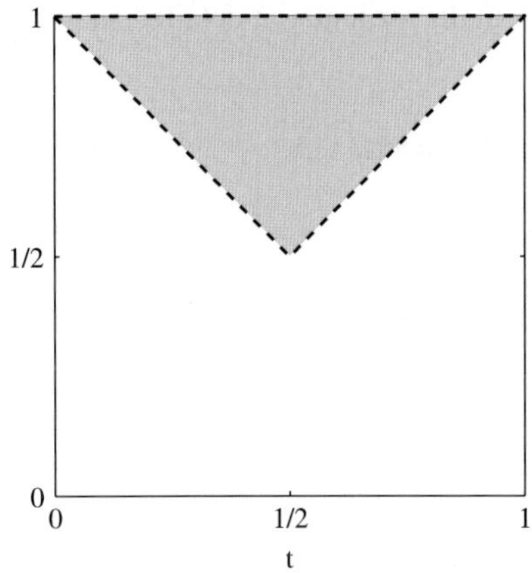

Figure 5.1. Admissible region (*shaded*) for the dependence function A

ILLUSTRATION 5.7. ▶

The functions A shown below satisfy the constraints of Theorem 5.7.

1. Let

$$A(t) = 1 \tag{5.30}$$

   for $t \in \mathbb{I}$; then $\mathbf{C}^*(u, v) = \mathbf{\Pi}_2(u, v)$.
2. Let

$$A(t) = \max\{t, 1-t\} \tag{5.31}$$

   for $t \in \mathbb{I}$; then $\mathbf{C}^*(u, v) = \mathbf{M}_2(u, v)$.
3. Let

$$A(t) = 1 - \min\{\beta t, \alpha(1-t)\}, \tag{5.32}$$

   with $t \in \mathbb{I}$ and $\alpha, \beta \in \mathbb{I}$; then Eq. (5.29) yields the Marshall-Olkin family of copulas (see Section C.13).
4. Let

$$A(t) = \left(t^\theta + (1-t)^\theta\right)^{1/\theta}, \tag{5.33}$$

   with $t \in \mathbb{I}$ and $\theta \geq 1$; then Eq. (5.29) yields the Gumbel-Hougaard family of copulas (see Section C.2), which reduces to $\mathbf{\Pi}_2$ as $\theta \to 1$.

Note how the constraints (1)–(2) correspond to limiting cases: the first involves the upper border of the admissibility region for the dependence function A, whereas the second relates to the lower border (see Figure 5.1). In particular, a detailed analysis of case (2) yields the following result:

$$A(1/2) = 1/2 \iff \mathbf{C}^* = \mathbf{M}_2, \tag{5.34}$$

which provides a characterization of $\mathbf{M}_2$ in terms of A. ◀

A further interesting construction is as follows [207].

ILLUSTRATION 5.8. ▶
Let the function A be given by

$$A(t) = 1 - t(1-t)\left(\alpha t + \beta(1-t)\right), \tag{5.35}$$

with $t \in \mathbb{I}$. From Figure 5.1, it is clear that $A'(0^+) \in [-1, 0]$, and $A'(1^-) \in [0, 1]$. As a consequence, α and β must belong to $\mathbb{I}$. Also, A will be convex when $\alpha \leq 2\beta$ and $\beta \leq 2\alpha$. This latter region is plotted in Figure 5.2.

Thus, if (α, β) lies in the shaded region of Figure 5.2, then A may generate an EVC through Eq. (5.29). ◀

The representation in terms of the dependence function A provides alternative ways to find the CDA of a given copula [2].

THEOREM 5.8. *Let* $\mathbf{C}^*$ *be an EVC with dependence function A. Then the copula* $\mathbf{C} \in CDA(\mathbf{C}^*)$ *if, and only if,*

$$\lim_{u,v\to 0} \frac{1 - \mathbf{C}(1-u, 1-v)}{u+v} = A(t) \tag{5.36}$$

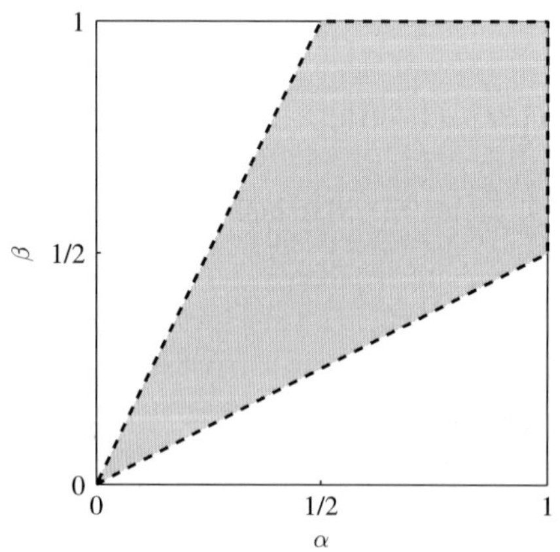

Figure 5.2. Admissible region (*shaded*) for the parameters (α, β) in Eq. (5.35)

holds for any sequence $(u, v) \in (0,1)^2$ *such that, for given* $t \in \mathbb{I}$,

$$\lim_{u,v\to 0} \frac{v}{u+v} = t.$$

As a consequence of the above theorem, a further important characterization of the CDA follows [2].

THEOREM 5.9. *Let* $\mathbf{C}^*$ *be an EVC with dependence function A. Then the copula* $\mathbf{C} \in CDA(\mathbf{C}^*)$ *if, and only if,*

$$\lim_{u\to 0} \frac{1-\mathbf{C}\left((1-u)^{1-t}, (1-u)^{t}\right)}{u} = A(t) \tag{5.37}$$

holds for all $t \in \mathbb{I}$.

An interesting point is as follows. If A is *symmetric* with respect to 1/2, i.e. if

$$A(t) = A(1-t) \tag{5.38}$$

holds for all $t \in \mathbb{I}$, then $\mathbf{C}^*$ is *exchangeable*. An example is the Gumbel-Hougaard EVC calculated in (4) in Illustration 5.7. Thus, the construction of A via Eq. (5.29) may provide an easy way to generate *asymmetric* families of 2-copulas. Note that, in applications, exchangeability may not always be a realistic assumption. In [111] an interesting procedure to generate new families of dependence functions is shown. In particular, an "asymmetrization" algorithm is introduced, as outlined in the following proposition.

PROPOSITION 5.4. *Let A, B be two dependence functions.*

1. ***(Convex combination)*** *If* $0 \le \lambda \le 1$, *then*

$$E(t) = \lambda A(t) + (1-\lambda)B(t), \tag{5.39}$$

   where $t \in \mathbb{I}$, *is a dependence function.*
2. ***(Asymmetrization)*** *If* $0 < \lambda, \kappa < 1$, *then*

$$\begin{aligned} E(t) = (\kappa t + \lambda(1-t))\, A\left(\frac{\kappa t}{\kappa t + \lambda(1-t)}\right) & \\ + ((1-\kappa)t + (1-\lambda)(1-t)) & \\ B\left(\frac{(1-\kappa)t}{(1-\kappa)t + (1-\lambda)(1-t)}\right), & \end{aligned} \tag{5.40}$$

   where $t \in \mathbb{I}$, *is a dependence function.*

NOTE 5.6. Clearly, (1) is a special case of (2), corresponding to $\kappa = \lambda$. Note how case (2) concerns the same ideas as outlined in Subsection C.15.2 for generating asymmetric copulas — see, in particular, Propositions C.2–C.4 and Note C.1.

Below we illustrate the issue by introducing a well known family of asymmetric 2-copulas.

ILLUSTRATION 5.9 (Asymmetric Logistic family). ►

Consider the dependence function B given by Eq. (5.33), generating the Gumbel-Hougaard family of copulas. According to Proposition 5.4, this family can be "enlarged" via Eq. (5.40): for instance, the independence copula $\mathbf{\Pi}_2$, generated by the dependence function $A \equiv 1$, can be used for this purpose. The resulting asymmetric model is a three-parameter EV family of copulas, whose dependence functions are of the form

$$E(t) = (\kappa t + \lambda(1-t)) + \left(\kappa^\theta t^\theta + \lambda^\theta (1-t)^\theta\right)^{1/\theta}, \tag{5.41}$$

where $t \in \mathbb{I}$, $0 < \lambda, \kappa < 1$, and $\theta \geq 1$. According to [280], this is known as the Asymmetric Logistic family. ◄

Another interesting point is the possibility to express two well-known measures of association, i.e. the *Kendall's* τ_K and the *Spearman's* ρ_S (see Section B.2), in terms of the dependence function A [35]:

$$\tau_K(A) = \int_0^1 \frac{t(1-t)}{A(t)} \, \mathrm{d}A'(t), \tag{5.42a}$$

and

$$\rho_S(A) = 12 \int_0^1 \frac{1}{(1+A(t))^2} \, \mathrm{d}t - 3. \tag{5.42b}$$

In addition, as shown in [117], also the function $K_{\mathbf{C}^*}$ (see Theorem 3.4 and the ensuing discussion) can be calculated by using A [117]. In fact, if $\mathbf{C}^*$ is generated via Eq. (5.29), then

$$K_{\mathbf{C}^*}(t) = t - (1 - \tau_K(A))\, t \ln(t), \tag{5.43}$$

where $t \in \mathbb{I}$, and $\tau_K(A)$ is given by Eq. (5.42a).

The introduction of the dependence function in a d-dimensional environment, $d > 2$, is more complex than in the bivariate case. An interesting approach is due to Pickands [217], who uses an appropriate finite measure on a suitable simplex, as already discussed in Section 2.2 when stating Theorem 2.1. However, we shall not pursue the issue here.

5.3. TAIL DEPENDENCE

As discussed in Section 3.4, the notion of *tail dependence* offers a useful tool for investigating the dependence of extremes in multivariate distributions: we know that extreme values mainly depend on the tails of a distribution. We now show how to proceed in a d-dimensional context, $d > 2$, generalizing the bivariate approach outlined in Section 3.4 — see also [42, 96, 260].

DEFINITION 5.6 (Tail dependence (d-dimensional case)). *The tail dependence coefficients of a d-dimensional vector $\boldsymbol{X} = (X_1, \ldots, X_d)$, $d > 2$, are defined as those of its bivariate marginals. In particular, provided that the limits exist, we have, for all $1 \le i \ne j \le d$:*

$$\lambda_U^{(i,j)} = \lim_{t \to 1^-} \mathbb{P}\left\{X_j > F_j^{[-1]}(t) \mid X_i > F_i^{[-1]}(t)\right\}, \tag{5.44}$$

and

$$\lambda_L^{(i,j)} = \lim_{t \to 0^+} \mathbb{P}\left\{X_j \le F_j^{[-1]}(t) \mid X_i \le F_i^{[-1]}(t)\right\}, \tag{5.45}$$

called, respectively, upper tail dependence coefficients and lower tail dependence coefficients.

Clearly, the copula formulations given by Eqs. (3.81)–(3.83) automatically hold. As a consequence,

$$\lambda_U^{(i,j)} = \lambda_U^{(j,i)} \quad \text{and} \quad \lambda_L^{(i,j)} = \lambda_L^{(j,i)} \tag{5.46}$$

for all $1 \le i \ne j \le d$. Thus, only $\binom{d}{2} = d(d-1)/2$ upper and lower dependence coefficients influence the asymptotic behavior of $\boldsymbol{X}$. In addition, following the discussion in Note 3.4, we now concentrate on the upper tail dependence coefficients only, and use the simpler notation λ_{ij} for $\lambda_U^{(i,j)}$.

As shown in Illustration 5.1, the d-dimensional independence copula $\mathbf{\Pi}_d$, joining independent r.v.'s, is also an EVC. The interesting point is that it is the only case (see also Proposition 2.6).

PROPOSITION 5.5. *Let $\mathbf{C}^*$ be a d-dimensional EVC for which $\lambda_{ij} = 0$ for all $1 \le i \ne j \le d$. Then $\mathbf{C}^* = \mathbf{\Pi}_d$.*

As anticipated at the end of Section 3.4, if $\mathbf{C}^*$ is the limiting EVC of a given copula $\mathbf{C}$, then it has the same tail dependence coefficients as $\mathbf{C}$, as follows.

PROPOSITION 5.6. *Let $\mathbf{C} \in CDA(\mathbf{C}^*)$ be a d-dimensional copula with tail dependence coefficients λ_{ij}, $1 \le i < j \le d$. Then*

$$\lambda_{ij}^* = \lambda_{ij}, \tag{5.47}$$

where the λ_{ij}^*'s represent the tail dependence coefficients of $\mathbf{C}^*$.

As an important consequence, the CDA of a copula $\mathbf{C}$ linking pairwise asymptotically independent r.v.'s is that of the independence copula $\mathbf{\Pi}_d$, as follows.

ILLUSTRATION 5.10. ►

As shown in Section C.3 and Illustration 4.5, the expression

$$\mathbf{C}_\theta(\boldsymbol{u}) = \left(u_1^{-\theta} + \cdots + u_d^{-\theta} - (d-1)\right)^{-1/\theta},$$

with $\theta > 0$, identifies a subfamily of copulas of the Clayton family. Using Theorem 5.2 it is not difficult to show that

$$\lim_{t\to\infty} t\left(1 - \mathbf{C}_\theta\left(\boldsymbol{u}^{1/t}\right)\right) = \lim_{t\to\infty} t\left(1 - \left(u_1^{-\theta/t} + \cdots + u_d^{-\theta/t} - (d-1)\right)^{-1/\theta}\right)$$
$$= -\sum_{i=1}^{d} \ln u_i.$$

Thus, the EVC of $\mathbf{C}_\theta$ is the independence copula $\mathbf{\Pi}_d$.

Alternatively, consider the bivariate marginals of $\mathbf{C}_\theta$, that are again Clayton copulas. The direct calculation of the upper tail dependence coefficients λ_{ij}'s yields

$$\lambda_{ij} = \lim_{t\to 1^-} \frac{1 - 2t + \mathbf{C}_\theta(t,t)}{1-t} = \lim_{t\to 1^-} \frac{1 - 2t + \left(t^{-\theta} + t^{-\theta} - 1\right)^{-1/\theta}}{1-t} = 0$$

for all $1 \le i \ne j \le d$. This result should be compared with the outcome of Illustration 3.7. ◄

As already mentioned in Section 3.4, the tail dependence coefficients of a convex linear combination of copulas are simply the convex linear combinations of those of the mixing copulas. Using Eq. (5.16), it is easy to calculate the EVC of a convex linear combination of copulas, and, in turn, the corresponding tail dependence coefficients. In fact, from Proposition 5.6, we may write

$$\lambda_{ij}^* = \sum_{l=1}^{k} \lambda_l \lambda_{ij}^{(l)}, \tag{5.48}$$

for all $1 \le i \ne j \le d$, where the index l identifies the l-th copula in the mixture.

A final important point [97] concerns the relationship between the dependence function A (for the bivariate case) and the tail dependence coefficients of $\mathbf{C}^*$ (see Section 3.4). In fact, by considering the upper tail dependence coefficient λ_U, it follows that

$$\lambda_U(\mathbf{C}^*) = 2 - 2A(1/2). \tag{5.49}$$

Also, by taking into account the dependence function related to the survival copula, this result can be extended to the lower tail dependence coefficient λ_L.

APPENDIX A

SIMULATION OF COPULAS

Copulas have primary and direct applications in the simulation of dependent variables. We now present general procedures to simulate bivariate, as well as multivariate, dependent variables. Other algorithms can be found in many of the exercises proposed by [207], as well as in Appendix C.

The mathematical kernel for simulating copulas is provided by the formulas in Eq. (3.26), where conditional probabilities are written in terms of the partial derivatives of copulas. Firstly we need to introduce the notion of a *quasi-inverse* of a distribution function.

DEFINITION A.1 (Quasi-inverse). *Let F be a univariate distribution function. A quasi-inverse of F is any function $F^{[-1]}: \mathbb{I} \to \overline{\mathbb{R}}$ such that:*

1. *if $t \in \text{Ran}(F)$, then $F^{[-1]}(t)$ is any number $x \in \overline{\mathbb{R}}$ such that $F(x) = t$, i.e., for all $t \in \text{Ran}(F)$,*

$$F\left(F^{[-1]}(t)\right) = t; \tag{A.1a}$$

2. *if $t \notin \text{Ran}(F)$, then*

$$F^{[-1]}(t) = \inf\{x : F(x) \geq t\} = \sup\{x : F(x) \leq t\}. \tag{A.1b}$$

Clearly, if F is strictly increasing it has a single quasi-inverse, which equals the (ordinary) inverse function F^{-1} (or, sometimes, $F^{(-1)}$).

A.1. THE 2-DIMENSIONAL CASE

A general algorithm for generating observations (x, y) from a pair of r.v.'s (X, Y) with marginals F_X, F_Y, joint distribution F_{XY}, and 2-copula $\mathbf{C}$ is as follows. By virtue of Sklar's Theorem (see Theorem 3.1), we need only to generate a pair (u, v) of observations of r.v.'s (U, V), Uniform on $\mathbb{I}$ and having the 2-copula $\mathbf{C}$. Then, using the *Probability Integral Transform*, we transform (u, v) into (x, y), i.e.

$$\begin{cases} x = F_X^{[-1]}(u) \\ y = F_Y^{[-1]}(v) \end{cases}. \tag{A.2}$$

In order to generate the pair (u, v) we may consider the conditional distribution of V given the event $\{U = u\}$:

$$c_u(v) = \mathbb{P}\{V \le v \mid U = u\} = \frac{\partial}{\partial u}\mathbf{C}(u, v). \tag{A.3}$$

A possible algorithm is as follows.

1. Generate independent variates u, t Uniform on $\mathbb{I}$.
2. Set $v = c_u^{[-1]}(t)$.

The desired pair is then (u, v). For other algorithms see [71, 156].

ILLUSTRATION A.1. ►

The Frank and the Gumbel-Hougaard families of 2-copulas are widely used in applications (e.g., in hydrology — see [66, 253, 254, 90, 67, 255]). As clarified in Appendix C (see Section C.1 and Section C.2, respectively), Frank's copulas may model both negatively and positively associated r.v.'s, whereas Gumbel-Hougaard's copulas only describe positive forms of association. In addition, Frank's copulas do not feature tail dependence, as opposed to Gumbel-Hougaard's copulas.

We show shortly several comparisons between simulated (U, V) samples extracted from Frank's and Gumbel-Hougaard's 2-copulas, where both U and V are Uniform on $\mathbb{I}$. In all cases the sample size is $N = 200$, and the random generator is the same for both samples, as well as the value of Kendall's τ_K. This gives us the possibility of checking how differently these copulas affect the joint behavior of r.v.'s subjected to their action.

The first example is given in Figure A.1. Here $\tau_K \approx 0.001$, i.e. U and V are very weakly positively associated (practically, they are independent, that is $\mathbf{C} \approx \mathbf{\Pi}_2$ for both families). Indeed, the two plots are almost identical, and the points are uniformly sparse within the unit square.

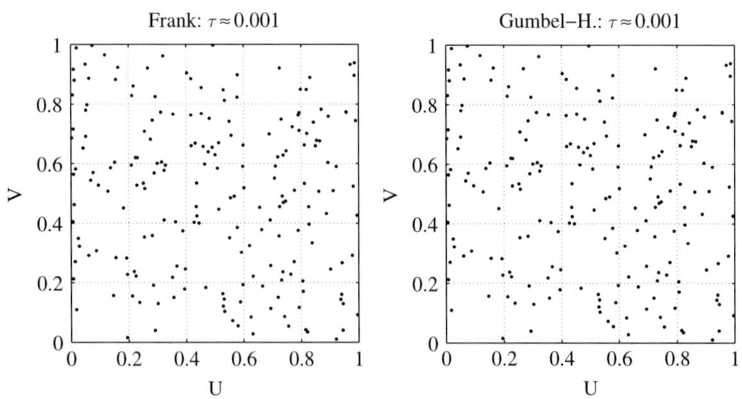

Figure A.1. Comparison between simulated samples extracted from Frank's and Gumbel-Hougaard's 2-copulas. Here $\tau_K \approx 0.001$

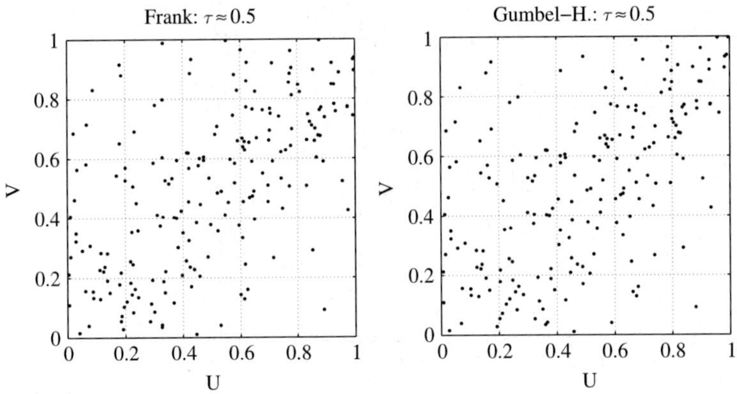

Figure A.2. Comparison between simulated samples extracted from Frank's and Gumbel-Hougaard's 2-copulas. Here $\tau_K \approx 0.5$

The second example is given in Figure A.2. Here $\tau_K \approx 0.5$, i.e. U and V are moderately positively associated. Actually, the points tend to dispose themselves along the main diagonal. The two plots are still quite similar. The main differences are evident only in the upper right corner, where the Gumbel-Hougaard copula tends to organize the points in a different way from that of the Frank copula.

The third example is given in Figure A.3. Here $\tau_K \approx 0.95$, i.e. U and V are strongly associated positively. Actually, the points clearly tend to dispose themselves along the main diagonal. The two plots are still quite similar. The main differences are evident only in the extreme upper right corner: the Gumbel-Hougaard copula tends to concentrate the points, whereas the Frank copula seems to make the points more sparse.

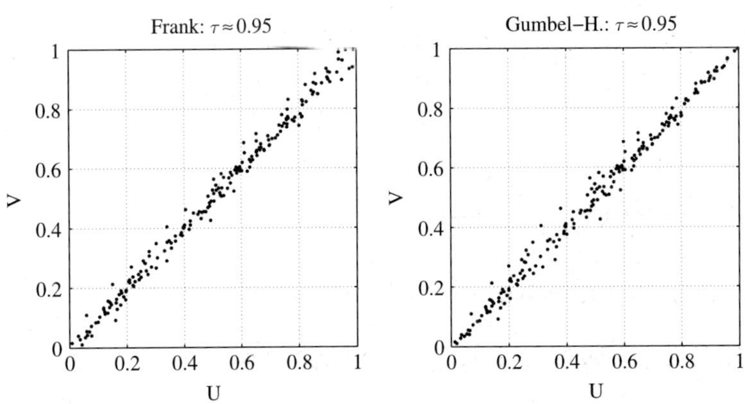

Figure A.3. Comparison between simulated samples extracted from Frank's and Gumbel-Hougaard's 2-copulas. Here $\tau_K \approx 0.95$

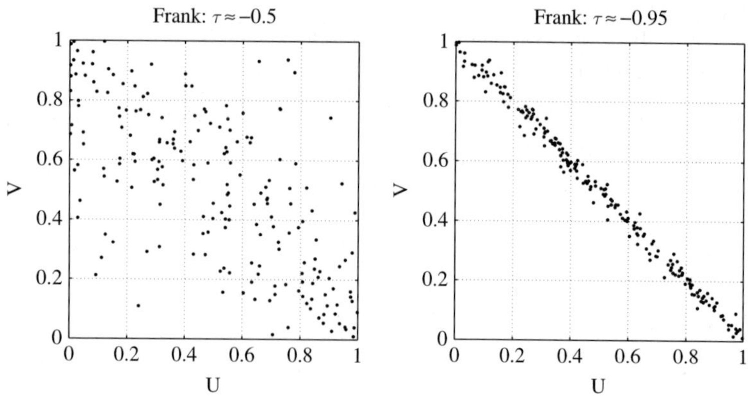

Figure A.4. Simulated samples extracted from Frank's 2-copulas: (*left*) $\tau_K \approx -0.5$; (*right*) $\tau_K \approx -0.95$

The last example concerns the negative association, and thus only samples extracted from the Frank family can be shown. In in Figure A.4 we show two simulations where $\tau_K \approx -0.5$ (corresponding to a moderate negative association), and $\tau_K \approx -0.95$ (corresponding to a strong negative association). In both cases the points clearly tend to disperse along the secondary diagonal of the unit square. As τ_K decreases, the observations get more concentrated along this line.

In Figures A.1–A.4 we show how the behavior of a bivariate sample may change when the degree of association between the variables considered (measured, e.g., by Kendall's τ_K) ranges over its domain, i.e. $[-1, +1]$. As an important conclusion of practical relevance, we must stress that visual comparisons are not sufficiente to decide which copula best describes the behavior of the available data. For instance, the plots for $\tau_K > 0$ show that Frank's and Gumbel-Hougaard's 2-copulas apparently behave in a similar way. Unfortunately, this is also a widespread practice as seen in the literature. On the contrary, only specific statistical tests (e.g., concerning tail dependence) may help in deciding whether or not a family of copulas should be considered for a given application. ◀

A.2. THE GENERAL CASE

Let F be a multivariate distribution with continuous marginals $F_1, \ldots, F_d$, and suppose that F can be expressed in a unique way via a d-copula $\mathbf{C}$ by virtue of Sklar's Theorem (see Theorem 4.2). In order to simulate a vector $(X_1, \ldots, X_d) \sim F$, it is sufficient to simulate a vector $(U_1, \ldots, U_d) \sim \mathbf{C}$, where the r.v.'s U_i's are Uniform on $\mathbb{I}$. By using Sklar's Theorem and the *Probability Integral Transform*

$$U_i = F_i(X_i) \Longleftrightarrow X_i = F_i^{[-1]}(U_i), \tag{A.4}$$

where $i = 1, \ldots, d$, the r.v.'s X_i's have marginal distributions F_i's, and joint distribution F. We now show how to simulate a sample extracted from $\mathbf{C}$. For the sake of simplicity, we assume that $\mathbf{C}$ is absolutely continuous.

1. To simulate the first variable U_1, it suffices to sample from a r.v. U_1' Uniform on $\mathbb{I}$. Let us call u_1 the simulated sample.
2. To obtain a sample u_2 from U_2, consistent with the previously sampled u_1, we need to know the distribution of U_2 conditional on the event $\{U_1 = u_1\}$. Let us denote this law by $\mathrm{G}_2(\cdot \mid u_1)$, given by:

$$\begin{aligned} \mathrm{G}_2(u_2 \mid u_1) &= \mathbf{P}(U_2 \le u_2 \mid U_1 = u_1) \\ &= \frac{\partial_{u_1} \mathbf{C}\,(u_1, u_2, 1, \ldots, 1)}{\partial_{u_1} \mathbf{C}\,(u_1, 1, \ldots, 1)} \\ &= \partial_{u_1} \mathbf{C}\,(u_1, u_2, 1, \ldots, 1). \end{aligned} \tag{A.5}$$

   Then we take $u_2 = \mathrm{G}_2^{-1}(u_2' \mid u_1)$, where u_2' is the realization of a r.v. U_2' Uniform on $\mathbb{I}$, that is independent of U_1'.
3. In general, to simulate a sample u_k from U_k, consistent with the previously sampled $u_1, \ldots, u_{k-1}$, we need to know the distribution of U_k conditional on the events $\{U_1 = u_1, \ldots, U_{k-1} = u_{k-1}\}$. Let us denote this law by $\mathrm{G}_k(\cdot \mid u_1, \ldots, u_{k-1})$, given by:

$$\begin{aligned} &\mathrm{G}_k(u_k \mid u_1, \ldots, u_{k-1}) \\ &\quad = \mathbb{P}\{U_k \le u_k \mid U_1 = u_1, \ldots, U_{k-1} = u_{k-1}\} \\ &\quad = \frac{\partial_{u_1,\ldots,u_{k-1}} \mathbf{C}\,(u_1, \ldots, u_k, 1, \ldots, 1)}{\partial_{u_1,\ldots,u_{k-1}} \mathbf{C}\,(u_1, \ldots, u_{k-1}, 1, \ldots, 1)}. \end{aligned} \tag{A.6}$$

   Then we take $u_k = \mathrm{G}_k^{-1}(u_k' \mid u_1, \ldots, u_{k-1})$, where u_k' is the realization of a r.v. U_k' Uniform on $\mathbb{I}$, that is independent of $U_1', \ldots, U_{k-1}'$.

Using the *Probability Integral Transform*, it is easy to generate the sample $(x_1, \ldots, x_d)$ extracted from F:

$$(x_1, \ldots, x_d) = \left(F_1^{-1}(u_1), \ldots, F_d^{-1}(u_d)\right). \tag{A.7}$$

Below we show how the "conditional" approach to the construction of multivariate copulas, introduced in Section 4.3, is also well suited for the simulation of multivariate vectors, for this task can be made simply by calculating partial derivatives of the copulas of interest.

ILLUSTRATION A.2 (Sea storm simulation). ►

As shown in Illustration 4.9, in [68] there is a characterization of the sea storm dynamics involving four variables: the significant wave height H, the storm duration D, the waiting time I between two successive "active" storm phases, and the storm wave direction A.

In order to simulate the full sea-state dynamics, one needs the 4-copula $\mathbf{C}_{HDIA}$ associated with F_{HDIA}. The algorithm explained in this Section for simulating multivariate data using copulas has a "nested" structure. An initial variable (e.g., H) is first simulated. Then, by using the information provided by $\mathbf{C}_{HD}$, the variable D can be simulated consistently with the previous realization of H. As the next step, the variable I can be simulated by exploiting $\mathbf{C}_{HDI}$, and the pair (H, D) just simulated. Note that all the three 2-copulas $\mathbf{C}_{HD}, \mathbf{C}_{HI}, \mathbf{C}_{DI}$ are required to carry out this step. Finally, the variable A can be simulated by using $\mathbf{C}_{HDIA}$, and the triple (H, D, A) just simulated. Here, all the six 2-copulas linking the four variables H, D, I, A are needed.

If $\mathbf{C}_{HDIA}$ is constructed as explained in Subsections 4.3.1–4.3.2, then an important result follows. In fact, only the knowledge of the four one-dimensional distributions F_H, F_D, F_I, F_A, and of the six 2-copulas $\mathbf{C}_{HD}$, $\mathbf{C}_{HI}$, $\mathbf{C}_{HA}$, $\mathbf{C}_{DI}$, $\mathbf{C}_{DA}$, $\mathbf{C}_{IA}$, is required to carry out the simulation. Clearly, this may represent a great advantage with respect to the estimation of parameters. In addition, the calculations greatly simplify, and the integral representations disappear: eventually, only trivial composite functions of partial derivatives of 2-copulas need to be evaluated. In turn, the numerical simulation of sea-states is quite fast.

Below we outline a step-by-step procedure to simulate a sequence of sea-states, assuming that the construction of the underlying model follows the "conditional" approach outlined above. We simplify and clarify the presentation at the expense of some abuse of mathematical notation. Here D_1 and D_2 denote, respectively, the partial derivatives with respect to the first and the second component. An obvious point is as follows: if the variables were simulated in a different order, then only the corresponding copulas should be changed, while the algorithm remains the same.

1. **Simulation of H.**
   Let U_1 be Uniform on $\mathbb{I}$. In order to simulate H set
   $$H = F_H^{-1}(U_1). \tag{A.8}$$

2. **Simulation of D.**
   Let U_2 be Uniform on $\mathbb{I}$ and independent of U_1. In order to simulate D consistently with H, the function
   $$\mathrm{G}_2(u_2 \mid u_1) = \partial_{u_1} \mathbf{C}_{HD}(u_1, u_2) \tag{A.9}$$
   must be calculated first. Then set
   $$D = F_D^{-1}(\mathrm{G}_2^{-1}(U_2 \mid U_1)). \tag{A.10}$$

3. **Simulation of I.**
Let U_3 be Uniform on $\mathbb{I}$ and independent of $\{U_1, U_2\}$. In order to simulate I consistently with (H, D), the function

$$G_3(u_3 \mid u_1, u_2) = \frac{\partial_{u_1,u_2} \mathbf{C}_{HDI}(u_1, u_2, u_3)}{\partial_{u_1,u_2} \mathbf{C}_{HD}(u_1, u_2)} \tag{A.11}$$

must be calculated first, where the numerator ϕ equals

$$\phi = \partial_{u_1} \mathbf{C}_{HI}\left(D_2 \mathbf{C}_{HD}(u_1, u_2), D_1 \mathbf{C}_{DI}(u_2, u_3)\right). \tag{A.12}$$

In fact, according to Eq. (4.22), ϕ can be written as

$$\begin{aligned} \partial_{u_1,u_2} \mathbf{C}_{HDI}(u_1, u_2, u_3) &= \partial_{u_1,u_2} \int_0^{u_2} \mathbf{C}_{HI}(\ldots, \ldots)\, dx \\ &= \partial_{u_1}\left(\partial_{u_2} \int_0^{u_2} \mathbf{C}_{HI}(\ldots, \ldots)\, dx\right), \end{aligned} \tag{A.13}$$

leading to the expression of ϕ shown in Eq. (A.12). Moreover, by taking the partial derivative with respect to u_1, we have the following simplification:

$$\begin{aligned} \phi = D_1 \mathbf{C}_{HI}\left(D_2 \mathbf{C}_{HD}(u_1, u_2), D_1 \mathbf{C}_{DI}(u_2, u_3)\right) \cdot \\ \cdot D_{12} \mathbf{C}_{HD}(u_1, u_2), \end{aligned} \tag{A.14}$$

and the right-most term equals the denominator in Eq. (A.11), which then cancels out. Then set

$$I = F_I^{-1}(G_3^{-1}(U_3 \mid U_1, U_2)). \tag{A.15}$$

4. **Simulation of A.**
Let U_4 be Uniform on $\mathbb{I}$ and independent of $\{U_1, U_2, U_3\}$. In order to simulate A consistently with (H, D, I), the function

$$G_4(u_4 \mid u_1, u_2, u_3) = \frac{\partial_{u_1,u_2,u_3} \mathbf{C}_{HDIA}(u_1, u_2, u_3, u_4)}{\partial_{u_1,u_2,u_3} \mathbf{C}_{HDI}(u_1, u_2, u_3)} \tag{A.16}$$

must be calculated first. Here the denominator equals $\partial_{u_3}\phi$, whereas the numerator ψ can be calculated as in Eq. (A.13), and is given by

$$\psi = D_1 \mathbf{C}_{HA}(\psi_1, \psi_2) \cdot \partial_{u_3}\phi, \tag{A.17}$$

where

$$\psi_1 = D_2 \mathbf{C}_{HI}\left(D_2 \mathbf{C}_{HD}(u_1, u_2), D_1 \mathbf{C}_{DI}(u_2, u_3)\right), \tag{A.18a}$$

$$\psi_2 = \mathrm{D}_2\,\mathbf{C}_{AI}\left(\mathrm{D}_2\,\mathbf{C}_{AD}\left(u_4, u_2\right), \mathrm{D}_1\,\mathbf{C}_{DI}\left(u_2, u_3\right)\right). \tag{A.18b}$$

As in the simulation of I, cancellations occur, and only a simplified version of the numerator remains in the expression of G_4. Then set

$$A = F_A^{-1}(\mathrm{G}_4^{-1}(U_4 \mid U_1, U_2, U_3)). \tag{A.19}$$

Note how the whole simulation procedure simply reduces to the calculation of partial derivatives of 2-copulas. Also, in some cases, the inverse functions used above can be calculated analytically; otherwise, a simple numerical search can be performed.

As an illustration, in Figure A.5 we show the same comparisons as those presented in Figure 4.2, but using now a data set of about 20,000 simulated sea storms. As in Illustration 4.9, the three variables H, D, and A are considered, and the 2-copula used for the pair (H, D) belongs to the Ali-Mikhail-Haq family (see Section C.4), while those used for (H, A) and (D, A) belong to the Frank family (see Section C.1). As expected, the agreement is good in all cases.

As a further illustration, the same comparison as just given is shown in Figure A.6, where the variables D, I, and A are considered. The 2-copula used for the pairs (D, I) and (I, A) belong to the Gumbel-Hougaard family (see Section C.2), while that used for (D, A) belongs to the Frank family (see Section C.1). Again, the fit is good in all cases. These plots should be compared to those in Figure 4.3, where a much smaller data set is used.

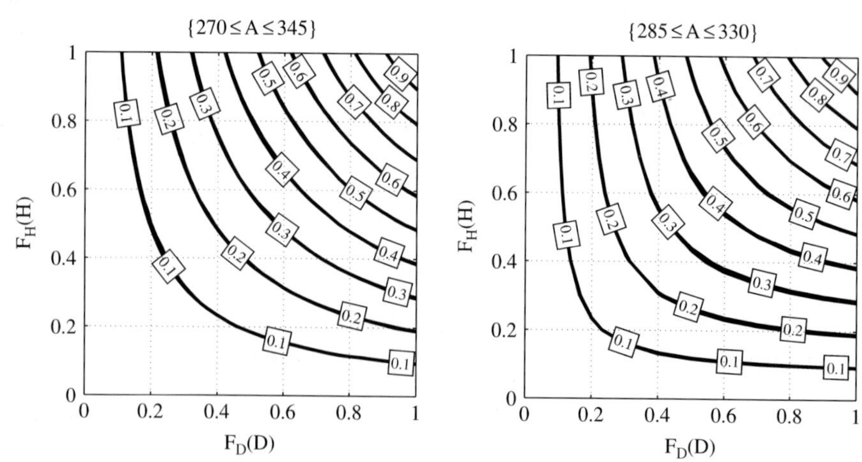

Figure A.5. Comparison between the level curves of the theoretical copulas fitted to the simulated (D, A, H) observations (*thin* lines), and those of the empirical copulas constructed using the same data (*thick* lines). The probability levels are as indicated (*labels*), as well as the conditioning events $\{\bullet \le A \le \bullet\}$

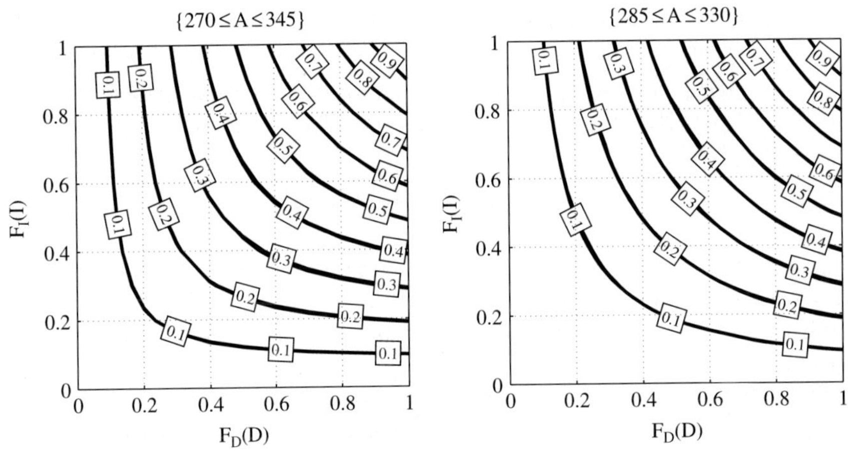

Figure A.6. Comparison between the level curves of the theoretical copulas fitted to the simulated (D, A, I) observations (*thin* lines), and those of the empirical copulas constructed using the same data (*thick* lines). The probability levels are as indicated (*labels*), as well as the conditioning events $\{\bullet \leq A \leq \bullet\}$

As a difference with the results shown in Illustration 4.9, the figures presented illustrate clearly how the situation may improve if sufficient data are made available when working with multivariate distributions. Indeed, in our examples the graphs of the theoretical copulas cannot be distinguished from those of the empirical ones. For these reasons, the possibility offered by copulas to simulate multidimensional vectors easily is invaluable in many practical applications. ◄

APPENDIX B

DEPENDENCE

(Written by Fabrizio Durante — Department of Knowledge-Based Mathematical Systems, Johannes Kepler University, Linz (Austria))

Dependence relations between random variables is one of the most studied subjects in Probability and Statistics. Unless specific assumptions are made about the dependence, no meaningful statistical model can be constructed. There are several ways to discuss and measure the dependence between random variables: valuable sources of information are [207, 155, 75], and references therein.

The aim of this Appendix is to collect and clarify the essential ideas about dependence, by emphasizing the role that copulas play in this context. In Section B.1 we present the basic concepts of dependence, and introduce the measures of dependence. In Section B.2 we present the measures of association for a pair of r.v.'s, and, in particular, we restrict ourselves to consider those measures (such as Kendall's τ_K and Spearman ρ_S) which provide information about a special form of dependence known as *concordance*.

B.1. BIVARIATE CONCEPTS OF DEPENDENCE

In this Section we review some bivariate dependence concepts for a pair of continuous r.v.'s X and Y. Wherever possible, we characterize these properties by using their copula $\mathbf{C}$. Analogous properties can be given in the d-variate case, $d \geq 3$. We refer the reader to [155, 75, 207].

B.1.1 Quadrant Dependence

The notion of *quadrant dependence* is important in applications, for risk assessment and reliability analysis.

DEFINITION B.1 (Quadrant dependence). *Let X and Y be a pair of continuous r.v.'s with joint distribution function F_{XY} and marginals F_X, F_Y.*

1. *X and Y are positively quadrant dependent (briefly, PQD) if*

$$\forall (x, y) \in \mathbb{R}^2 \quad \mathbb{P}\{X \leq x, Y \leq y\} \geq \mathbb{P}\{X \leq x\}\,\mathbb{P}\{Y \leq y\}. \tag{B.1}$$

2. *X and Y are negatively quadrant dependent (briefly, NQD) if*

$$\forall(x, y) \in \mathbb{R}^2 \quad \mathbb{P}\{X \leq x, Y \leq y\} \leq \mathbb{P}\{X \leq x\}\mathbb{P}\{Y \leq y\}. \tag{B.2}$$

Intuitively, X and Y are PQD if the probability that they are simultaneously large or simultaneously small is at least as great as it would be if they were independent.

In terms of distribution functions, the notion of quadrant dependence can be expressed as follows:

1. X and Y are positively quadrant dependent if

$$\forall(x, y) \in \mathbb{R}^2 \quad F_{XY}(x, y) \geq F_X(x)F_Y(y).$$

2. X and Y are negatively quadrant dependent if

$$\forall(x, y) \in \mathbb{R}^2 \quad F_{XY}(x, y) \leq F_X(x)F_Y(y).$$

Note that the quadrant dependence properties are invariant under strictly increasing transformations and, hence, they can be easily expressed in terms of the copula $\mathbf{C}$ of X and Y:

1. X and Y are PQD if, and only if,

$$\mathbf{C}(u, v) \geq \mathbf{\Pi}_2(u, v)$$

for all $(u, v) \in \mathbb{I}^2$.

2. X and Y are NQD if, and only if

$$\mathbf{C}(u, v) \leq \mathbf{\Pi}_2(u, v)$$

for all $(u, v) \in \mathbb{I}^2$.

Intuitively, X and Y are PQD if the graph of their copula $\mathbf{C}$ lies on or above the graph of the independence copula $\mathbf{\Pi}_2$. By using Proposition 3.2 and Proposition 3.3, we can easily prove that, if (X, Y) is PQD, then $(-X, -Y)$ is PQD, and $(-X, Y)$ and $(X, -Y)$ are NQD. The PQD property of X and Y can also be characterized in terms of covariance between X and Y (if it exists), as shown in [176] — see also Definition 2.5, where the stronger notion of (positively) associated r.v.'s was introduced, and Theorem 5.5.

PROPOSITION B.1. *Two r.v.'s X and Y are PQD if, and only if, the covariance* $\mathbb{C}(f(X), g(Y)) \geq 0$ *for all increasing functions f and g for which the expectations* $\mathbb{E}(f(X))$, $\mathbb{E}(g(Y))$, *and* $\mathbb{E}(f(X)g(Y))$ *exist.*

The notion of *orthant dependence* was introduced in Definition 2.4 as a multi-dimensional generalization of that of quadrant dependence. In terms of d-variate copulas we may rewrite Eqs. (2.27)–(2.28) as, respectively,

$$\mathbf{C}(u_1, \ldots, u_d) \geq \mathbf{\Pi}_d(u_1, \ldots, u_d) = u_1 \cdots u_d \tag{B.3}$$

and

$$\widetilde{\mathbf{C}}(u_1, \ldots, u_d) \geq (1-u_1) \cdots (1-u_d) \tag{B.4}$$

for all $\boldsymbol{u} \in \mathbb{I}^d$, where $\widetilde{\mathbf{C}}$ denotes the d-dimensional joint survival function corresponding to $\mathbf{C}$ (see also Theorem 5.6).

B.1.2 Tail Monotonicity

The notion of *tail monotonicity* is important in applications, for risk assessment and reliability analysis.

DEFINITION B.2 (Tail monotonicity). *Let X and Y be a pair of continuous r.v.'s.*

1. *Y is left tail decreasing in X (briefly, LTD($Y \mid X$)) if*

$$\mathbb{P}\{Y \leq y \mid X \leq x\} \quad \textit{is decreasing in } x \tag{B.5}$$

for all y.

2. *X is left tail decreasing in Y (briefly, LTD($X \mid Y$)) if*

$$\mathbb{P}\{X \leq x \mid Y \leq y\} \quad \textit{is decreasing in } y \tag{B.6}$$

for all x.

3. *Y is right tail increasing in X (briefly, RTI($Y \mid X$)) if*

$$\mathbb{P}\{Y > y \mid X > x\} \quad \textit{is increasing in } x \tag{B.7}$$

for all y.

4. *X is right tail increasing in Y (briefly, RTI($X \mid Y$)) if*

$$\mathbb{P}\{X > x \mid Y > y\} \quad \textit{is increasing in } y \tag{B.8}$$

for all x.

Intuitively, LTD($Y \mid X$) means that Y is more likely to take on smaller values when X decreases. Analogously, RTI($Y \mid X$) means that Y is more likely to take on larger values when X increases. Note that, if X and Y satisfy each of the four properties of tail monotonicity, then X and Y are PQD.

In terms of the copula $\mathbf{C}$ of X and Y, the above properties have the following characterization:

1. LTD($Y \mid X$) if, and only if, for every $v \in \mathbb{I}$,

$$u \mapsto \frac{\mathbf{C}(u,v)}{u} \quad \text{is decreasing.}$$

2. LTD($X \mid Y$) if, and only if, for every $u \in \mathbb{I}$,

$$v \mapsto \frac{\mathbf{C}(u,v)}{v} \quad \text{is decreasing.}$$

3. RTI($Y \mid X$) if, and only if, for every $v \in \mathbb{I}$,

$$u \mapsto \frac{v - \mathbf{C}(u,v)}{1-u} \quad \text{is decreasing.}$$

4. RTI($Y \mid X$) if, and only if, for every $u \in \mathbb{I}$,

$$v \mapsto \frac{u - \mathbf{C}(u,v)}{1-v} \quad \text{is decreasing.}$$

In terms of partial derivatives of $\mathbf{C}$, the above conditions can be also expressed in the following forms:

1. LTD($Y \mid X$) if, and only if, for every $v \in \mathbb{I}$,

$$\frac{\partial \mathbf{C}(u,v)}{\partial u} \leq \frac{\mathbf{C}(u,v)}{u}$$

for almost all $u \in \mathbb{I}$.

2. LTD($X \mid Y$) if, and only if, for every $u \in \mathbb{I}$,

$$\frac{\partial \mathbf{C}(u,v)}{\partial v} \leq \frac{\mathbf{C}(u,v)}{v}$$

for almost all $v \in \mathbb{I}$.

3. RTI($Y \mid X$) if, and only if, for every $v \in \mathbb{I}$,

$$\frac{\partial \mathbf{C}(u,v)}{\partial u} \geq \frac{v - \mathbf{C}(u,v)}{1-u}$$

for almost all $u \in \mathbb{I}$.

4. RTI($X \mid Y$) if, and only if, for every $u \in \mathbb{I}$,

$$\frac{\partial \mathbf{C}(u,v)}{\partial v} \geq \frac{u - \mathbf{C}(u,v)}{1-v}$$

for almost all $v \in \mathbb{I}$.

B.1.3 Stochastic Monotonicity

The notion of *stochastic monotonicity* is important in applications, for risk assessment and reliability analysis.

DEFINITION B.3 (Stochastic monotonicity). *Let X and Y be a pair of continuous r.v.'s.*

1. *Y is stochastically increasing in X (briefly, SI($Y \mid X$)) if, and only if,*

$$x \mapsto \mathbb{P}\{Y > y \mid X = x\} \tag{B.9}$$

is increasing for all y.

2. *X is stochastically increasing in Y (briefly, SI($X \mid Y$)) if, and only if,*

$$y \mapsto \mathbb{P}\{X > x \mid Y = y\} \tag{B.10}$$

is increasing for all x.

3. *Y is stochastically decreasing in X (briefly, SD($Y \mid X$)) if, and only if,*

$$x \mapsto \mathbb{P}\{Y > y \mid X = x\} \tag{B.11}$$

is decreasing for all y.

4. *X is stochastically decreasing in Y (briefly, SD($X \mid Y$)) if, and only if,*

$$y \mapsto \mathbb{P}\{X > x \mid Y = y\} \tag{B.12}$$

is decreasing for all x.

Intuitively, SI($Y \mid X$) means that Y is more likely to take on larger values as X increases. In terms of the copula $\mathbf{C}$ of X and Y, the above properties have the following characterization:

1. SI($Y \mid X$) if, and only if, $u \mapsto \mathbf{C}(u, v)$ is concave for every $v \in \mathbb{I}$.
2. SI($X \mid Y$) if, and only if, $v \mapsto \mathbf{C}(u, v)$ is concave for every $u \in \mathbb{I}$.
3. SD(Y|X) if, and only if, $u \mapsto \mathbf{C}(u, v)$ is convex for every $v \in \mathbb{I}$.
4. SD(X|Y) if, and only if, $v \mapsto \mathbf{C}(u, v)$ is convex for every $u \in \mathbb{I}$.

Note that, if X and Y are r.v.'s such that SI($Y \mid X$), then LTD($Y \mid X$) and RTI($Y \mid X$) follow. Analogously, SI($X \mid Y$) implies LTD($X \mid Y$) and RTI($X \mid Y$). In particular, if X and Y are r.v.'s such that their joint distribution function H is a bivariate EV distribution, then SI($Y \mid X$) and SI($X \mid Y$) follow [106].

B.1.4 Corner Set Monotonicity

The notion of *corner set monotonicity* is important in applications, for risk assessment and reliability analysis.

DEFINITION B.4 (Corner set monotonicity). *Let X and Y be a pair of continuous r.v.'s.*

1. *X and Y are left corner set decreasing (briefly, LCSD(X, Y)) if, and only if, for all x and y,*

$$\mathbb{P}\{X \leq x, Y \leq y \mid X \leqslant x', Y \leq y'\} \tag{B.13}$$

is decreasing in x' and in y'.

2. *X and Y are left corner set increasing (briefly, LCSI(X, Y)) if, and only if, for all x and y,*

$$\mathbb{P}\{X \leq x, Y \leq y \mid X \leq x', Y \leq y'\} \tag{B.14}$$

is increasing in x' and in y'.

3. *X and Y are right corner set increasing (briefly, RCSI(X, Y)) if, and only if, for all x and y,*

$$\mathbb{P}\{X > x, Y > y \mid X > x', Y > y'\} \tag{B.15}$$

is increasing in x' and in y'.

4. *X and Y are right corner set decreasing (briefly, RCSD(X, Y)) if, and only if, for all x and y,*

$$\mathbb{P}\{X > x, Y > y \mid X > x', Y > y'\} \tag{B.16}$$

is decreasing in x' and in y'.

Note that, if LCSD(X, Y), then LTD$(Y \mid X)$ and LTD$(X \mid Y)$ follow. Analogously, if RCSI(X, Y), then RTI$(Y \mid X)$ and RTI$(X \mid Y)$ follow. In terms of the copula $\mathbf{C}$ of X and Y, and of the corresponding survival copula $\overline{\mathbf{C}}$, these properties have the following characterization:

1. LCSD(X, Y) if, and only if, $\mathbf{C}$ is TP$_2$, i.e. for every u, u', v, v' in $\mathbb{I}$, $u \leq u'$, $v \leq v'$,

$$\mathbf{C}(u, v)\,\mathbf{C}(u', v') \geq \mathbf{C}(u, v')\,\mathbf{C}(u', v);$$

2. RCSI(X, Y) if, and only if, $\overline{\mathbf{C}}$ is TP$_2$, i.e. for every u, u', v, v' in $\mathbb{I}$, $u \leq u'$, $v \leq v'$,

$$\overline{\mathbf{C}}(u, v)\,\overline{\mathbf{C}}(u', v') \geq \overline{\mathbf{C}}(u, v')\,\overline{\mathbf{C}}(u', v).$$

In Table B.1 we summarize the relationships between the positive dependence concepts illustrated above [207].

Table B.1. Relationships between positive dependence concepts

SI($Y \mid X$)	$\Longrightarrow$	RTI($Y \mid X$)	$\Longleftarrow$	RCSI(X, Y)
$\Downarrow$		$\Downarrow$		$\Downarrow$
LTD($Y \mid X$)	$\Longrightarrow$	PQD(X, Y)	$\Longleftarrow$	RTI($X \mid Y$)
$\Uparrow$		$\Uparrow$		$\Uparrow$
LCSD(X, Y)	$\Longrightarrow$	LTD($X \mid Y$)	$\Longleftarrow$	SI($X \mid Y$)

B.1.5 Dependence Orderings

After the introduction of some dependence concepts, it is natural to ask whether one bivariate distribution function is more dependent than another, according to some prescribed dependence concept. Comparisons of this type are made by introducing a partial ordering in the set of all bivariate distribution functions having the same marginals (and, hence, involving the concept of copula). The most common dependence ordering is the *concordance ordering* (also called *PQD ordering*).

DEFINITION B.5 (Concordance ordering). *Let H and H' be continuous bivariate distribution functions with copulas $\mathbf{C}$ and $\mathbf{C}'$, respectively, and the same marginals F and G. H' is said to be more concordant (or more PQD) than H if*

$$\forall (x, y) \in \mathbb{R}^2 \qquad H(x, y) \leq H'(x, y), \tag{B.17a}$$

or, equivalently,

$$\forall (u, v) \in \mathbb{I}^2 \qquad \mathbf{C}(u, v) \leq \mathbf{C}'(u, v). \tag{B.17b}$$

If H' is more concordant than H, we write $H \prec_{\mathbf{C}} H'$ (or simply $H \prec H'$).

If we consider a family of copulas $\{\mathbf{C}_\theta\}$, indexed by a parameter θ belonging to an interval of $\mathbb{R}$, we say that $\mathbf{C}_\theta$ is *positively ordered* if $\theta_1 \leq \theta_2$ implies $\mathbf{C}_{\theta_1} \prec \mathbf{C}_{\theta_2}$ in the concordance ordering.

A dependence ordering related to LTD and RTI concepts is given in [7]. An ordering based on SI concepts is presented in [155], where an ordering based on the TP_2 notion is also discussed.

B.1.6 Measure of Dependence

There are several ways to discuss and measure the dependence between random variables. Intuitively, a measure of dependence indicates how closely two r.v.'s X and Y are related, with extremes at mutual independence and (monotone) dependence. Most importantly, some of these measures are *scale-invariant*, i.e. they remain unchanged under strictly increasing transformations of the variables of interest. Thus, from Proposition 3.2, they are expressible in terms of the copula linking these variables [265].

Practitioners should primarily consider dependence measures that depend only upon the copula of the underlying random vector. Unfortunately, this is not true for the often used Pearson's linear correlation coefficient ρ_P, that strongly depends upon the marginal laws (especially outside the framework of elliptically contoured distributions — see Section C.11; for a discussion about ρ_P see [85]).

In 1959, A. Rényi [231] proposed the following set of axioms for a *measure of dependence*. Here we outline a slightly modified version of them.

DEFINITION B.6 (Measure of dependence). *A numerical measure δ between two continuous r.v.'s X and Y with copula $\mathbf{C}$ is a measure of dependence if it satisfies the following properties:*

1. *δ is defined for every pair (X, Y) of continuous r.v.'s;*
2. *$0 \leq \delta_{X,Y} \leq 1$;*
3. *$\delta_{X,Y} = \delta_{Y,X}$;*
4. *$\delta_{X,Y} = 0$ if, and only if, X and Y are independent;*
5. *$\delta_{X,Y} = 1$ if, and only if, each of X and Y is almost surely a strictly monotone function of the other;*
6. *if α and β are almost surely strictly monotone functions on* Ran X *and* Ran Y, *respectively, then $\delta_{\alpha(X),\beta(Y)} = \delta_{X,Y}$;*
7. *if $\{(X_n, Y_n)\}$, $n \in \mathbb{N}$, is a sequence of continuous r.v.'s with copula $\mathbf{C}_n$, and if $\{\mathbf{C}_n\}$ converges pointwise to a copula $\mathbf{C}$, then*

$$\lim_{n\to\infty} \delta_{X_n,Y_n} = \delta_{X,Y}.$$

Note that, in the above sense, Pearson's linear correlation coefficient ρ_P is not a measure of dependence: in fact, $\rho_P(X, Y) = 0$ does not imply that X and Y are independent. A measure of dependence is given, instead, by the *maximal correlation coefficient* ρ_P^* defined by:

$$\rho_P^* = \sup_{f,g} \rho_P(f(X), g(Y)), \tag{B.18}$$

where the supremum is taken over all Borel functions f and g for which the correlation $\rho_P(f(X), g(Y))$ is well defined. However, this measure is too often equal to one, and cannot be effectively calculated.

Another example is given by the *Schweizer-Wolff measure of dependence* [265] defined, for continuous r.v.'s X and Y with copula $\mathbf{C}$, by

$$\delta_{\mathbf{C}} = 12 \iint_{\mathbb{I}^2} | \mathbf{C}(u, v) - uv | \, du\, dv. \tag{B.19}$$

We anticipate here that, if X and Y are PQD, then $\delta_{\mathbf{C}} = \rho_S$, and if X and Y are NQD, then $\delta_{\mathbf{C}} = -\rho_S$, where ρ_S denotes the Spearman's coefficient (see Subsection B.2.3). More details can be found in [207].

B.2. MEASURES OF ASSOCIATION

A numerical *measure of association* is a statistical summary of the degree of relationship between variables. For the ease of comparison, coefficients of association are usually constructed to vary between -1 and $+1$. Their values increase as the strength of the relationship increases, with a $+1$ (or -1) value when there is perfect positive (or negative) association. Each coefficient of association measures a special type of relationship: for instance, Pearson's product-moment correlation coefficient ρ_P measures the amount of linear relationship.

The most widely known (and used), scale-invariant, measures of association are the *Kendall's* τ_K and the *Spearman's* ρ_S, both of which measure a form of dependence known as *concordance*. These two measures also play an important role in applications, since the practical fit of a copula to the available data is often carried out via the estimate of τ_K or ρ_S (see Chapter 3 and Appendix C). Note that both τ_K and ρ_S always exist (being based on the ranks), whereas the existence of other standard measures (such as ρ_P) may depend upon that of the second-order moments of the variables of interest — and is not guaranteed, e.g., for heavy tailed r.v.'s (see, e.g., the discussion in [67, 255]).

B.2.1 Measures of Concordance

Roughly speaking, two r.v.'s are *concordant* if small values of one are likely to be associated with small values of the other, and large values of one are likely to be associated with large values of the other.

More precisely, let (x_i, y_i) and (x_j, y_j) be two observations from a vector (X, Y) of continuous r.v.'s. Then, (x_i, y_i) and (x_j, y_j) are concordant if

$$(x_i - x_j)(y_i - y_j) > 0, \tag{B.20a}$$

and discordant if

$$(x_i - x_j)(y_i - y_j) < 0. \tag{B.20b}$$

A mathematical definition of a *measure of concordance* is as follows [258].

DEFINITION B.7 (Measure of concordance). *A numeric measure of association κ between two continuous r.v.'s X and Y with copula **C** is a measure of concordance if it satisfies the following properties:*

1. *κ is defined for every pair (X, Y) of continuous r.v.'s;*
2. $-1 \le \kappa_{X,Y} \le 1$, $\kappa_{X,X} = 1$, *and* $\kappa_{X,-X} = -1$;
3. $\kappa_{X,Y} = \kappa_{Y,X}$;
4. *if X and Y are independent then* $\kappa_{X,Y} = 0$;
5. $\kappa_{-X,Y} = \kappa_{X,-Y} = -\kappa_{X,Y}$;

6. *if* (X_1, Y_1) *and* (X_2, Y_2) *are random vectors with copulas* $\mathbf{C}_1$ *and* $\mathbf{C}_2$*, respectively, such that* $\mathbf{C}_1 \prec \mathbf{C}_2$*, then* $\kappa_{X_1,Y_1} \leq \kappa_{X_2,Y_2}$*;*
7. *if* $\{(X_n, Y_n)\}$*,* $n \in \mathbb{N}$*, is a sequence of continuous r.v.'s with copula* $\mathbf{C}_n$*, and if* $\{\mathbf{C}_n\}$ *converges pointwise to a copula* $\mathbf{C}$*, then*

$$\lim_{n\to\infty} \kappa_{\mathbf{C}_n} = \kappa_{\mathbf{C}}.$$

We anticipate here that, as a consequence of the above definition, both Kendall's τ_K and Spearman's ρ_S turn out to be measures of concordance. We need now to introduce the *concordance function*.

DEFINITION B.8 (Concordance function). *Let* $(X_1, Y_1), (X_2, Y_2)$ *be independent vectors of continuous r.v.'s with marginals* F_X, F_Y *and copulas* $\mathbf{C}_1, \mathbf{C}_2$*. The difference*

$$Q = \mathbb{P}\{(X_1 - X_2)(Y_1 - Y_2) > 0\} - \mathbb{P}\{(X_1 - X_2)(Y_1 - Y_2) < 0\} \tag{B.21}$$

defines the concordance function Q*.*

The important point is that Q depends upon the distributions of (X_1, Y_1) and (X_2, Y_2) only through their copulas $\mathbf{C}_1$ and $\mathbf{C}_2$. In fact, it can be shown that

$$Q = Q(\mathbf{C}_1, \mathbf{C}_2) = 4 \iint_{\mathbb{I}^2} \mathbf{C}_2(u, v)\, d\mathbf{C}_1(u, v) - 1. \tag{B.22}$$

Note that Q is symmetric in its arguments [207].

B.2.2 Kendall's τ_K

The population version of Kendall's τ_K [160, 170, 239, 207] is defined as the difference between the probability of concordance and the probability of discordance.

DEFINITION B.9 (Kendall's τ_K). *Let* (X_1, Y_1) *and* (X_2, Y_2) *be i.i.d. vectors of continuous r.v.'s. The difference*

$$\tau_K = \mathbb{P}\{(X_1 - X_2)(Y_1 - Y_2) > 0\} - \mathbb{P}\{(X_1 - X_2)(Y_1 - Y_2) < 0\} \tag{B.23}$$

defines the population version of Kendall's τ_K*.*

Evidently, if $\mathbf{C}$ is the copula of X and Y, then

$$\tau_K^{X,Y} = \tau_K^{\mathbf{C}} = Q(\mathbf{C}, \mathbf{C}) = 4 \iint_{\mathbb{I}^2} \mathbf{C}(u, v)\, d\mathbf{C}(u, v) - 1. \tag{B.24}$$

Note that the integral defining Q corresponds to the expected value of the r.v. $W = \mathbf{C}(U, V)$ introduced in Proposition 3.5, i.e.

$$\tau_K^{\mathbf{C}} = 4\,\mathbb{E}\,(\mathbf{C}\,(U, V)) - 1. \tag{B.25}$$

In particular, in the Archimedean case, Kendall's τ_K can be expressed [207] as a function of the generator γ of $\mathbf{C}$:

$$\tau_K^{\mathbf{C}} = 1 + 4\int_0^1 \frac{\gamma(t)}{\gamma'(t)}\,\mathrm{d}t. \tag{B.26}$$

Note that, for a copula $\mathbf{C}$ with a singular component, the expression in Eq. (B.24) for $\tau_K^{\mathbf{C}}$ may be difficult to calculate, and can be replaced by the following formula:

$$\tau_K^{\mathbf{C}} = 1 - 4\iint_{\mathbb{I}^2} \left(\frac{\partial \mathbf{C}\,(u, v)}{\partial u}\right) \cdot \left(\frac{\partial \mathbf{C}\,(u, v)}{\partial v}\right) \mathrm{d}u\,\mathrm{d}v. \tag{B.27}$$

Note that, if X and Y are PQD, then $\tau_K^{X,Y} \geq 0$, and, analogously, if X and Y are NQD, then $\tau_K^{X,Y} \leq 0$. In particular, τ_K is increasing with respect to the concordance ordering introduced in Definition B.5. Possible multivariate extensions of τ_K are discussed in [152, 204].

The sample version t of Kendall's τ_K is easy to calculate:

$$t = \frac{c - d}{c + d}, \tag{B.28}$$

where c (d) represent the number of concordant (discordant) pairs in a sample of size n from a vector of continuous r.v.'s (X, Y). Unfortunately, in applications it often happens that continuous variables are "discretely" sampled, due to a finite instrumental resolution. For instance, the rainfall depth could be returned as an integer multiple of 0.1 mm, or the storm duration could be expressed in hours and rounded to an integer value (see, e.g., [66, 253]). Clearly, this procedure introduces repetitions in the observed values (called *ties* in statistics), which may adversely affect the estimation of τ_K. However, corrections to Eq. (B.28) are specified for solving the problem (see, e.g., the formulas in [223]).

ILLUSTRATION B.1 (Storm intensity–duration (cont.)). ►

In [255] a Frank's 2-copula is used to model the relationship between the (average) *storm intensity* I and the (wet) *storm duration* W — see also Illustration 4.4. The Authors carry out a seasonal analysis, and investigate the "strength" of the association between the r.v.'s I and W. In particular, they consider a sequence of increasing thresholds $\nu_1 < \cdots < \nu_n$ of the *storm volume* $V = IW$, and estimate the values of Kendall's τ_K for all those pairs (I, W) satisfying $IW > \nu_i$, $i = 1, \ldots, n$. In Figure B.1 we plot the results obtained; for the sake of comparison, the corresponding values of Pearson's linear correlation coefficient ρ_P are also shown.

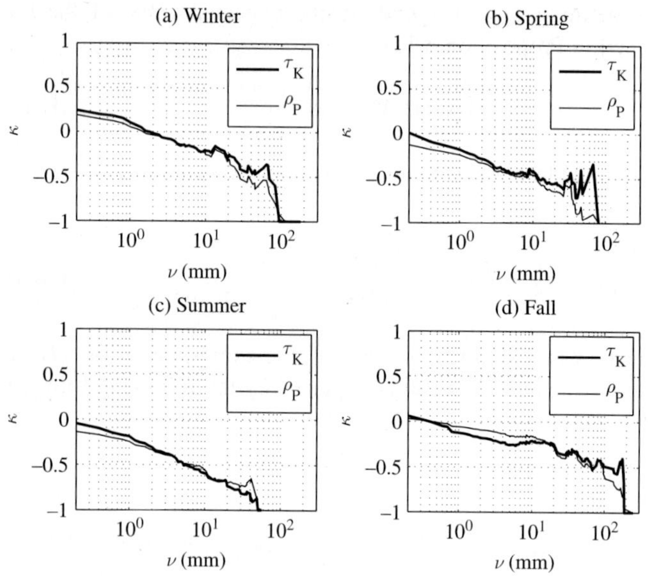

Figure B.1. Plot of the measures of association τ_K and ρ_P as a function of the volume threshold ν by considering the pairs (I, W) for which $V = IW > \nu$, for the data analysed by [255] in each season

Figure B.1 deserves several comments. The analysis of the behavior of τ_K is distribution-free, since there is no need to estimate the marginal laws of I and W. On the contrary, these marginals must be calculated before estimating ρ_P, since the existence of Pearson's linear correlation coefficient may depend upon such distributions, and must be proved in advance.

Except for very small values of the storm volume V (at most, $\nu = 2$ mm in Winter), I and W are always negatively associated, and such a link becomes stronger considering more and more extreme storms (i.e., for larger and larger values of V): apparently, the rate of "increase" of the association strength towards the limit value -1 is logarithmic. Note that the points in the upper tail are scattered simply because in that region ($\nu \gg 1$) very few storms are present, which affects the corresponding statistical analysis.

Overall, a negative association between I and W has to be expected (see, e.g., the discussion in [66]), for in the real world we usually observe that the strongest intensities are associated with the shortest durations, and the longest durations with the weakest intensities. ◄

B.2.3 Spearman's ρ_S

As with Kendall's τ_K, also the population version of Spearman's ρ_S [170, 239, 207] is based on concordance and discordance.

DEFINITION B.10 (Spearman's ρ_S). *Let (X_1, Y_1), (X_2, Y_2) and (X_3, Y_3) be three independent random vectors with a common joint continuous distribution. The difference*

$$\rho_S = 3\left(\mathbb{P}\{(X_1 - X_2)(Y_1 - Y_3) > 0\} - \mathbb{P}\{(X_1 - X_2)(Y_1 - Y_3) < 0\}\right) \tag{B.29}$$

defines the population version of Spearman's ρ_S.

Evidently, if $\mathbf{C}$ is the copula of (X, Y), then $(X_1, Y_1) \sim \mathbf{C}$, but $(X_2, Y_3) \sim \mathbf{\Pi}_2$, since X_2 and Y_3 are independent. As a consequence,

$$\rho_S^{X,Y} = \rho_S^{\mathbf{C}} = 3Q(\mathbf{C}, \mathbf{\Pi}_2). \tag{B.30}$$

Also Spearman's ρ_S can be written in terms of a suitable expectation:

$$\rho_S^{\mathbf{C}} = 12 \iint_{\mathbb{I}^2} uv \, \mathrm{d}\mathbf{C}(u, v) - 3 = 12\,\mathbb{E}(UV) - 3. \tag{B.31}$$

A practical interpretation of Spearman's ρ_S arises from rewriting the above formula as

$$\rho_S^{\mathbf{C}} = 12 \iint_{\mathbb{I}^2} [\mathbf{C}(u, v) - uv] \, \mathrm{d}u \, \mathrm{d}v. \tag{B.32}$$

Thus, $\rho_S^{\mathbf{C}}$ is proportional to the signed volume between the graphs of $\mathbf{C}$ and the independence copula $\mathbf{\Pi}_2$. Roughly, $\rho_S^{\mathbf{C}}$ measures the "average distance" between the joint distribution of X and Y (as represented by $\mathbf{C}$), and independence (given by $\mathbf{\Pi}_2$). Note that, if X and Y are PQD, then $\rho_S^{X,Y} \geq 0$, and, analogously, if X and Y are NQD, then $\rho_S^{X,Y} \leq 0$. In particular, ρ_S is increasing with respect to the concordance ordering introduced in Definition B.5. Possible multivariate extensions of ρ_S are discussed in [152, 204].

The sample version r of Spearman's ρ_S is easy to calculate:

$$r = 1 - \frac{6 \sum_{i=1}^{n} (R_i - S_i)^2}{n^3 - n}, \tag{B.33}$$

where $R_i = \mathrm{Rank}(x_i)$, $S_i = \mathrm{Rank}(y_i)$, and n is the sample size. As already mentioned before, instrumental limitations may adversely affect the estimation of ρ_S in practice due to the presence of ties. However, corrections to Eq. (B.33) are specified for solving the problem (see, e.g., the formulas in [223]).

ILLUSTRATION B.2 (Storm intensity–duration (cont.)). ►

As discussed in Illustration B.1, in [255] a Frank's 2-copula is used to model the relation between the (average) *storm intensity* I and the (wet) *storm duration*

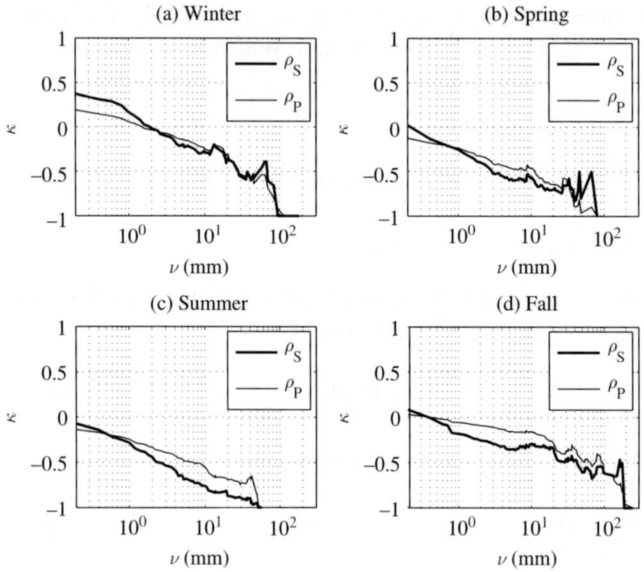

Figure B.2. Plot of the measures of association ρ_S and ρ_P as a function of the volume threshold ν by considering the pairs (I, W) for which $V = IW > \nu$, for the data analysed by [255] in each season

W. The Authors estimate the values of Spearman's ρ_S for all those pairs (I, W) satisfying $IW > \nu_i$, $i = 1, \ldots, n$, by considering a sequence of increasing thresholds $\nu_1 < \cdots < \nu_n$ of the *storm volume* $V = IW$. In Figure B.2 we plot the results obtained; for the sake of comparison, the corresponding values of Pearson's linear correlation coefficient ρ_P are also shown.

The same comments as in Illustration B.1 hold in the present case. ◀

APPENDIX C

FAMILIES OF COPULAS

(Written by Fabrizio Durante — Department of Knowledge-Based Mathematical Systems, Johannes Kepler University, Linz (Austria))

Research on copulas seems to generate new formulas endlessly. A full list of well known copulas includes several types. It is not feasible to include all of these in this book. Instead, we summarize a few families that find numerous applications in practice. For a more extensive list see [145, 155, 207], where additional mathematical properties are also found.

C.1. THE FRANK FAMILY

One of the possible equivalent expressions for members of this family is

$$\mathbf{C}_\theta(u,v) = \frac{1}{\ln\theta}\ln\left[1+\frac{(\theta^u-1)(\theta^v-1)}{\theta-1}\right], \tag{C.1}$$

where $u, v \in \mathbb{I}$, and $\theta \geq 0$ is a dependence parameter [99, 203, 108].

If U and V are r.v.'s with copula $\mathbf{C}_\theta$, then they are PQD for $0 \leq \theta < 1$, and NQD for $\theta > 1$. The limiting case $\theta = 1$ occurs when U and V are independent r.v.'s, i.e. $\mathbf{C}_1 = \mathbf{\Pi}_2$. Moreover, $\mathbf{C}_0 = \mathbf{M}_2$ and $\mathbf{C}_\infty = \mathbf{W}_2$, and thus the Frank family is comprehensive. Also, since $\theta_1 \leq \theta_2$ implies $\mathbf{C}_{\theta_1} \prec \mathbf{C}_{\theta_2}$, this family is positively ordered. Further mathematical properties can be found in [207].

Every copula of the Frank family is absolutely continuous, and its density has a simple expression given by

$$\mathbf{c}_\theta(u,v) = -\frac{(\theta-1)\theta^{u+v}\ln\theta}{\theta^{u+v}-\theta^u-\theta^v+\theta}. \tag{C.2}$$

Copulas belonging to the Frank family are strict Archimedean. The expression of the generator is, for $t \in \mathbb{I}$,

$$\gamma(t) = -\ln\left(\frac{\theta^t-1}{\theta-1}\right). \tag{C.3}$$

These 2-copulas are the only Archimedean ones that satisfy the functional equation $\mathbf{C} = \overline{\mathbf{C}}$ for radial symmetry.

Two useful relationships exist between θ and, respectively, Kendall's τ_K and Spearman's ρ_S:

$$\tau_K(\theta) = 1 - 4\,\frac{D_1(-\ln\theta) - 1}{\ln\theta}, \tag{C.4a}$$

$$\rho_S(\theta) = 1 - 12\,\frac{D_2(-\ln\theta) - D_1(-\ln\theta)}{\ln\theta}, \tag{C.4b}$$

where D_1 and D_2 are, respectively, the *Debye* functions of order 1 and 2 [181]. For a discussion on the estimate of θ using τ_K and ρ_S see [108].

The lower and upper tail dependence coefficients for the members of this family are equal to 0.

As an illustration, we plot the Frank 2-copula and the corresponding level curves in Figures C.1–C.3, for different values of θ.

As a comparison with Figure 3.1 and Figure 3.4, note how in Figure C.1 $\mathbf{C}$ approximates $\mathbf{M}_2$, for $\theta \approx 0$, while in Figure C.2 $\mathbf{C}$ approximates $\mathbf{W}_2$, for $\theta \gg 1$. In Figure C.3 we show one of the Frank 2-copulas used in [66]: here a negative association is modeled.

A general algorithm for generating observations (u, v) from a pair of r.v.'s (U, V) Uniform on $\mathbb{I}$, and having a Frank 2-copula $\mathbf{C}_\theta$, can be constructed by using the method outlined in Section A.1.

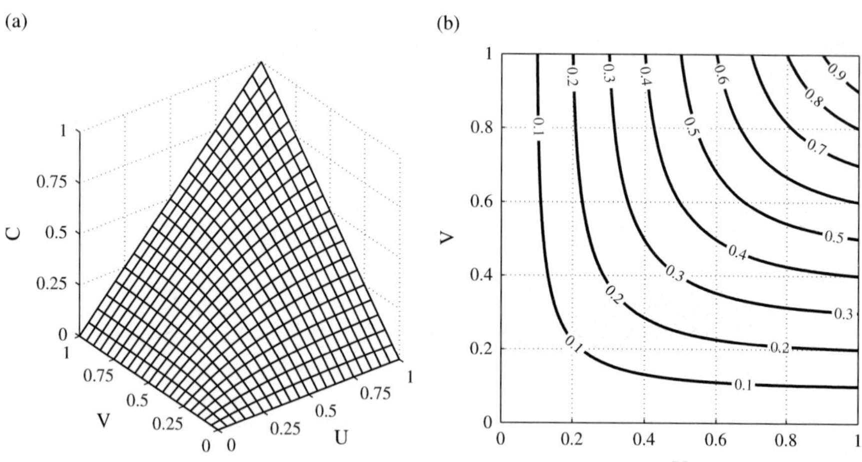

Figure C.1. The Frank 2-copula and the corresponding level curves. Here the parameter is $\theta = 0.02$. The probability levels are as indicated

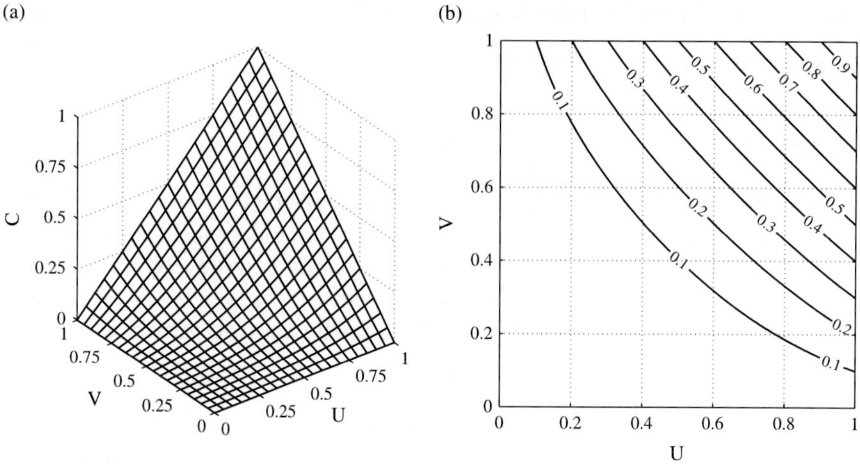

Figure C.2. The Frank 2-copula and the corresponding level curves. Here the parameter is $\theta = 50$. The probability levels are as indicated

Due to Proposition 4.1, the Frank family can be extended to the d-dimensional case, $d \geq 3$, if we restrict the range of the parameter θ to the interval $(0, 1)$, where γ^{-1} is completely monotonic. Its generalization is given by

$$\mathbf{C}_{\theta}(\boldsymbol{u}) = \frac{1}{\ln \theta} \ln \left[1 + \frac{(\theta^{u_1} - 1) \cdots (\theta^{u_d} - 1)}{(\theta - 1)^{d-1}} \right], \tag{C.5}$$

where $\theta \in (0, 1)$.

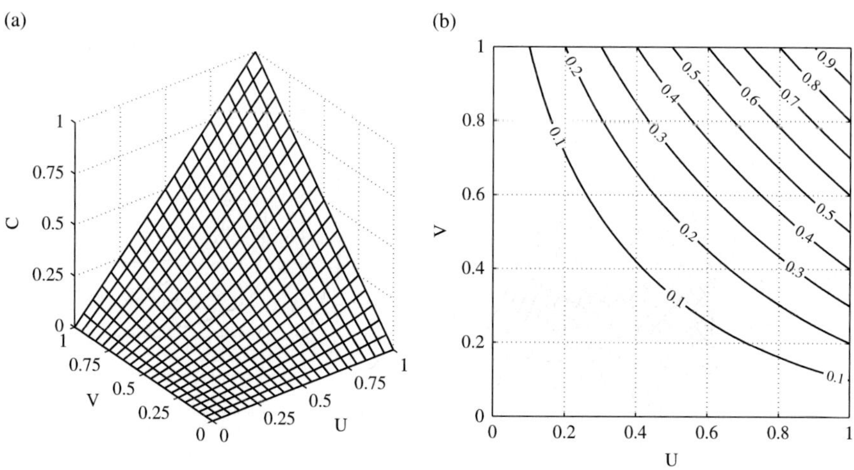

Figure C.3. The Frank 2-copula and the corresponding level curves. Here the parameter is $\theta \approx 12.1825$, as used in [66]. The probability levels are as indicated

C.2. THE GUMBEL-HOUGAARD FAMILY

The standard expression for members of this family is

$$\mathbf{C}_\theta(u,v) = e^{-\left[(-\ln u)^\theta + (-\ln v)^\theta\right]^{1/\theta}}, \tag{C.6}$$

where $u, v \in \mathbb{I}$, and $\theta \geq 1$ is a dependence parameter [207].

If U and V are r.v.'s with copula $\mathbf{C}_\theta$, then they are independent for $\theta = 1$, i.e. $\mathbf{C}_1 = \mathbf{\Pi}_2$, and PQD for $\theta > 1$. In particular, this family is positively ordered, with $\mathbf{C}_\infty = \mathbf{M}_2$, and its members are absolutely continuous.

Copulas belonging to the Gumbel-Hougaard family are strict Archimedean. The expression of the generator is, for $t \in \mathbb{I}$,

$$\gamma(t) = (-\ln t)^\theta. \tag{C.7}$$

The following relationship exists between θ and Kendall's τ_K:

$$\tau_K(\theta) = \frac{\theta - 1}{\theta}, \tag{C.8}$$

which may provide a way to fit a Gumbel-Hougaard 2-copula to the available data.

The lower and upper tail dependence coefficients for the members of this family are given by, respectively, $\lambda_L = 0$ and $\lambda_U = 2 - 2^{1/\theta}$.

As an illustration, we plot the Gumbel-Hougaard 2-copula and the corresponding level curves in Figures C.4–C.5, for different values of θ.

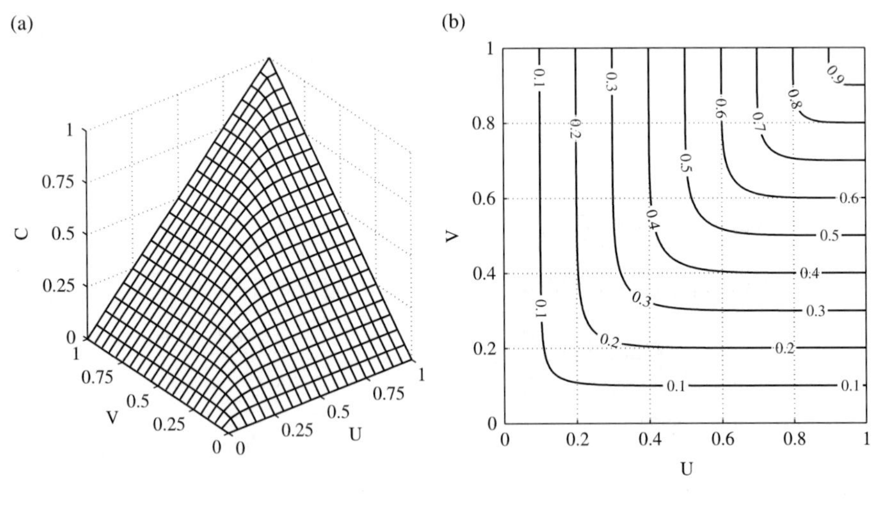

Figure C.4. The Gumbel-Hougaard 2-copula and the corresponding level curves. Here the parameter is $\theta = 5$. The probability levels are as indicated

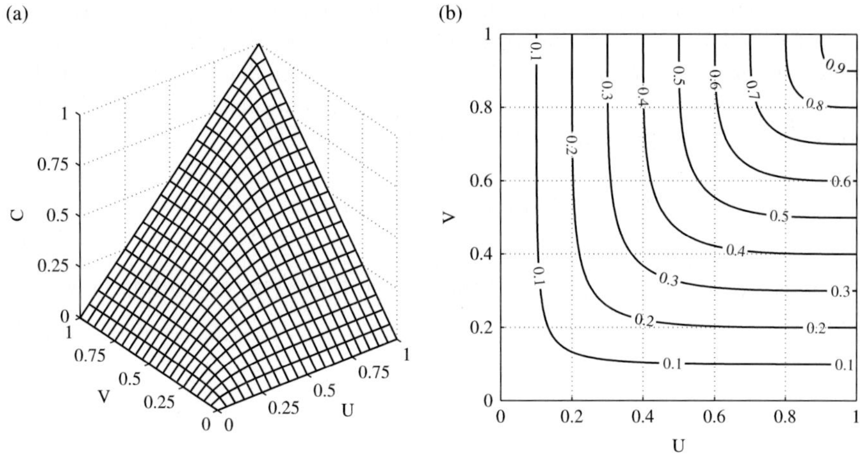

Figure C.5. The Gumbel-Hougaard 2-copula and the corresponding level curves. Here the parameter is $\theta \approx 3.055$, as used in [67]. The probability levels are as indicated

As a comparison with Figure 3.1 and Figure 3.4, note how in Figure C.4 $\mathbf{C}$ approximates $\mathbf{M}_2$, for sufficiently large θ. In Figure C.5 we show the Gumbel-Hougaard 2-copula used in [67]: here a weaker association is modeled.

A general algorithm for generating observations (u, v) from a pair of r.v.'s (U, V) Uniform on $\mathbb{I}$, and having a Gumbel-Hougaard 2-copula $\mathbf{C}_\theta$, can be constructed by using the method outlined in Section A.1.

Due to Proposition 4.1, the Gumbel-Hougaard family can be extended to the d-dimensional case, $d \geq 3$, and its expression is given by

$$\mathbf{C}_\theta(\boldsymbol{u}) = e^{-\left[(-\ln u_1)^\theta + \cdots + (-\ln u_d)^\theta\right]^{1/\theta}}. \tag{C.9}$$

Every member of this class is a MEV copula. As shown in [115], if H is a MEV distribution whose copula $\mathbf{C}$ is Archimedean, then $\mathbf{C}$ belongs to the Gumbel-Hougaard family (see Illustration 5.1).

C.3. THE CLAYTON FAMILY

The standard expression for members of this family is

$$\mathbf{C}_\theta(u, v) = \left(\max\left\{(u^{-\theta} + v^{-\theta} - 1), 0\right\}\right)^{-1/\theta}, \tag{C.10}$$

where $u, v \in \mathbb{I}$, and $\theta \geq -1$ is a dependence parameter [207].

If U and V are r.v.'s with copula $\mathbf{C}_\theta$, then they are PQD for $\theta > 0$, and NQD for $-1 \leq \theta < 0$. The limiting case $\theta = 0$ corresponds to the independent case, i.e. $\mathbf{C}_0 = \mathbf{\Pi}_2$. In particular, this family is positively ordered, and its members are absolutely continuous for $\theta > 0$.

The Clayton family is the only "truncation invariant family" [211], in the sense that, if U and V are r.v.'s with copula $\mathbf{C}$, then, given $u_0, v_0 \in (0, 1)$, the copula of the conditional r.v.'s U, given that $U \leq u_0$, and V, given that $V \leq v_0$, is again $\mathbf{C}$.

Copulas belonging to the Clayton family are Archimedean, and they are strict when $\theta > 0$. The expression of the generator is, for $t \in \mathbb{I}$,

$$\gamma(t) = \frac{1}{\theta}(t^{-\theta} - 1). \tag{C.11}$$

Using Eq. (B.26), it is possible to derive the following relationship between θ and Kendall's τ_K:

$$\tau_K(\theta) = \frac{\theta}{\theta + 2}, \tag{C.12}$$

which may provide a way to fit a Clayton 2-copula to the available data.

The lower and upper tail dependence coefficients for the members of this family are given by, respectively, $\lambda_L = 2^{-1/\theta}$ and $\lambda_U = 0$, for $\theta \geq 0$.

As an illustration, we plot the Clayton 2-copula and the corresponding level curves in Figures C.6–C.8, for different values of θ.

As a comparison with Figure 3.1 and Figure 3.4, note how in Figure C.6 $\mathbf{C}$ approximates $\mathbf{M}_2$, for sufficiently large θ, while in Figure C.7 $\mathbf{C}$ approximates $\mathbf{W}_2$, for sufficiently small θ. In Figure C.8 we show the Clayton 2-copula used in [109]: here a weak positive association is modeled.

A general algorithm for generating observations (u, v) from a pair of r.v.'s (U, V) Uniform on $\mathbb{I}$, and having a Clayton 2-copula $\mathbf{C}_\theta$, is as follows:

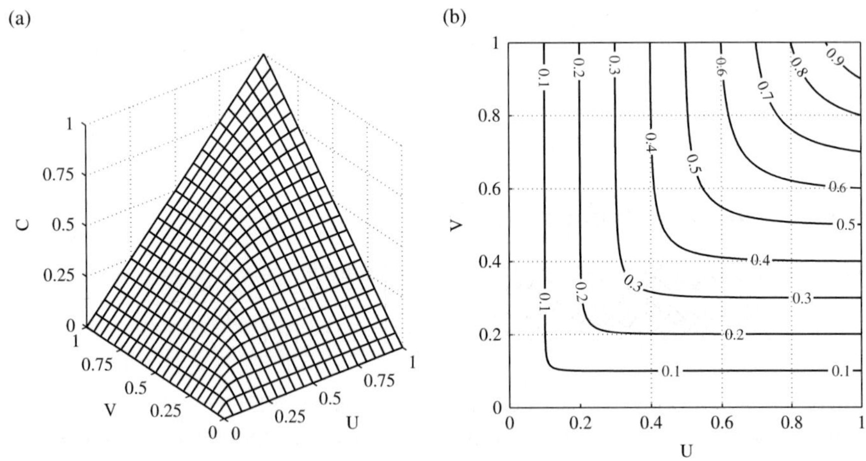

Figure C.6. The Clayton 2-copula and the corresponding level curves. Here the parameter is $\theta = 5$. The probability levels are as indicated

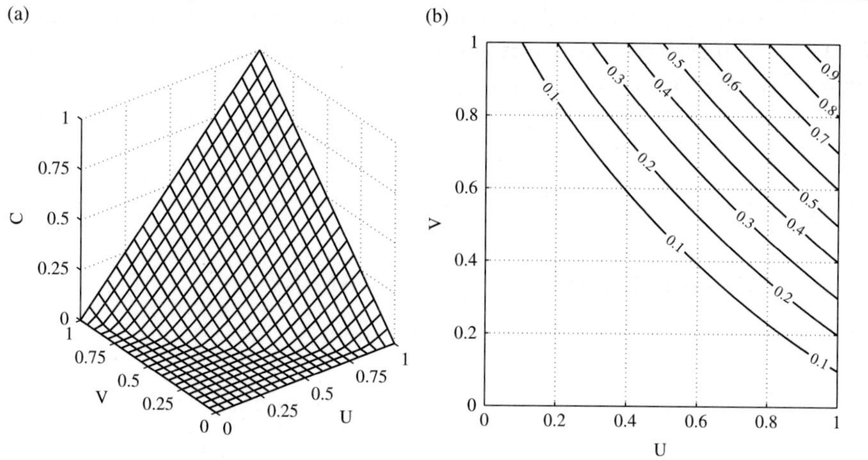

Figure C.7. The Clayton 2-copula and the corresponding level curves. Here the parameter is $\theta = -0.75$. The probability levels are as indicated

1. Generate independent variates x and y with a standard Exponential distribution.
2. Generate a variate z, independent of x and y, with Gamma distribution $\Gamma(\theta, 1)$.
3. Set $u = [1 + (x/z)]^{-\theta}$ and $v = [1 + (y/z)]^{-\theta}$.

The desired pair is then (u, v). For more details, see [71].

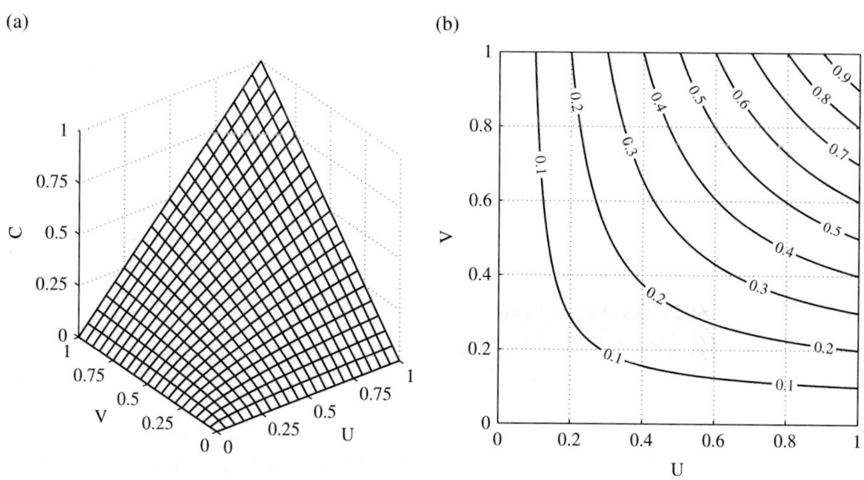

Figure C.8. The Clayton 2-copula and the corresponding level curves. Here the parameter is $\theta \approx 0.449$, as used in [109]. The probability levels are as indicated

Due to Proposition 4.1, the Clayton family can be extended to the d-dimensional case, $d \geq 3$, if we consider the parameter range $\theta > 0$, where γ^{-1} is completely monotonic. Its generalization is given by

$$\mathbf{C}_\theta(\boldsymbol{u}) = (u_1^{-\theta} + \cdots + u_d^{-\theta} - d + 1)^{-1/\theta}, \tag{C.13}$$

where $\theta > 0$.

C.4. THE ALI-MIKHAIL-HAQ (AMH) FAMILY

The standard expression for members of this family is

$$\mathbf{C}_\theta(u, v) = \frac{uv}{1 - \theta(1-u)(1-v)}, \tag{C.14}$$

where $u, v \in \mathbb{I}$, and $-1 \leq \theta \leq 1$ is a dependence parameter [207].

If U and V are r.v.'s with copula $\mathbf{C}_\theta$, then they are independent for $\theta = 0$, i.e. $\mathbf{C}_0 = \mathbf{\Pi}_2$, PQD for $\theta > 0$, and NQD for $\theta < 0$. In particular, this family is positively ordered.

If $\theta = 1$, then $\mathbf{C}_1(u, v) = uv/(u + v - uv)$: this copula belongs to many families of Archimedean copulas, as noted in [206]. For instance, it belongs to the Clayton family by taking $\theta = 1$ in Eq. (C.10).

The harmonic mean of two Ali-Mikhail-Haq 2-copulas is again an Ali-Mikhail-Haq 2-copula, i.e. if $\mathbf{C}_{\theta_1}$ and $\mathbf{C}_{\theta_2}$ are given by Eq. (C.14), then their harmonic mean is $\mathbf{C}_{(\theta_1+\theta_2)/2}$. In addition, each Ali-Mikhail-Haq 2-copula can be written as a weighted harmonic mean of the two extreme members of the family, i.e.

$$\mathbf{C}_\theta(u, v) = \frac{1}{\frac{1-\theta}{2}\frac{1}{\mathbf{C}_{-1}(u,v)} + \frac{1+\theta}{2}\frac{1}{\mathbf{C}_1(u,v)}} \tag{C.15}$$

for all $\theta \in [-1, 1]$.

Copulas belonging to the Ali-Mikhail-Haq family are strict Archimedean. The expression of the generator is, for $t \in \mathbb{I}$,

$$\gamma(t) = \ln \frac{1 - \theta(1-t)}{t}. \tag{C.16}$$

A useful relationship exists between θ and Kendall's τ_K:

$$\tau_K(\theta) = 1 - \frac{2(\theta + (1-\theta)^2 \ln(1-\theta))}{3\theta^2}, \tag{C.17}$$

which may provide a way to fit a Ali-Mikhail-Haq 2-copula to the available data. Copulas of this family show a limited range of dependence, which restricts their use in applications: in fact, Kendall's τ_K only ranges from ≈ -0.1817 to $1/3$, as θ goes from -1 to 1.

The lower and upper tail dependence coefficients for the members of this family are equal to 0, for $\theta \in (-1, 1)$.

As an illustration, we plot the Ali-Mikhail-Haq 2-copula and the corresponding level curves in Figures C.9–C.11, for different values of θ.

As a comparison with Figure 3.1 and Figure 3.4, note how in Figure C.9 $\mathbf{C}$ fails to approximate $\mathbf{M}_2$ sufficiently, although $\theta \approx 1$, and how in Figure C.10 $\mathbf{C}$ fails to approximate $\mathbf{W}_2$ sufficiently, although $\theta \approx -1$. In Figure C.11 we show the Ali-Mikhail-Haq 2-copula used in [68]: here a positive association is modeled.

A general algorithm for generating observations (u, v) from a pair of r.v.'s (U, V) Uniform on $\mathbb{I}$, and having an Ali-Mikhail-Haq 2-copula $\mathbf{C}_\theta$, is as follows:

1. Generate independent variates u, t Uniform on $\mathbb{I}$.
2. Set $a = 1 - u$.
3. Set $b = -\theta(2at + 1) + 2\theta^2 a^2 t + 1$ and $c = \theta^2(4a^2 t - 4at + 1) + \theta(4at - 4a + 2) + 1$.
4. Set $v = \frac{2t(a\theta-1)^2}{b+\sqrt{c}}$.

The desired pair is then (u, v). For more details, see [156].

Due to Proposition 4.1, this family can be extended for $d \geq 3$, if we consider the parameter range $\theta > 0$, where γ^{-1} is completely monotonic. Then, the generalization of the Ali-Mikhail-Haq family in d-dimensions is given by

$$\mathbf{C}_\theta(\boldsymbol{u}) = \frac{(1-\theta)\prod_{i=1}^d u_i}{\prod_{i=1}^d [1-\theta(1-u_i)] - \theta \prod_{i=1}^d u_i}, \tag{C.18}$$

where $\theta > 0$.

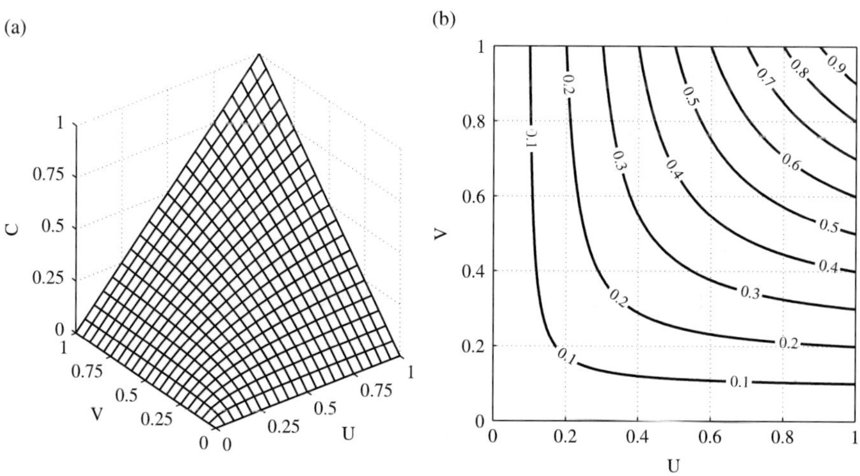

Figure C.9. The Ali-Mikhail-Haq 2-copula and the corresponding level curves. Here the parameter is $\theta = 0.99$. The probability levels are as indicated

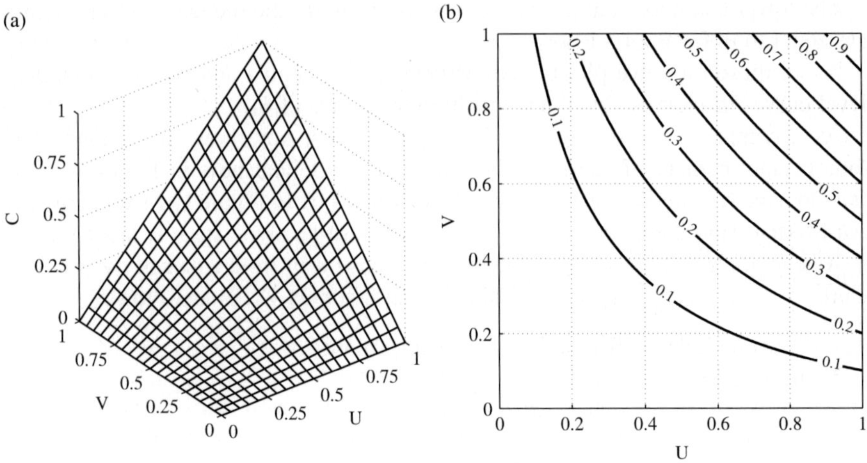

Figure C.10. The Ali-Mikhail-Haq 2-copula and the corresponding level curves. Here the parameter is $\theta = -0.99$. The probability levels are as indicated

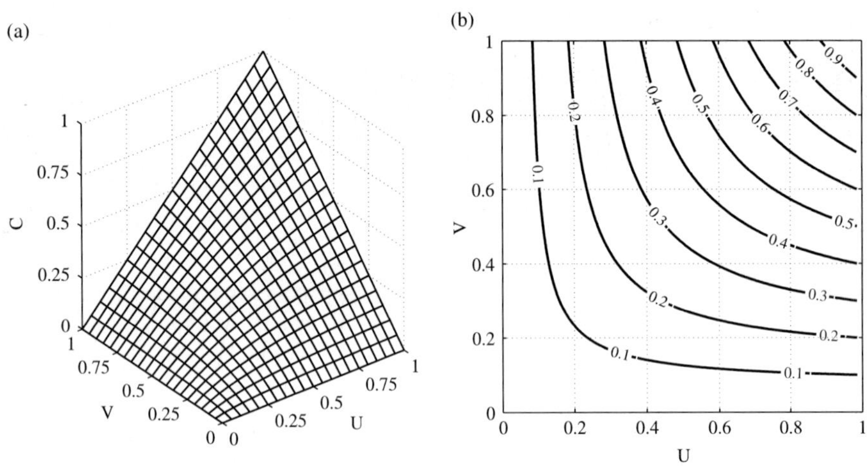

Figure C.11. The Ali-Mikhail-Haq 2-copula and the corresponding level curves. Here the parameter is $\theta \approx 0.829$, as used in [68]. The probability levels are as indicated

C.5. THE JOE FAMILY

The standard expression for members of this family is

$$\mathbf{C}_\theta (u, v) = 1 - \left[(1-u)^\theta + (1-v)^\theta - (1-u)^\theta (1-v)^\theta\right]^{1/\theta}, \tag{C.19}$$

where $u, v \in \mathbb{I}$, and $\theta \geq 1$ is a dependence parameter [153, 207].

If U and V are r.v.'s with copula $\mathbf{C}_\theta$, then they are independent for $\theta = 1$, i.e. $\mathbf{C}_1 = \mathbf{\Pi}_2$, and PQD for $\theta > 1$. In particular, this family is positively ordered, with $\mathbf{C}_\infty = \mathbf{M}_2$, and its members are absolutely continuous.

Copulas belonging to the Joe family are strict Archimedean. The expression of the generator is, for $t \in \mathbb{I}$,

$$\gamma(t) = -\ln\left[1-(1-t)^\theta\right]. \tag{C.20}$$

Using Eq. (B.26), it is possible to calculate the expression of Kendall's τ_K numerically.

The lower and upper tail dependence coefficients for the members of this family are given by, respectively, $\lambda_L = 0$ and $\lambda_U = 2 - 2^{1/\theta}$.

As an illustration, in Figures C.12–C.13 we plot the Joe 2-copula and the corresponding level curves, for different values of θ.

As a comparison with Figure 3.1 and Figure 3.4, note how in Figure C.12 $\mathbf{C}$ approximates $\mathbf{\Pi}_2$, for $\theta \approx 1$, while in Figure C.13 $\mathbf{C}$ approximates $\mathbf{M}_2$, for sufficiently large θ.

A general algorithm for generating observations (u, v) from a pair of r.v.'s (U, V) Uniform on $\mathbb{I}$, and having a Joe 2-copula $\mathbf{C}_\theta$, can be constructed by using the method outlined in Section A.1.

Due to Proposition 4.1, this family can be extended for $d \geq 3$, if we consider the parameter range $\theta \geq 1$, where γ^{-1} is completely monotonic. Then, the generalization of the Joe family in d-dimensions is given by

$$\mathbf{C}_\theta(\boldsymbol{u}) = 1 - \left[1 - \prod_{i=1}^{d}\left(1-(1-u_i)^\theta\right)\right]^{1/\theta}, \tag{C.21}$$

where $\theta \geq 1$.

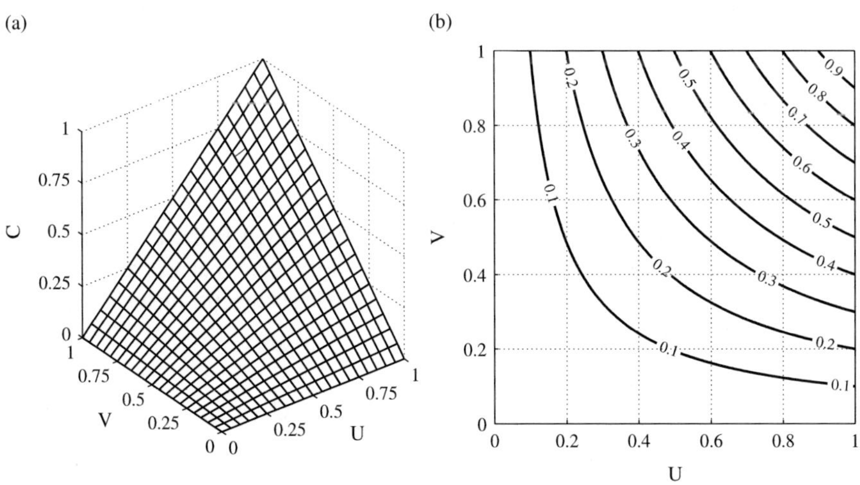

Figure C.12. The Joe 2-copula and the corresponding level curves. Here the parameter is $\theta = 1.05$. The probability levels are as indicated

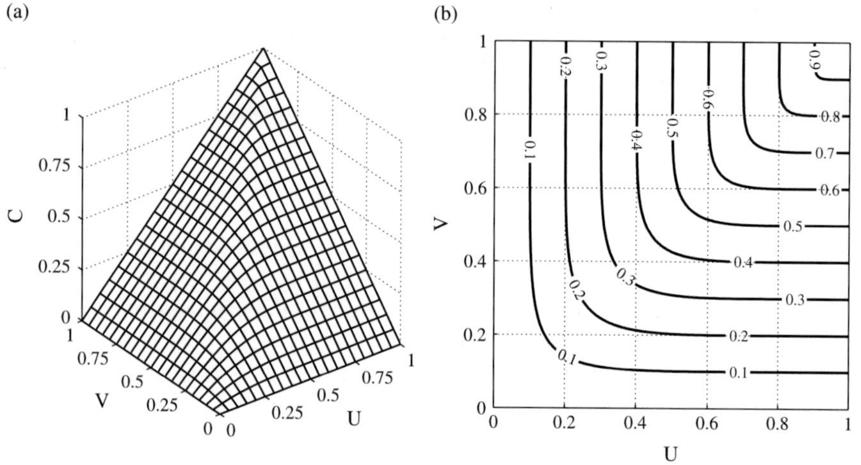

Figure C.13. The Joe 2-copula and the corresponding level curves. Here the parameter is $\theta = 7$. The probability levels are as indicated

C.6. THE FARLIE-GUMBEL-MORGENSTERN (FGM) FAMILY

The standard expression for members of this family is

$$\mathbf{C}_\theta(u, v) = uv + \theta uv(1-u)(1-v). \tag{C.22}$$

where $u, v \in \mathbb{I}$, and $\theta \in [-1, 1]$ is a dependence parameter [207].

If U and V are r.v.'s with copula $\mathbf{C}_\theta$, then they are independent for $\theta = 0$, i.e. $\mathbf{C}_0 = \mathbf{\Pi}_2$, PQD for $\theta > 0$ and NQD for $\theta < 0$. In particular, this family is positively ordered.

Every member of the FGM family is absolutely continuous, and its density has a simple expression given by

$$\mathbf{c}_\theta(u, v) = 1 + \theta(1-2u)(1-2v). \tag{C.23}$$

The arithmetic mean of two FGM 2-copulas is again an FGM 2-copula, i.e. if $\mathbf{C}_{\theta_1}$ and $\mathbf{C}_{\theta_2}$ are given by Eq. (C.22), then their arithmetic mean is $\mathbf{C}_{(\theta_1+\theta_2)/2}$. In addition, each FGM 2-copula can be written as the arithmetic mean of the two extreme members of the family, i.e.

$$\mathbf{C}_\theta(u, v) = \frac{1-\theta}{2}\mathbf{C}_{-1}(u, v) + \frac{1+\theta}{2}\mathbf{C}_1(u, v) \tag{C.24}$$

for all $\theta \in [-1, 1]$.

The FGM 2-copulas satisfy the functional equation $\mathbf{C} = \overline{\mathbf{C}}$ for radial symmetry.

Two useful relationships exist between θ and, respectively, Kendall's τ_K and Spearman's ρ_S:

$$\tau_K(\theta) = \frac{2\theta}{9}, \tag{C.25a}$$

$$\rho_S(\theta) = \frac{\theta}{3}. \tag{C.25b}$$

Therefore, the values of $\tau_K(\theta)$ have a range of $[-2/9, 2/9]$, and the values of $\rho_S(\theta)$ have a range of $[-1/3, 1/3]$. These limited intervals restrict the usefulness of this family for modeling.

The lower and upper tail dependence coefficients for the members of this family are equal to 0.

As an illustration, in Figures C.14–C.15 we plot the Farlie-Gumbel-Morgenstern 2-copula and the corresponding level curves, for different values of θ.

As a comparison with Figure 3.1 and Figure 3.4, note how in Figure C.14 $\mathbf{C}$ fails to approximate $\mathbf{M}_2$ sufficiently, although $\theta \approx 1$, and how in Figure C.15 $\mathbf{C}$ fails to approximate $\mathbf{W}_2$ sufficiently, although $\theta \approx -1$.

A general algorithm for generating observations (u, v) from a pair of r.v.'s (U, V) Uniform on $\mathbb{I}$, and having a Farlie-Gumbel-Morgenstern 2-copula $\mathbf{C}_\theta$, is as follows:

1. Generate independent variates u, t Uniform on $\mathbb{I}$.
2. Set $a = \theta(1-2u) - 1$ and $b = \sqrt{a^2 - 4(a+1)t}$.
3. Set $v = 2t/(b-a)$.

The desired pair is then (u, v). For more details, see [156].

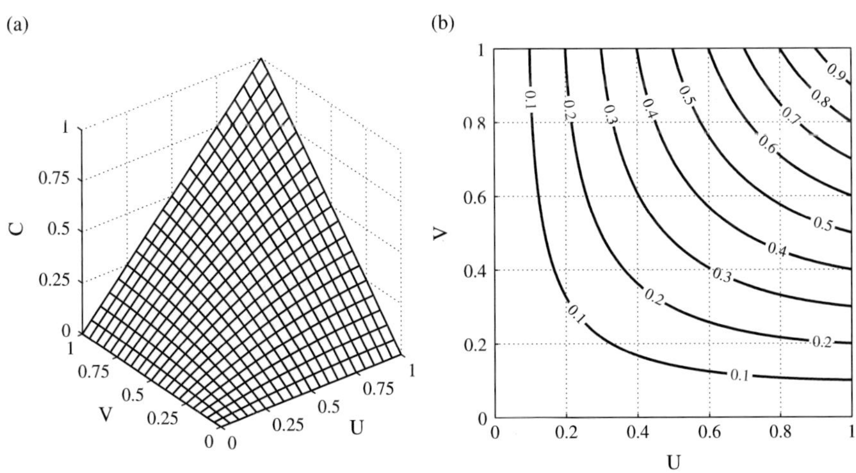

Figure C.14. The Farlie-Gumbel-Morgenstern 2-copula and the corresponding level curves. Here the parameter is $\theta = 0.99$. The probability levels are as indicated

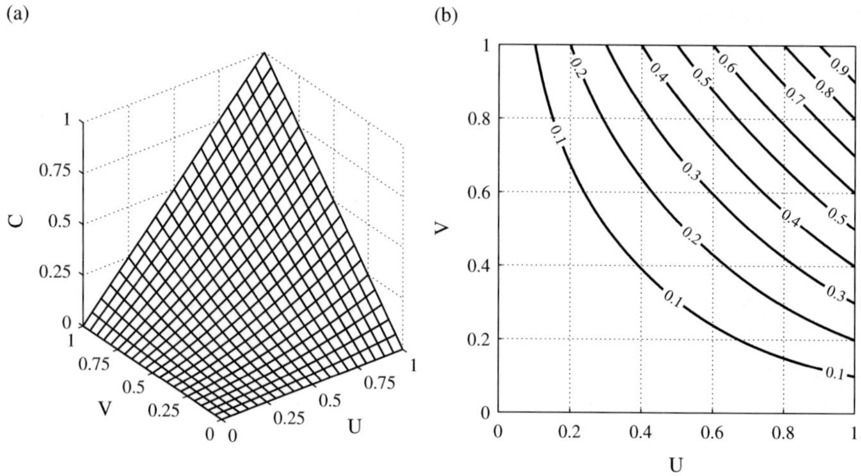

Figure C.15. The Farlie-Gumbel-Morgenstern 2-copula and the corresponding level curves. Here the parameter is $\theta = -0.99$. The probability levels are as indicated

In order to extend the range of dependence of the FGM family, several generalizations are proposed [75]. In particular, the following family of copulas is introduced in [238]:

$$\mathbf{C}_{f,g}(u, v) = uv + f(u)g(v), \tag{C.26}$$

where f and g are two real functions defined on $\mathbb{I}$, with $f(0) = g(0) = 0$ and $f(1) = g(1) = 0$, which satisfy the conditions:

$$|f(u_1) - f(u_2)| \leq \mathrm{M}|u_1 - u_2|, \tag{C.27a}$$

$$|g(u_1) - g(u_2)| \leq \frac{1}{\mathrm{M}}|u_1 - u_2|, \tag{C.27b}$$

for all $u_1, u_2 \in \mathbb{I}$, with $\mathrm{M} > 0$. Note that, by taking $f(u) = \theta u(1-u)$ and $g(v) = v(1-v)$, Eq. (C.26) describes the FGM family. Instead, by taking $f(u) = \theta u^\alpha (1-u)^\beta$ and $g(v) = v^\alpha (1-v)^\beta$, Eq. (C.26) describes the family given in [174].

The FGM family can be extended to the d-dimensional case, $d \geq 3$ — see [75, 207] for more details. Such an extension depends upon $2^d - d - 1$ parameters, and has the following form:

$$\mathbf{C}_\theta(\boldsymbol{u}) = \left(\prod_{i=1}^d u_i\right) \cdot \left[1 + \sum_{k=2}^d \sum_{1 \leqslant j_1 < \cdots < j_k \leqslant d} \theta_{j_1 \cdots j_k} (1 - u_{j_1}) \cdots (1 - u_{j_k})\right], \tag{C.28}$$

for suitable values of the parameters. In particular, for $d = 3$, four parameters are required, yielding the expression

$$\begin{aligned}\mathbf{C}_\theta(\boldsymbol{u}) = u_1u_2u_3[1 + \theta_{12}(1-u_1)(1-u_2) + \theta_{13}(1-u_1)(1-u_3) \\ + \theta_{23}(1-u_2)(1-u_3) + \theta_{123}(1-u_1)(1-u_2)(1-u_3)],\end{aligned} \tag{C.29}$$

for suitable values of the θ's.

C.7. THE PLACKETT FAMILY

The standard expression for members of this family is

$$\begin{aligned}\mathbf{C}_\theta(u, v) = \frac{1 + (\theta-1)(u+v)}{2(\theta-1)} \\ - \frac{\sqrt{[1 + (\theta-1)(u+v)]^2 - 4uv\theta(\theta-1)}}{2(\theta-1)},\end{aligned} \tag{C.30}$$

where $u, v \in \mathbb{I}$, and $\theta \geq 0$ is a dependence parameter [219, 145, 207].

If U and V are r.v.'s with copula $\mathbf{C}_\theta$, then they are PQD for $\theta > 1$, and NQD for $\theta < 1$. The limiting case $\theta = 1$ correspondes to the independent case, i.e. $\mathbf{C}_1 = \mathbf{\Pi}_2$. Moreover, $\mathbf{C}_0 = \mathbf{W}_2$ and $\mathbf{C}_\infty = \mathbf{M}_2$, and thus the Plackett family is comprehensive. In particular, this family is positively ordered. Every Plackett copula is absolutely continuous, and satisfies the functional equation $\mathbf{C} = \overline{\mathbf{C}}$ for radial symmetry.

No closed form exists for the value of Kendall's τ_K. Instead, Spearman's ρ_S is given by:

$$\rho_S(\theta) = \frac{\theta+1}{\theta-1} - \frac{2\theta \ln\theta}{(\theta-1)^2} \tag{C.31}$$

for $\theta \neq 1$.

The lower and upper tail dependence coefficients for the members of this family are equal to 0.

As an illustration, we plot the Plackett 2-copula and the corresponding level curves in Figures C.16–C.17, for different values of θ.

As a comparison with Figure 3.1 and Figure 3.4, note how in Figure C.16 $\mathbf{C}$ approximates $\mathbf{M}_2$, for $\theta \gg 1$, while in Figure C.17 $\mathbf{C}$ approximates $\mathbf{W}_2$, for $\theta \approx 0$.

A general algorithm for generating observations (u, v) from a pair of r.v.'s (U, V), Uniform on $\mathbb{I}$ and having a Plackett 2-copula $\mathbf{C}_\theta$ is as follows:

1. Generate independent variates u, t Uniform on $\mathbb{I}$.
2. Set $a = t(1-t)$.
3. Set $b = \theta + a(\theta-1)^2$, $c = 2a(u\theta^2 + 1 - u) + \theta(1-2a)$, and $d = \sqrt{\theta \cdot [\theta + 4au(1-u)(1-\theta)^2]}$.
4. Set $v = \frac{c-(1-2t)d}{2b}$.

The desired pair is then (u, v). For more details, see [156].

A multivariate extension of this family is presented in [196].

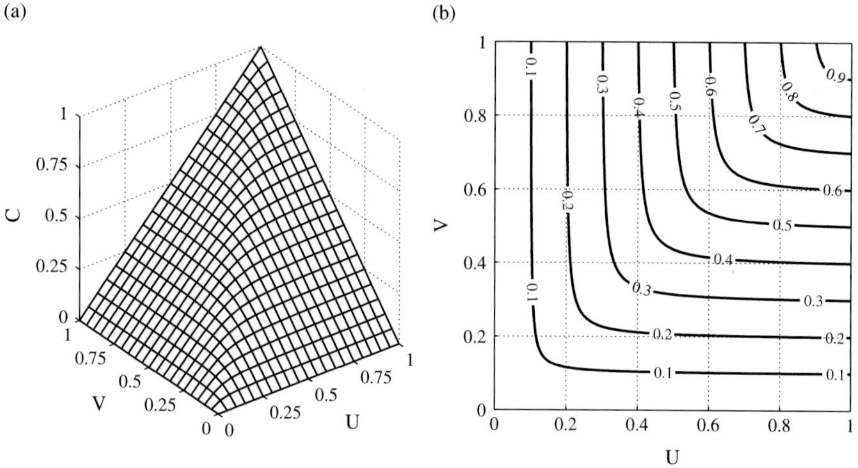

Figure C.16. The Plackett 2-copula and the corresponding level curves. Here the parameter is $\theta = 50$. The probability levels are as indicated

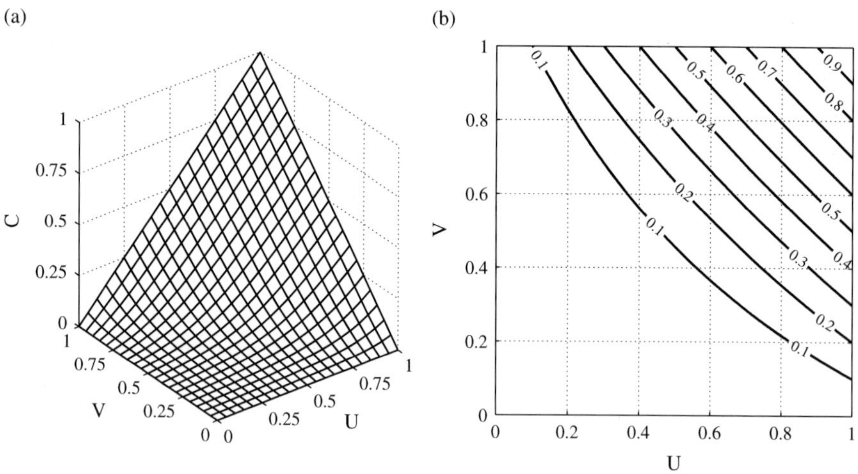

Figure C.17. The Plackett 2-copula and the corresponding level curves. Here the parameter is $\theta = 0.1$. The probability levels are as indicated

C.8. THE RAFTERY FAMILY

The standard expression for members of this family is

$$\mathbf{C}_\theta(u, v) = \mathbf{M}_2(u, v) + \frac{1-\theta}{1+\theta}(uv)^{1/(1-\theta)}\left[1 - (\max\{u, v\})^{-(1+\theta)/(1-\theta)}\right], \tag{C.32}$$

where $u, v \in \mathbb{I}$, and $\theta \in \mathbb{I}$ is a dependence parameter [227, 207].

If U and V are r.v.'s with copula $\mathbf{C}_\theta$, then they are independent for $\theta = 0$, i.e. $\mathbf{C}_0 = \mathbf{\Pi}_2$, and PQD for $\theta > 0$, with $\mathbf{C}_1 = \mathbf{M}_2$. Moreover, this family is positively ordered, and its members are absolutely continuous.

Two useful relationships exist between θ and, respectively, Kendall's τ_K and Spearman's ρ_S:

$$\tau_K(\theta) = \frac{2\theta}{3-\theta}, \tag{C.33a}$$

$$\rho_S(\theta) = \frac{\theta(4-3\theta)}{(2-\theta)^2}, \tag{C.33b}$$

which may provide a way to fit a Raftery 2-copula to a sample.

The lower and upper tail dependence coefficients for the members of this family are given by, respectively, $\lambda_L = 2\theta/(\theta+1)$ and $\lambda_U = 0$.

As an illustration, we plot the Raftery 2-copula and the corresponding level curves in Figures C.18–C.19, for different values of θ.

As a comparison with Figure 3.1 and Figure 3.4, note how in Figure C.18 $\mathbf{C}$ approximates $\mathbf{M}_2$, for $\theta \approx 1$, while in Figure C.19 $\mathbf{C}$ approximates $\mathbf{\Pi}_2$, for $\theta \approx 0$.

A general algorithm for generating observations (u, v) from a pair of r.v.'s (U, V) Uniform on $\mathbb{I}$, and having a Raftery 2-copula $\mathbf{C}_\theta$, can be constructed by using the method outlined in Section A.1.

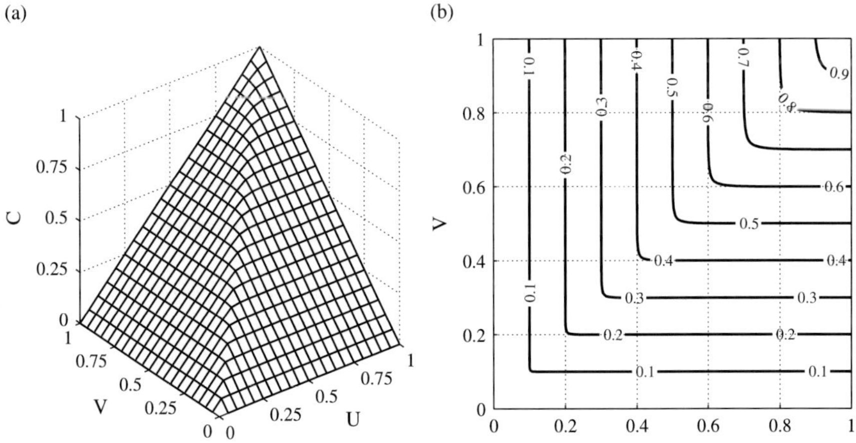

Figure C.18. The Raftery 2-copula and the corresponding level curves. Here the parameter is $\theta = 0.95$. The probability levels are as indicated.

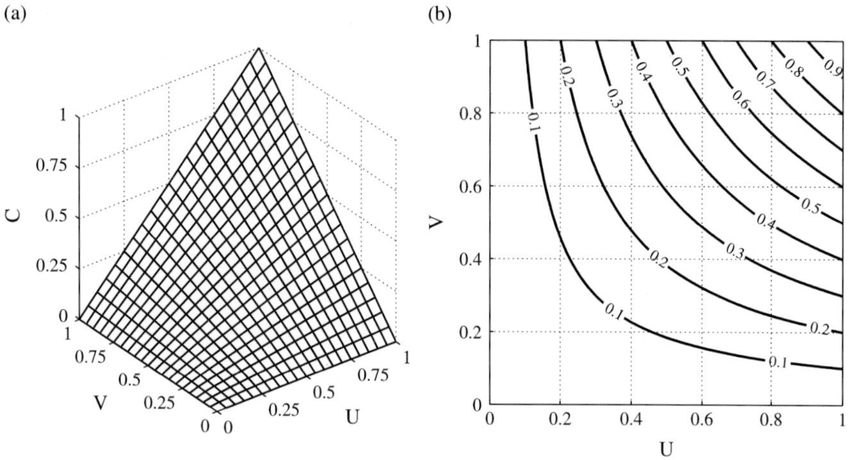

Figure C.19. The Raftery 2-copula and the corresponding level curves. Here the parameter is $\theta = 0.05$. The probability levels are as indicated.

C.9. THE GALAMBOS FAMILY

The standard expression for members of this family is

$$\mathbf{C}_\theta(u, v) = uv \exp\left(\left[(-\ln u)^{-\theta} + (-\ln v)^{-\theta}\right]^{-1/\theta}\right), \tag{C.34}$$

where $u, v \in \mathbb{I}$, and $\theta \geq 0$ is a dependence parameter [104, 153].

If U and V are r.v.'s with copula $\mathbf{C}_\theta$, then they are independent for $\theta = 0$, i.e. $\mathbf{C}_0 = \mathbf{\Pi}_2$, and PQD for $\theta > 0$. In particular, this family is positively ordered, with $\mathbf{C}_\infty = \mathbf{M}_2$, and its members are absolutely continuous. Most importantly, copulas belonging to the Galambos family are EV copulas.

Using Eq. (B.26) it is possible to calculate the expression of Kendall's τ_K numerically.

The lower and upper tail dependence coefficients for the members of this family are given by, respectively, $\lambda_L = 0$ and $\lambda_U = 2^{-1/\theta}$.

As an illustration, we plot the Galambos 2-copula and the corresponding level curves in Figures C.20–C.22, for different values of θ.

As a comparison with Figure 3.1 and Figure 3.4, note how in Figure C.20 $\mathbf{C}$ approximates $\mathbf{M}_2$, for θ large enough, while in Figure C.21 $\mathbf{C}$ approximates $\mathbf{\Pi}_2$, for $\theta \approx 0$. In Figure C.22 we show the Galambos 2-copula used in [109]: here a low positive association is found to a large extent.

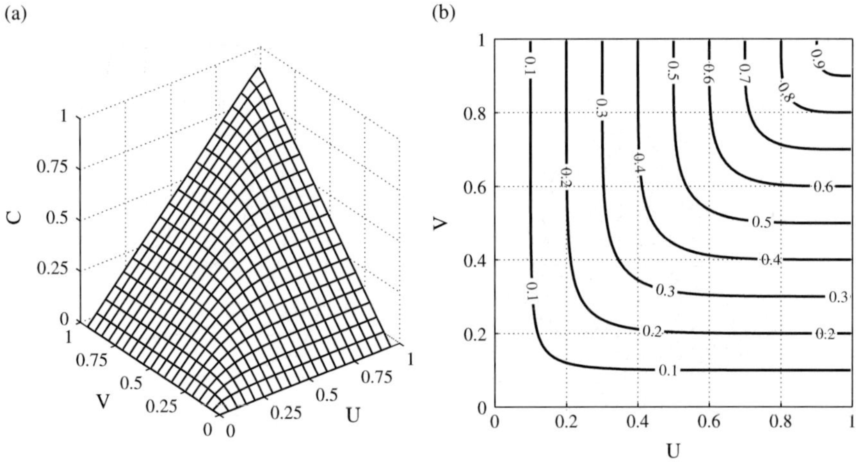

Figure C.20. The Galambos 2-copula and the corresponding level curves. Here the parameter is $\theta = 3$. The probability levels are as indicated

A general algorithm for generating observations (u, v) from a pair of r.v.'s (U, V) Uniform on $\mathbb{I}$, and having a Galambos 2-copula $\mathbf{C}_\theta$, can be constructed by using the method outlined in Section A.1.

A multivariate (partial) extension of this family to the d-dimensional case, $d \geq 3$, is presented in [155].

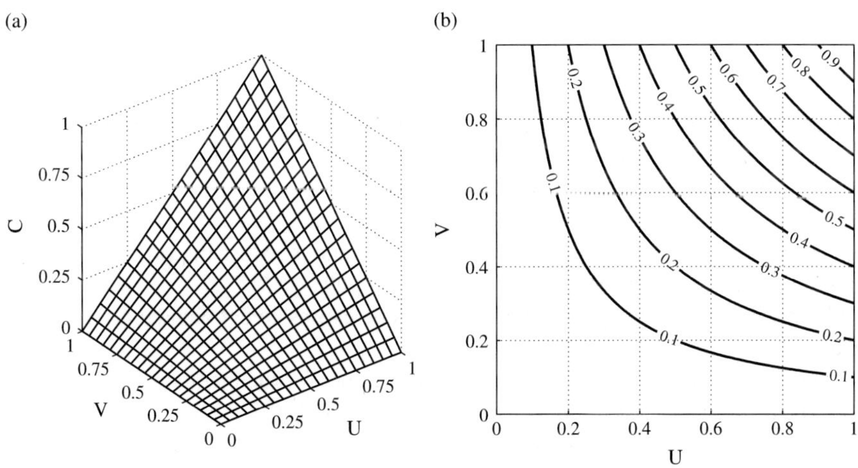

Figure C.21. The Galambos 2-copula and the corresponding level curves. Here the parameter is $\theta = 0.01$. The probability levels are as indicated

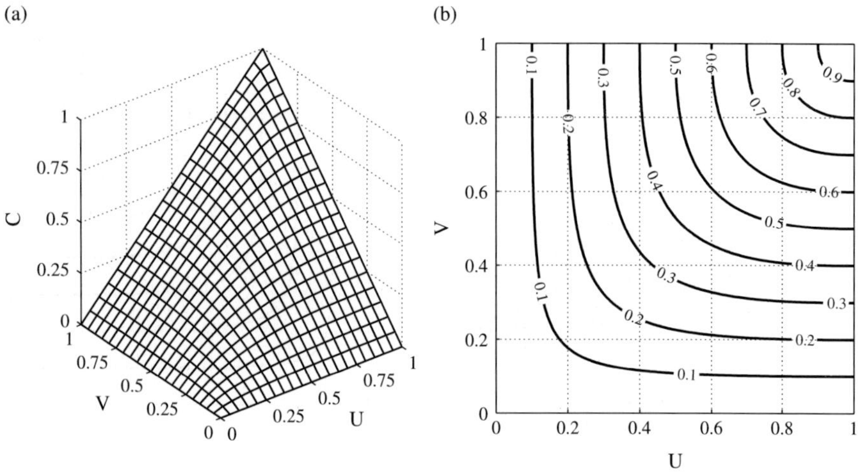

Figure C.22. The Galambos 2-copula and the corresponding level curves. Here the parameter is $\theta \approx 1.464$, as used in [109]. The probability levels are as indicated.

C.10. THE HÜSLER-REISS FAMILY

The standard expression for members of this family is

$$\mathbf{C}_\theta(u,v) = \exp\left[(\ln u)\,\Phi\left(\frac{1}{\theta}+\frac{\theta}{2}\ln\left(\frac{\ln u}{\ln v}\right)\right) + (\ln v)\,\Phi\left(\frac{1}{\theta}+\frac{\theta}{2}\ln\left(\frac{\ln v}{\ln u}\right)\right)\right], \tag{C.35}$$

where $u, v \in \mathbb{I}$, $\theta \geq 0$ is a dependence parameter, and Φ is the univariate standard Normal distribution [144, 153].

If U and V are r.v.'s with copula $\mathbf{C}_\theta$, then they are independent for $\theta = 0$, i.e. $\mathbf{C}_0 = \mathbf{\Pi}_2$, and PQD for $\theta > 0$. In particular, this family is positively ordered, with $\mathbf{C}_\infty = \mathbf{M}_2$, and its members are absolutely continuous. Most importantly, copulas belonging to the Hüsler-Reiss family are EV copulas.

Using Eq. (B.26) it is possible to calculate the expression of Kendall's τ_K numerically.

The lower and upper tail dependence coefficients for the members of this family are given by, respectively, $\lambda_L = 0$ and $\lambda_U = 2 - 2\Phi(1/\theta)$.

As an illustration, we plot the Hüsler-Reiss 2-copula and the corresponding level curves in Figures C.23–C.25, for different values of θ.

As a comparison with Figure 3.1 and Figure 3.4, note how in Figure C.23 $\mathbf{C}$ approximates $\mathbf{M}_2$, for sufficiently large θ, while in Figure C.24 $\mathbf{C}$ approximates $\mathbf{\Pi}_2$, for sufficiently small θ. In Figure C.25 we show the Hüsler-Reiss 2-copula used in [109]: here a low positive association is seen to a large extent.

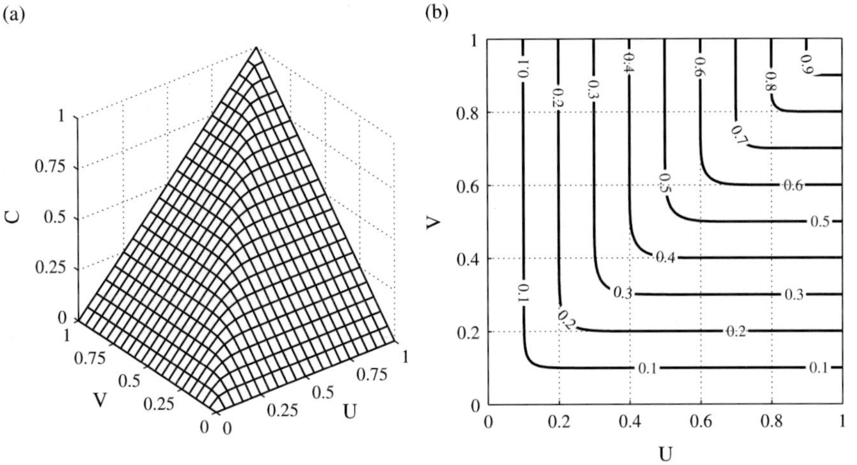

Figure C.23. The Hüsler-Reiss 2-copula and the corresponding level curves. Here the parameter is $\theta = 10$. The probability levels are as indicated.

A general algorithm for generating observations (u, v) from a pair of r.v.'s (U, V) Uniform on $\mathbb{I}$, and having a Hüsler-Reiss 2-copula $\mathbf{C}_\theta$, can be constructed by using the method outlined in Section A.1.

A multivariate (and partial) extension of this family to the d-dimensional case, $d \geq 3$, is presented in [155].

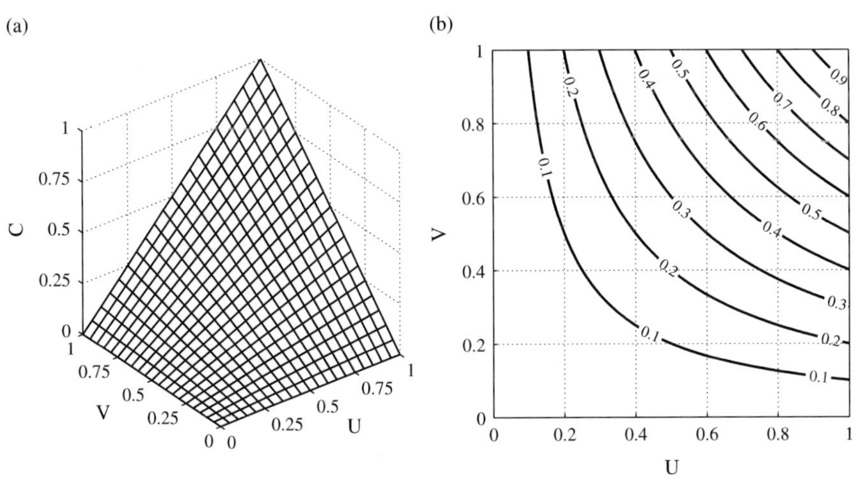

Figure C.24. The Hüsler-Reiss 2-copula and the corresponding level curves. Here the parameter is $\theta = 0.05$. The probability levels are as indicated.

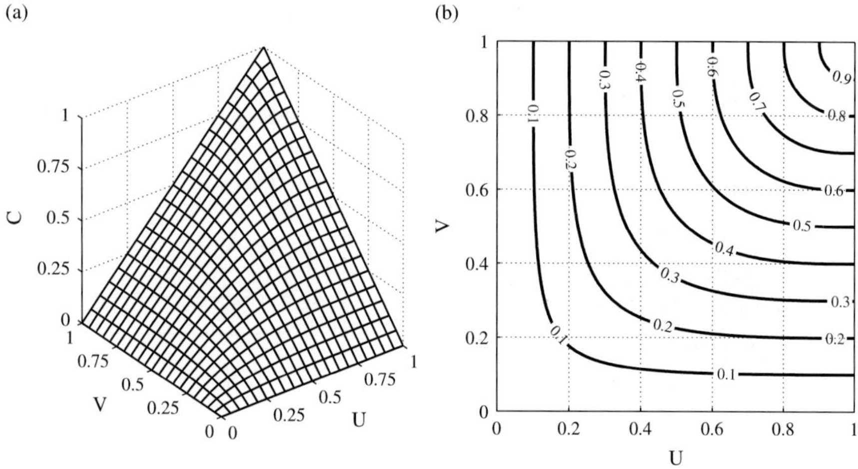

Figure C.25. The Hüsler-Reiss 2-copula and the corresponding level curves. Here the parameter is $\theta \approx 2.027$, as used in [109]. The probability levels are as indicated.

C.11. THE ELLIPTICAL FAMILY

A random vector $\boldsymbol{X} \in \mathbb{R}^d$ is said to have an *elliptical distribution* if it admits the stochastic representation

$$\boldsymbol{X} = \boldsymbol{\mu} + R\boldsymbol{A}\boldsymbol{U}, \tag{C.36}$$

where $\boldsymbol{\mu} \in \mathbb{R}^d$, R is a positive r.v. independent of $\boldsymbol{U}$, $\boldsymbol{U}$ is a random vector Uniform on the unit sphere in $\mathbb{R}^d$, and $\boldsymbol{A}$ is a fixed $d \times d$ matrix such that $\Sigma = \boldsymbol{A}\boldsymbol{A}^T$ is non-singular.

The density function of an elliptical distribution (if it exists) is given by, for $\boldsymbol{x} \in \mathbb{R}^d$,

$$f(\boldsymbol{x}) = |\Sigma|^{-1/2} g((\boldsymbol{x} - \boldsymbol{\mu})^T \Sigma^{-1} (\boldsymbol{x} - \boldsymbol{\mu})) \tag{C.37}$$

for some function $g : \mathbb{R} \to \mathbb{R}_+$, called a *density generator*, uniquely determined by the distribution of R (for more details, see [89, 88, 84]). For instance, the function

$$g(t) = C \exp(-t/2) \tag{C.38}$$

generates the multivariate Normal distribution, with a suitable normalizing constant C. Similarly,

$$g(t) = C\,(1 + t/m)^{-(d+m)/2} \tag{C.39}$$

generates the multivariate t-Student distribution, with a suitable normalizing constant C and $m \in \mathbb{N}$.

All the marginals of an elliptical distribution are the same, and are elliptically distributed. The unique copula associated to an elliptical distribution is called *elliptical copula*, and can be obtained by means of the inversion method (see, e.g., Corollary 3.1 or Corollary 4.1). If H is a d-variate elliptical distribution, with univariate marginal distribution functions F, then the corresponding elliptical d-copula is given by

$$\mathbf{C}(\boldsymbol{u}) = H(F^{-1}(u_1), \ldots, F^{-1}(u_d)). \tag{C.40}$$

The copula of the bivariate Normal distribution, also called *Gaussian* copula, is given by

$$\mathbf{C}_\theta(u, v) = \int_{-\infty}^{\Phi^{-1}(u)} \int_{-\infty}^{\Phi^{-1}(v)} \frac{1}{2\pi\sqrt{1-\theta^2}} \exp\left(-\frac{s^2 - 2\theta st + t^2}{2(1-\theta^2)}\right) \mathrm{d}s\, \mathrm{d}t, \tag{C.41}$$

where $\theta \in [-1, 1]$, and Φ^{-1} denotes the inverse of the univariate Normal distribution.

If U and V are r.v.'s with copula $\mathbf{C}_\theta$, then they are PQD for $\theta > 0$, and NQD for $\theta < 0$. The limiting case $\theta = 0$ corresponds to the independent case, i.e. $\mathbf{C}_0 = \mathbf{\Pi}_2$. This family is positively ordered, and also comprehensive, since $\mathbf{C}_{-1} = \mathbf{W}_2$ and $\mathbf{C}_1 = \mathbf{M}_2$. Every Gaussian 2-copula satisfies the functional equation $\mathbf{C} = \widehat{\mathbf{C}}$ for radial symmetry.

Two useful relationships exist between θ and, respectively, Kendall's τ_K and Spearman's ρ_S:

$$\tau_K(\theta) = \frac{2}{\pi} \arcsin\theta, \tag{C.42a}$$

$$\rho_S(\theta) = \frac{6}{\pi} \arcsin\frac{\theta}{2}, \tag{C.42b}$$

which may provide a way to fit a Gaussian 2-copula to the available data.

Gaussian copulas have lower and upper tail dependence coefficients equal to 0.

The copula of the bivariate t-Student distribution with $\nu > 2$ degrees of freedom, also called the *t-Student* copula, is given by

$$\mathbf{C}_\theta(u, v) = \int_{-\infty}^{t_\nu^{-1}(u)} \int_{-\infty}^{t_\nu^{-1}(v)} \frac{1}{2\pi\sqrt{1-\theta^2}} \left(1 + \frac{s^2 - 2\theta st + t^2}{\nu(1-\theta^2)}\right)^{-(\nu+2)/2} \mathrm{d}s\, \mathrm{d}t, \tag{C.43}$$

where $\theta \in [-1, 1]$, and t_ν^{-1} denotes the inverse of the univariate t-distribution.

The expressions of Kendall's τ_K and Spearman's ρ_S of a t-copula are the same as of those in Eqs. (C.42).

The lower and upper tail dependence coefficients are equal, and are given by

$$\lambda_U = 2t_{\nu+1}\left(-\frac{\sqrt{\nu+1}\,\sqrt{1-\theta}}{\sqrt{1+\theta}}\right). \tag{C.44}$$

Therefore, the tail dependence coefficient is increasing in θ, and decreasing in ν, and tends to zero as $\nu \to \infty$ if $\theta < 1$ (see also [261]).

A general algorithm for generating observations (u, v) from a pair of r.v.'s (U, V) Uniform on $\mathbb{I}$, and having a 2-copula either of Gaussian or t-Student type, is outlined in [84].

The d-variate extensions, $d \geq 3$, of the Gaussian and the t-Student copulas can be constructed in an obvious way by using, respectively, the multivariate version of Normal and t-Student distributions.

In applications, multivariate elliptical copulas are sometimes used together with different types of univariate marginals (not necessarily elliptically distributed), in order to obtain new classes of multivariate distributions functions called *meta-elliptical* (see [88, 1] and also [110]).

C.12. THE FRÉCHET FAMILY

The standard expression for members of this family is

$$\mathbf{C}_{\alpha,\beta}(u, v) = \alpha\mathbf{M}_2(u, v) + (1-\alpha-\beta)\mathbf{\Pi}_2(u, v) + \beta\mathbf{W}_2(u, v), \tag{C.45}$$

where $u, v \in \mathbb{I}$, and α, β are dependence parameters, with $\alpha, \beta \geq 0$ and $\alpha+\beta \leq 1$ [207]. Therefore, the copulas of this family are *convex combinations* of $\mathbf{\Pi}_2$, $\mathbf{W}_2$, $\mathbf{M}_2$. In fact, $\mathbf{C}_{1,0} = \mathbf{M}_2$, $\mathbf{C}_{0,0} = \mathbf{\Pi}_2$ and $\mathbf{C}_{0,1} = \mathbf{W}_2$. In turn, this family is comprehensive. When $\alpha, \beta \neq 0$, $\mathbf{C}_{\alpha,\beta}$ has a singular component, and hence it is not absolutely continuous.

Two useful relationships exist between α and β and, respectively, Kendall's τ_K and Spearman's ρ_S:

$$\tau_K(\alpha, \beta) = \frac{(\alpha-\beta)(\alpha+\beta+2)}{3}, \tag{C.46a}$$

$$\rho_S(\alpha, \beta) = \alpha - \beta, \tag{C.46b}$$

which may provide a method of fitting a Fréchet 2-copula to the available data.

The lower and upper tail dependence coefficients for the members of this family are equal to α.

As an illustration, in Figures C.26–C.27 we plot the Fréchet 2-copula and the corresponding level curves, for different values of the parameters α, β.

Figures C.26–C.27 should be compared to Figure 3.1 and Figure 3.4.

Although the Fréchet family does not seem to have desirable properties, it was recently rediscovered in applications. In fact, as shown in [306], every 2-copula can be approximated in a unique way by a member of the Fréchet family, and the error bound can be estimated. Accordingly, practitioners can deal with a complicated copula by focusing on its approximation by means of a suitable Fréchet copula.

A slight modification of this family is the so-called *Linear Spearman* copula (see [155, family B11] and [142]), given by, for $\theta \in \mathbb{I}$,

$$\mathbf{C}_{\theta}(u, v) = (1-\theta)\mathbf{\Pi}_2(u, v) + \theta\mathbf{M}_2(u, v). \tag{C.47}$$

This family is positively ordered, and the dependence parameter θ is equal to the value of Spearman's ρ_S, i.e. $\rho_S(\theta) = \theta$.

The Linear Spearman copula can be extended to the d-dimensional case, $d \geq 3$, in an obvious way. The corresponding expression is given by

$$\mathbf{C}_{\theta}(\boldsymbol{u}) = (1-\theta)\mathbf{\Pi}_d(\boldsymbol{u}) + \theta\mathbf{M}_d(\boldsymbol{u}), \tag{C.48}$$

for $\theta \in \mathbb{I}$. Note that, instead, the Fréchet family cannot be extended in the same manner: in fact, $\mathbf{W}_d$ is not a copula for $d \geq 3$ (see Illustration 4.1).

C.13. THE MARSHALL-OLKIN FAMILY

The standard expression for members of this family is

$$\mathbf{C}_{\alpha,\beta}(u, v) = \min\left\{u^{1-\alpha}v, uv^{1-\beta}\right\}, \tag{C.49}$$

where $u, v \in \mathbb{I}$, and α, β are dependence parameters, with $0 \leq \alpha, \beta \leq 1$ [185, 186, 207]. Note that $\mathbf{C}_{\alpha,0} = \mathbf{C}_{0,\beta} = \mathbf{\Pi}_2$ and $\mathbf{C}_{1,1} = \mathbf{M}_2$.

Copulas of this family have an absolutely continuous component, and a singular component given by, respectively,

$$A_{\mathrm{C}} = uv^{1-\beta} - \frac{\alpha\beta}{\alpha+\beta-\alpha\beta}\left(u^{\alpha}\right)^{(\alpha+\beta-\alpha\beta)/\alpha\beta} \tag{C.50a}$$

for $u^{\alpha} < v^{\beta}$, and

$$S_{\mathrm{C}} = \frac{\alpha\beta}{\alpha+\beta-\alpha\beta}\left(\min\left\{u^{\alpha}, v^{\beta}\right\}\right)^{(\alpha+\beta-\alpha\beta)/\alpha\beta}. \tag{C.50b}$$

Two useful relationships exist between the parameters α, β and, respectively, Kendall's τ_K and Spearman's ρ_S:

$$\tau_K(\alpha, \beta) = \frac{\alpha\beta}{\alpha - \alpha\beta + \beta}, \tag{C.51a}$$

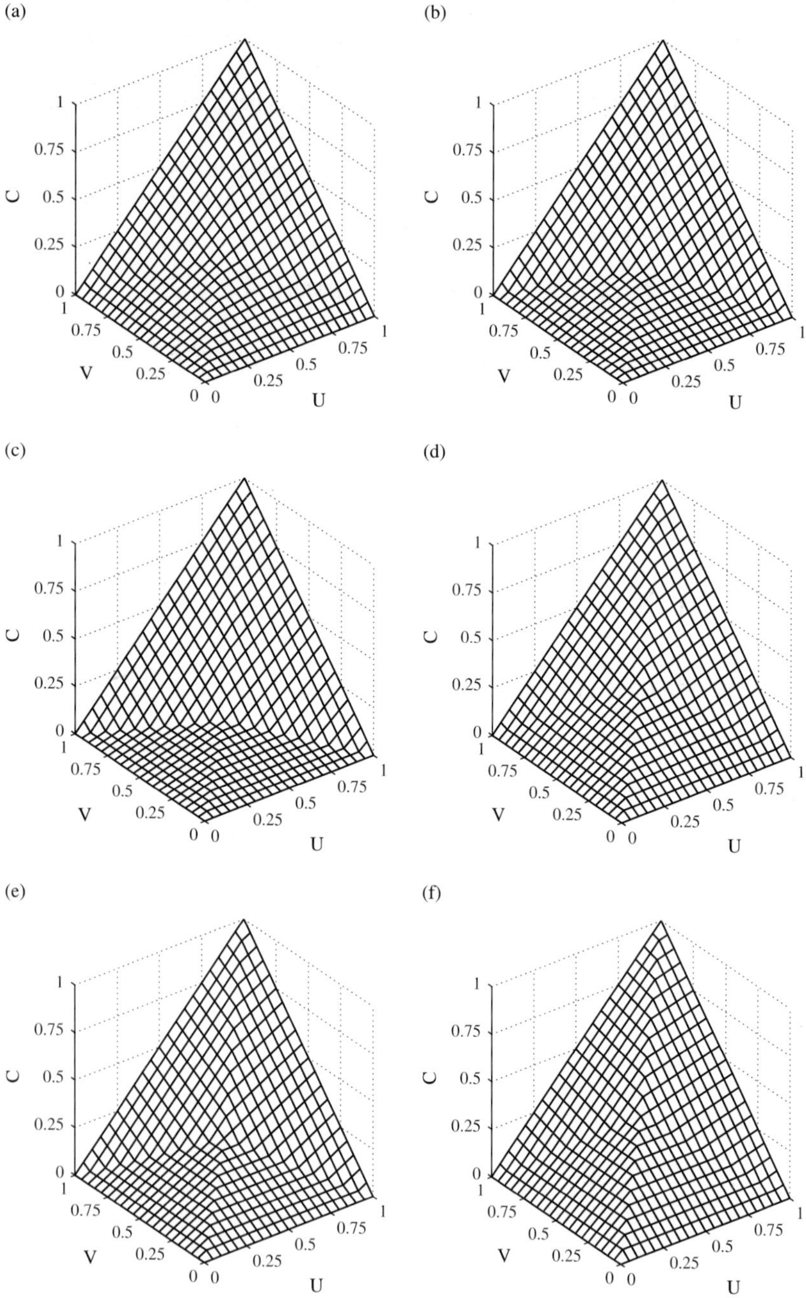

Figure C.26. The Fréchet 2-copula for several values of the parameters α, β: (*a*) $\alpha = 0.25, \beta = 0.25$, (*b*) $\alpha = 0.25, \beta = 0.5$, (*c*) $\alpha = 0.25, \beta = 0.75$, (*d*) $\alpha = 0.5, \beta = 0.25$, (*e*) $\alpha = 0.5, \beta = 0.5$, (*f*) $\alpha = 0.75, \beta = 0.25$.

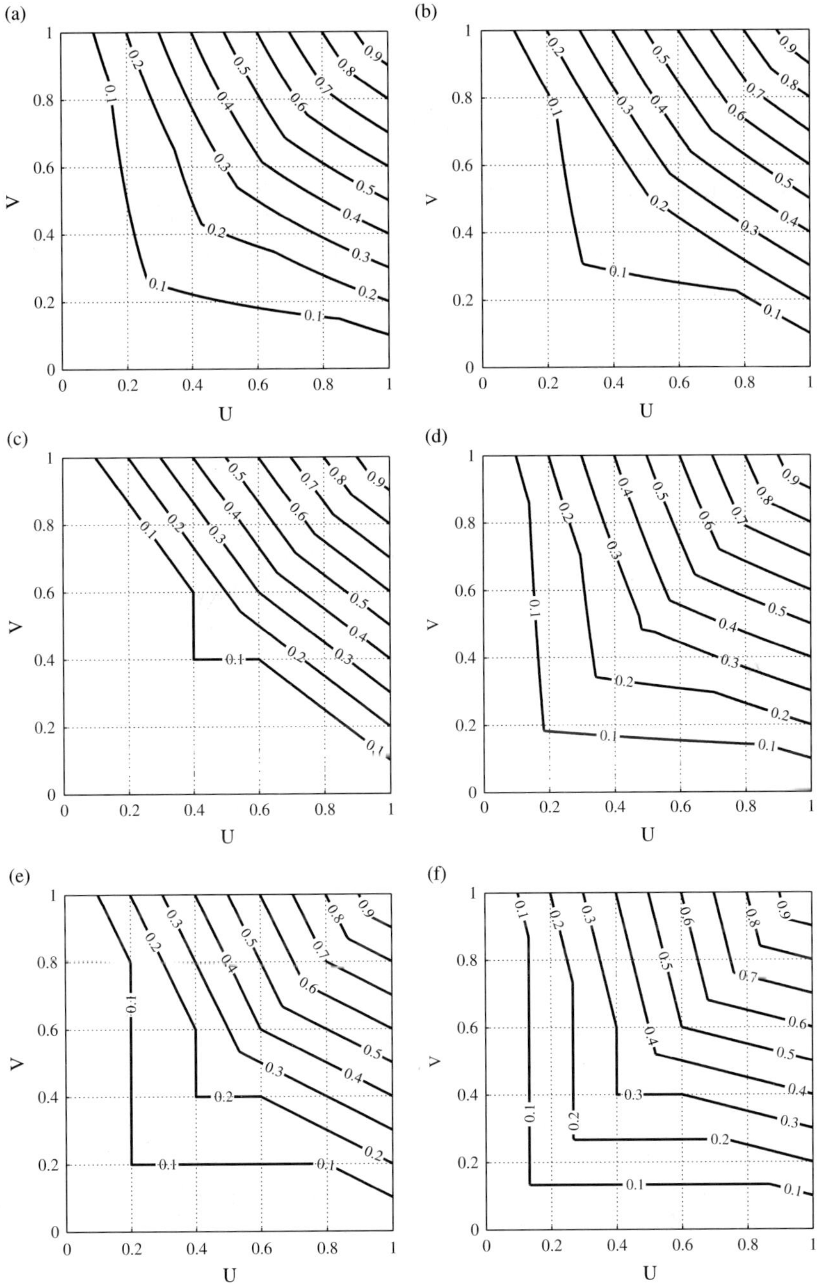

Figure C.27. The level curves of the Fréchet 2-copula for several values of the parameters α, β: (*a*) $\alpha = 0.25, \beta = 0.25$, (*b*) $\alpha = 0.25, \beta = 0.5$, (*c*) $\alpha = 0.25, \beta = 0.75$, (*d*) $\alpha = 0.5, \beta = 0.25$, (*e*) $\alpha = 0.5, \beta = 0.5$, (*f*) $\alpha = 0.75, \beta = 0.25$. The probability levels are as indicated.

$$\rho_S(\alpha, \beta) = \frac{3\alpha\beta}{2\alpha - \alpha\beta + 2\beta}, \tag{C.51b}$$

which may provide a way to fit a Marshall-Olkin 2-copula to the available data.

The lower and upper tail dependence coefficients for the members of this family are given by, respectively, $\lambda_L = 0$ and $\lambda_U = \min\{\alpha, \beta\}$.

As an illustration, we plot the Marshall-Olkin 2-copula and the corresponding level curves in Figures C.28–C.30, for different values of the parameters α, β.

Figures C.28–C.30 should be compared to Figure 3.1 and Figure 3.4.

A general algorithm for generating observations (u, v) from a pair of r.v.'s (U, V) Uniform on $\mathbb{I}$, and having a Marshall-Olkin 2-copula $\mathbf{C}_{\alpha,\beta}$, is as follows:

1. Generate independent variates r, s, and t Uniform on $\mathbb{I}$.
2. For any $\lambda_{12} > 0$, set $\lambda_1 = \lambda_{12}(1-\alpha)/\alpha$ and $\lambda_2 = \lambda_{12}(1-\beta)/\beta$
3. Set $x = \min\left\{\frac{-\ln r}{\lambda_1}, \frac{-\ln t}{\lambda_{12}}\right\}$ and $y = \min\left\{\frac{-\ln s}{\lambda_2}, \frac{-\ln t}{\lambda_{12}}\right\}$.
4. Set $u = \exp[-(\lambda_1 + \lambda_{12})x]$ and $v = \exp[-(\lambda_2 + \lambda_{12})y]$.

The desired pair is then (u, v). For more details, see [71].

The subfamily of Marshall-Olkin copulas obtained by taking $\alpha = \beta$ is of particular interest. This is known as the *Cuadras-Augé* family, and its members are given by

$$\mathbf{C}_\theta(u, v) = (\mathbf{\Pi}_2(u, v))^{1-\theta}(\mathbf{M}_2(u, v))^\theta, \tag{C.52}$$

where $u, v \in \mathbb{I}$, and $\theta \in \mathbb{I}$ is a dependence parameter [55, 207].

If U and V are r.v.'s with copula $\mathbf{C}_\theta$, then they are independent for $\theta = 0$, i.e. $\mathbf{C}_0 = \mathbf{\Pi}_2$, and PQD for $\theta > 0$, with $\mathbf{C}_1 = \mathbf{M}_2$. In particular, this family is positively ordered. These 2-copulas satisfy the functional equation $\mathbf{C} = \overline{\mathbf{C}}$ for radial symmetry.

As an illustration, see Figure C.28*ab*, Figure C.29*cd*, and Figure C.30*ef*, where the parameters α, β of the corresponding Marshall-Olkin 2-copulas are equal to one another.

The Cuadras-Augé copulas can be extended for $d \geq 3$ in an obvious way. The corresponding expression is given by

$$\mathbf{C}_\theta(\boldsymbol{u}) = (\mathbf{\Pi}_d(\boldsymbol{u}))^{1-\theta}(\mathbf{M}_d(\boldsymbol{u}))^\theta \tag{C.53}$$

for $\theta \in \mathbb{I}$. For an example see Illustration 5.5.

The multivariate extension of Marshall-Olkin 2-copulas has a complicated form, even if it can be easily simulated due to its probabilistic interpretation [84]. In the particular case $d = 3$, the Marshall-Olkin 3-copula depends on a multidimensional parameter $\boldsymbol{\theta}$, having nine components, and is given by

$$\begin{aligned}\mathbf{C}_{\boldsymbol{\theta}}(\boldsymbol{u}) = & u_1 u_2 u_3 \min\left\{u_1^{-\theta_{12}^1}, u_2^{-\theta_{12}^2}\right\} \min\left\{u_1^{-\theta_{13}^1}, u_3^{-\theta_{13}^2}\right\} \\ & \min\left\{u_2^{-\theta_{23}^1}, u_3^{-\theta_{23}^2}\right\} \max\left\{u_1^{-\theta_{123}^1}, u_2^{-\theta_{123}^2}, u_3^{-\theta_{123}^3}\right\}.\end{aligned}$$

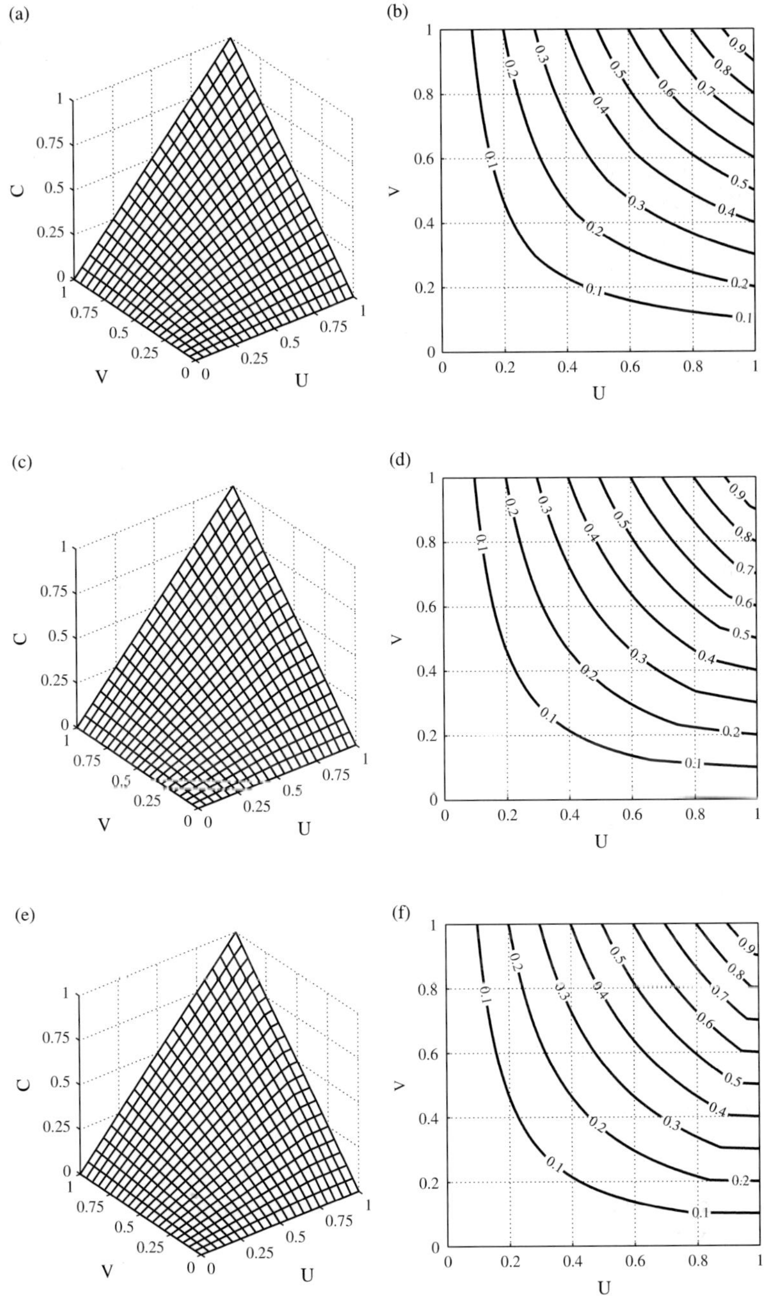

Figure C.28. The Marshall-Olkin 2-copula and the corresponding level curves for several values of the parameters α, β. Here $\alpha = 0.1$, and *(a,b)* $\beta = 0.1$, *(c,d)* $\beta = 0.5$, *(e,f)* $\beta = 0.9$. The probability levels are as indicated

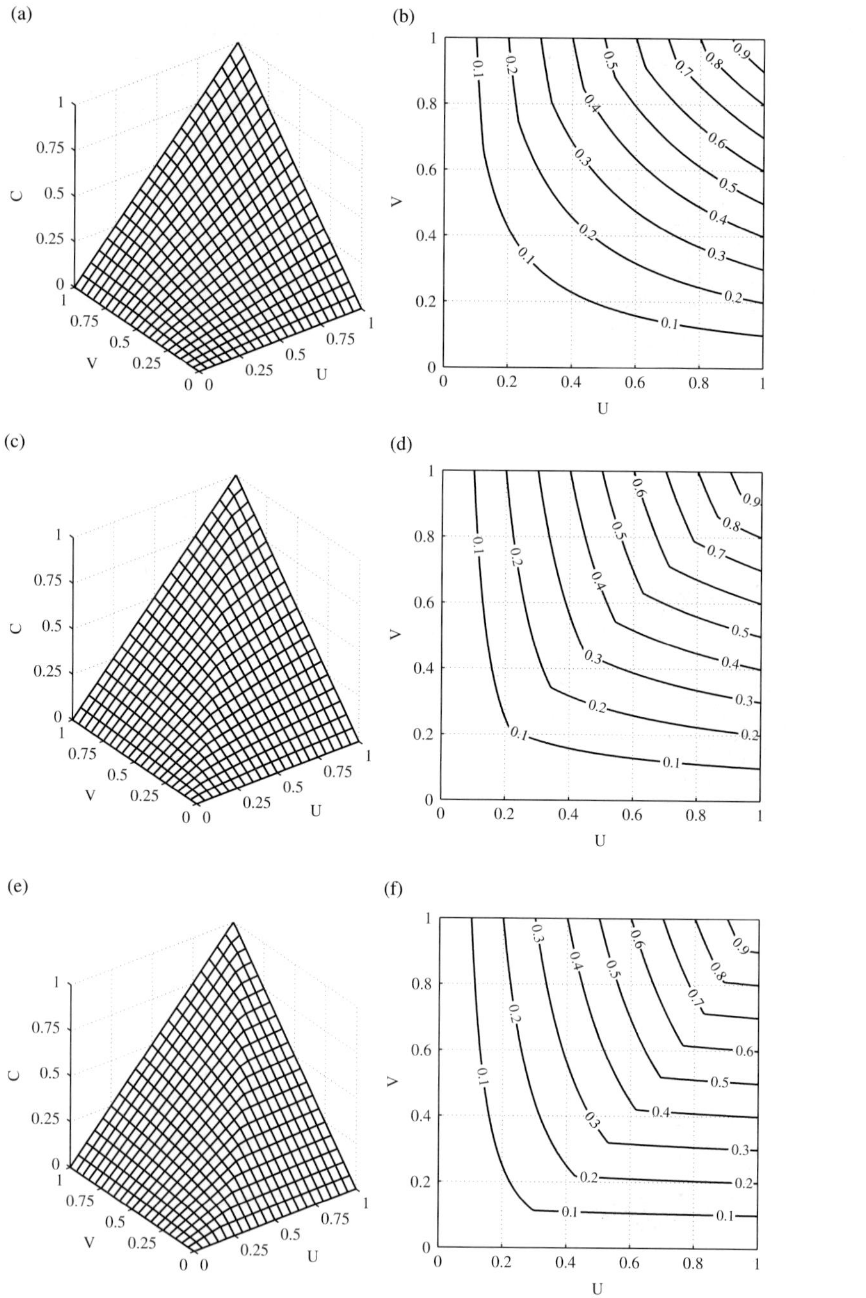

Figure C.29. The Marshall-Olkin 2-copula and the corresponding level curves for several values of the parameters α, β. Here $\alpha = 0.5$, and *(a,b)* $\beta = 0.1$, *(c,d)* $\beta = 0.5$, *(e,f)* $\beta = 0.9$. The probability levels are as indicated

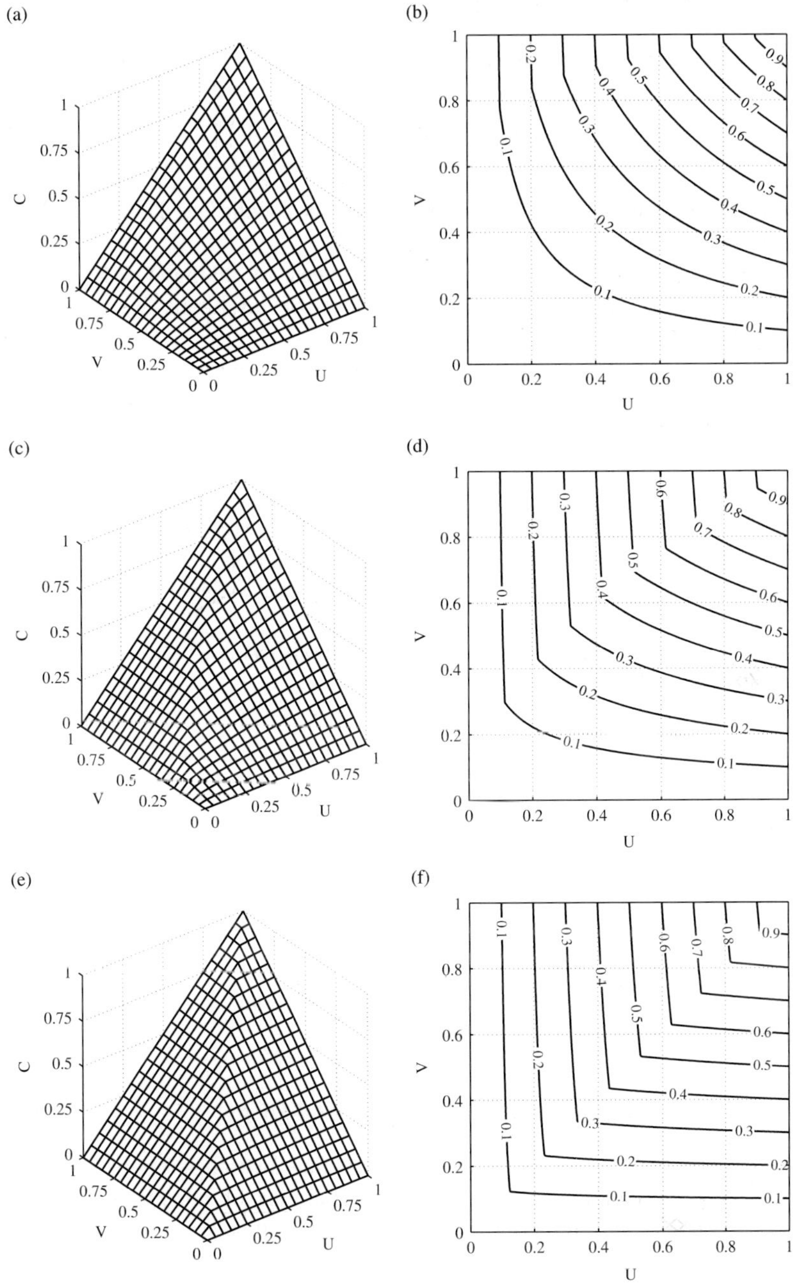

Figure C.30. The Marshall-Olkin 2-copula and the corresponding level curves for several values of the parameters α, β. Here $\alpha = 0.9$, and *(a,b)* $\beta = 0.1$, *(c,d)* $\beta = 0.5$, *(e,f)* $\beta = 0.9$. The probability levels are as indicated

C.14. THE ARCHIMAX FAMILY

The standard expression for members of this family is

$$\mathbf{C}_{\gamma,A}(u,v) = \gamma^{-1}\left[(\gamma(u)+\gamma(v))\,A\left(\frac{\gamma(u)}{\gamma(u)+\gamma(v)}\right)\right] \tag{C.54}$$

where $u, v \in \mathbb{I}$, γ is the generator of an Archimedean copula (see Definition 3.9), and A is a dependence function (see Definition 5.5).

This family includes both the Archimedean copulas, by taking $A = 1$, and the EV copulas, by taking $\gamma(t) = \ln(1/t)$. The members of this family are absolutely continuous if $\gamma(t)/\gamma'(t) \to 0$ as $t \to 0^+$.

Archimax copulas that are neither Archimedean nor extreme can be constructed at will. For instance, consider the dependence function

$$A(t) = \theta t^2 - \theta t + 1,$$

with $\theta \in \mathbb{I}$ (this dependence function is used in *Tawn's mixed model*, see [280]), and as γ use the generator of the Clayton family (see Section C.3).

For an Archimax copula $\mathbf{C}_{\gamma,A}$, the value of Kendall's τ_K is given by

$$\tau_K(\gamma, A) = \tau_K(A) + (1 - \tau_K(A))\tau_K(\gamma), \tag{C.55}$$

where

$$\tau_K(A) = \int_0^1 \frac{t(1-t)}{A(t)}\, \mathrm{d}A'(t), \tag{C.56a}$$

and

$$\tau_K(\gamma) = 1 + 4\int_0^1 \frac{\gamma(t)}{\gamma'(t)}\, \mathrm{d}t, \tag{C.56b}$$

are, respectively, the values of Kendall's τ_K of an EV copula generated by A, and of an Archimedean copula generated by γ.

C.15. CONSTRUCTION METHODS FOR COPULAS

In this section we present three methods for constructing 2-copulas. These constructions are very important in applications, where families of copulas with more parameters may sometimes provide significantly better fits than any monoparametric subfamily.

The first method is discussed in [80, 163, 164, 198], and includes, as special cases, the family of copulas BB1–BB7 presented in [155]. The second method arises from the ideas presented in [111], and has recently been generalized in [76]. Based on the fact that several families of 2-copulas are symmetric, i.e. $\mathbf{C}(u, v) = \mathbf{C}(v, u)$, this method provides a simple way to generate copulas that may fail this property. Asymmetric families of copulas sometimes improve the fit of a model (see, for instance, [109, 121]). The third method presents some constructions of copulas based on their diagonal section. Other procedures (e.g., geometric and algebraic methods, shuffles, and ordinal sums) can be found in [207].

C.15.1 Transformation of Copulas

Let us denote by Θ the set of continuous and strictly increasing functions $h : \mathbb{I} \to \mathbb{I}$, with $h(0) = 0$ and $h(1) = 1$. Given a 2-copula $\mathbf{C}$ and a function $h \in \Theta$, let $\mathbf{C}_h$ be the *h-transformation* of $\mathbf{C}$ defined by

$$\mathbf{C}_h(u, v) = h^{-1}(\mathbf{C}(h(u), h(v))). \tag{C.57}$$

Note that $\mathbf{C}_h$ need not be a copula. In fact, if $h(t) = t^2$ and $\mathbf{C} = \mathbf{W}_2$, then $(\mathbf{W}_2)_h$ is not 2-increasing. However, we have the following result.

PROPOSITION C.1. *For each $h \in \Theta$, the following statements are equivalent:*

1. *h is concave;*
2. *for every copula* $\mathbf{C}$*, the h-transformation of* $\mathbf{C}$ *given by Eq. (C.57) is a copula.*

We now give some examples.

ILLUSTRATION C.1. ►

Consider the strict generator $\gamma_\theta(t) = (t^{-\theta} - 1)/\theta$, $\theta > 0$, with inverse $\gamma_\theta^{-1}(t) = (1 + \theta t)^{-1/\theta}$, of an Archimedean copula belonging to the Clayton family $\mathbf{C}_\theta$ (see Section C.3). Set $h(t) = \exp(-\gamma(t))$. Then $h(t)$ satisfies the assumptions of Proposition C.1, and it is easy to verify that $(\mathbf{\Pi}_2)_h = \mathbf{C}_\theta$. Therefore, every Clayton copula can be obtained as a transformation of the copula $\mathbf{\Pi}_2$ with respect to a suitable function h. The same procedure can be applied to obtain any family of strict Archimedean copulas. ◄

ILLUSTRATION C.2. ►

Consider the strict generator $\gamma_\theta(t) = (-\ln t)^\theta$ of the Gumbel-Hougaard family $\mathbf{C}_\theta$ (see Section C.2). Set $h_\delta(t) = \exp\left(-(t^{-\delta} - 1)/\delta\right)$, with $\delta > 0$. Then $h(t)$ satisfies the assumptions of Proposition C.1, and it is easy to verify that the h_δ-transformation of $\mathbf{C}_\theta$, $(\mathbf{C}_\theta)_{h_\delta}$, is a member of the family BB3 given in [155].

In general, let $\gamma_{1,\theta}$ and $\gamma_{2,\delta}$, respectively, be the strict generators of the two Archimedean families of copulas $\mathbf{C}_\theta$ and $\mathbf{C}_\delta$. Set $h_{1,\theta}(t) = \exp(-\gamma_{1,\theta}(t))$ and $h_{2,\delta}(t) = \exp(-\gamma_{2,\delta}(t))$. Then $(\mathbf{C}_\theta)_{h_{2,\delta}}$ is a two-parameter Archimedean copula, with additive generator $\gamma_{1,\theta} \circ h_{2,\delta}$. Similarly, $(\mathbf{C}_\delta)_{h_{1,\theta}}$ is a two-parameter Archimedean copula with additive generator $\gamma_{2,\delta} \circ h_{1,\theta}$. In this way, we can obtain the families of copulas BB1–BB7 given in [155]. ◄

Note that, if H is a bivariate distribution function with marginals F and G, and h satisfies the assumptions of Proposition C.1, then $\widetilde{H}(x, y) = h(H(x, y))$ is a bivariate distribution function with marginals $h(F)$ and $h(\mathrm{G})$, and the copula of $\widetilde{H}$ is $\mathbf{C}_{h^{-1}}$. A generalization of this transformation to the d-dimensional case, $d \geqslant 3$, is considered in [198].

C.15.2 Composition of Copulas

Let us denote by Θ the set of continuous and strictly increasing functions $h : \mathbb{I} \to \mathbb{I}$, with $h(0) = 0$ and $h(1) = 1$. Let $H : \mathbb{I}^2 \to \mathbb{I}$ be increasing in each variable, with $H(0, 0) = 0$ and $H(1, 1) = 1$. Given f_1, f_2, g_1 and g_2 in Θ, and $\mathbf{A}$ and $\mathbf{B}$ 2-copulas, the function

$$F_{\mathbf{A},\mathbf{B}}(u, v) = H\left(\mathbf{A}(f_1(u), g_1(v)), \mathbf{B}(f_2(u), g_2(v))\right) \tag{C.58}$$

is called the *composition* of $\mathbf{A}$ and $\mathbf{B}$. In general, $F_{\mathbf{A},\mathbf{B}}$ is not a copula. In [76] some conditions on H, f_1, f_2, g_1 and g_2 are given in order to ensure that $F_{\mathbf{A},\mathbf{B}}$ is a copula. Here, we only present two special cases of composition [111].

PROPOSITION C.2. *Let* $\mathbf{A}$ *and* $\mathbf{B}$ *be 2-copulas. Then*

$$\mathbf{C}_{\alpha,\beta}(u, v) = \mathbf{A}(u^{\alpha}, v^{\beta})\,\mathbf{B}(u^{1-\alpha}, v^{1-\beta}). \tag{C.59}$$

defines a family of copulas $\mathbf{C}_{\alpha,\beta}$*, with parameters* $\alpha, \beta \in \mathbb{I}$.

In particular, if $\alpha = \beta = 1$, then $\mathbf{C}_{1,1} = \mathbf{A}$, and, if $\alpha = \beta = 0$, then $\mathbf{C}_{0,0} = \mathbf{B}$. For $\alpha \neq \beta$, the copula $\mathbf{C}$ in Eq. (C.59) is, in general, *asymmetric*, that is $\mathbf{C}(u, v) \neq \mathbf{C}(v, u)$ for some $(u, v) \in \mathbb{I}^2$.

An interesting statistical interpretation can be given for this family. Let U_1, V_1, U_2 and V_2 be r.v.'s Uniform on $\mathbb{I}$. If $\mathbf{A}$ is the copula of (U_1, U_2), and $\mathbf{B}$ is the copula of (V_1, V_2), and the pairs (U_1, U_2) and (V_1, V_2) are independent, then $\mathbf{C}_{\alpha,\beta}$ is the joint distribution function of

$$U = \max\left\{U_1^{1/\alpha}, V_1^{1/(1-\alpha)}\right\} \quad \text{and} \quad V = \max\left\{U_2^{1/\beta}, V_2^{1/(1-\beta)}\right\}. \tag{C.60}$$

In particular, this probabilistic interpretation yields an easy way to simulate a copula expressed by Eq. (C.59), provided that $\mathbf{A}$ and $\mathbf{B}$ can be easily simulated. A simpler way of constructing asymmetric copulas is based on the following result.

PROPOSITION C.3 (Khoudraji). *Let* $\mathbf{C}$ *be a symmetric copula,* $\mathbf{C} \neq \mathbf{\Pi}_2$*. A family of asymmetric copulas* $\mathbf{C}_{\alpha,\beta}$*, with parameters* $0 < \alpha, \beta < 1$, $\alpha \neq \beta$*, that includes* $\mathbf{C}$ *as a limiting case, is given by*

$$\mathbf{C}_{\alpha,\beta}(u, v) = u^{\alpha} v^{\beta} \cdot \mathbf{C}\left(u^{1-\alpha}, v^{1-\beta}\right). \tag{C.61}$$

ILLUSTRATION C.3. ►

Let $\mathbf{A}$ and $\mathbf{B}$ be Archimedean 2-copulas generated, respectively, by γ_1 and γ_2. In view of Proposition C.3, the following functions are copulas for all $\alpha, \beta \in \mathbb{I}$:

$$\mathbf{C}_{\alpha,\beta}(u, v) = \gamma_1^{-1}(\gamma_1(u^\alpha) + \gamma_1(v^\beta)) \cdot \gamma_2^{-1}(\gamma_2(u^{1-\alpha}) + \gamma_2(v^{1-\beta})). \tag{C.62}$$

In particular, if $\gamma_1(t) = \gamma_2(t) = (-\ln t)^\delta$, with $\delta \geq 1$, then $\mathbf{A}$ and $\mathbf{B}$ belong to the Gumbel-Hougaard family of copulas (see Section C.2).

By considering Eq. (C.59), we obtain a three-parameter family of asymmetric copulas

$$\mathbf{C}_{\alpha,\beta,\delta}(u, v) = \exp\Big(-\left[(-\alpha \ln u)^\delta + (-\beta \ln v)^\delta\right]^{1/\delta} - \left[(-\overline{\alpha} \ln u)^\delta + (-\overline{\beta} \ln v)^\delta\right]^{1/\delta}\Big), \tag{C.63}$$

where $\overline{\alpha} = 1 - \alpha$ and $\overline{\beta} = 1 - \beta$, representing a generalization of the Gumbel-Hougaard family (also obtained by the method outlined in [111]). ◄

The construction in Eq. (C.59) has the following generalization to the d-dimensional case, $d \geqslant 3$.

PROPOSITION C.4. *Let* $\mathbf{A}$ *and* $\mathbf{B}$ *be d-copulas. Then*

$$\mathbf{C}_{\alpha_1,\dots,\alpha_d}(\mathbf{u}) = \mathbf{A}(u_1^{\alpha_1}, \dots, u_d^{\alpha_d})\, \mathbf{B}(u_1^{1-\alpha_1}, \dots, u_d^{1-\alpha_d}) \tag{C.64}$$

defines a family of d-copulas $\mathbf{C}_{\alpha_1,\dots,\alpha_d}$ *with parameters* $\alpha_1, \dots, \alpha_d \in \mathbb{I}$.

NOTE C.1 (EV copulas). If $\mathbf{A}$ and $\mathbf{B}$ are EV copulas, then the copula $\mathbf{C}$ given by Eq. (C.64) is also an EV copula: this is a direct consequence of the fact that every EV copula is max-stable — see Section 5.1. In particular, in the bivariate case, we can calculate explicitly the dependence function [111] associated with $\mathbf{C}$ — see Section 5.2.

ILLUSTRATION C.4. ►

Let $\mathbf{A}_\theta$ be a 3-dimensional Cuadras-Augé copula (see Section C.13), with $\theta \in \mathbb{I}$, and let $\mathbf{B} = \mathbf{\Pi}_3$. Given $\alpha_1, \alpha_2, \alpha_3 \in \mathbb{I}$, let $\mathbf{C}$ be as in Eq. (C.64), i.e.

$$\mathbf{C}(\mathbf{u}) = u_1^{1-\theta\alpha_1} u_2^{1-\theta\alpha_2} u_3^{1-\theta\alpha_3} \min\left\{u_1^{\theta\alpha_1}, u_2^{\theta\alpha_2}, u_3^{\theta\alpha_3}\right\}. \tag{C.65}$$

Then, using Note C.1, $\mathbf{C}$ is an EV copula (see also Illustration 5.5). Moreover, the marginals $\mathbf{C}_{i,j}$ of $\mathbf{C}$, with $i, j = 1, 2, 3$ and $i \neq j$, are Marshall-Olkin bivariate copulas, with parameters $(\theta\alpha_i, \theta\alpha_j)$. ◄

C.15.3 Copulas with Given Diagonal Section

The problem of finding a copula $\mathbf{C}$ with a given *diagonal section* (see Definition 3.3) has been discussed in several papers [101, 208, 102, 79]. Its relevance stems from the fact that, if $\mathbf{C}$ is a copula associated with two r.v.'s U and V Uniform on $\mathbb{I}$, then the diagonal section of $\mathbf{C}$ contains some information about the behavior of the r.v.'s $\max\{U, V\}$ and $\min\{U, V\}$ (see also Illustration 3.3).

The diagonal section of a copula $\mathbf{C}$ is the function $\boldsymbol{\delta}_{\mathbf{C}} : \mathbb{I} \to \mathbb{I}$ given by $\boldsymbol{\delta}_{\mathbf{C}}(t) = \mathbf{C}(t, t)$, and satisfies the following properties:

1. $\boldsymbol{\delta}_{\mathbf{C}}(0) = 0$ and $\boldsymbol{\delta}_{\mathbf{C}}(1) = 1$;
2. $\boldsymbol{\delta}_{\mathbf{C}}(t) \le t$ for all $t \in \mathbb{I}$;
3. $\boldsymbol{\delta}_{\mathbf{C}}$ is increasing;
4. $\boldsymbol{\delta}_{\mathbf{C}}$ is 2-Lipschitz, i.e. $|\boldsymbol{\delta}_{\mathbf{C}}(t) - \boldsymbol{\delta}_{\mathbf{C}}(s)| \le 2|t - s|$ for all $s, t \in \mathbb{I}$.

We denote by $\mathcal{D}$ the set of all functions with properties (1)–(4), and a function in $\mathcal{D}$ is called a *diagonal*. The question naturally arising is whether, for each diagonal $\boldsymbol{\delta}$, there exists a copula whose diagonal section equals $\boldsymbol{\delta}$. Here we write $\boldsymbol{\delta}_t$ to indicate $\boldsymbol{\delta}(t)$.

The first example is the *Bertino* copula $\boldsymbol{B}_\delta : \mathbb{I}^2 \to \mathbb{I}$ given by

$$\boldsymbol{B}_\delta(u, v) = \min\{u, v\} - \min\left\{t - \boldsymbol{\delta}_t : t \in [\min\{u, v\}, \max\{u, v\}]\right\}. \tag{C.66}$$

The second example is the *Diagonal* copula $\boldsymbol{D}_\delta : \mathbb{I}^2 \to \mathbb{I}$ given by

$$\boldsymbol{D}_\delta(u, v) = \min\left\{\frac{\boldsymbol{\delta}_u + \boldsymbol{\delta}_v}{2}, u, v\right\}. \tag{C.67}$$

If $\mathbf{C}$ is a copula whose diagonal section is $\boldsymbol{\delta}$, then $\mathbf{C} \succ \boldsymbol{B}_\delta$. Moreover, if $\mathbf{C}$ is also symmetric, it follows that $\mathbf{C} \prec \boldsymbol{D}_\delta$.

The values of Kendall's τ_K for, respectively, Bertino and Diagonal copulas are given by:

$$\tau_K(\boldsymbol{B}_\delta) = 8 \cdot \int_0^1 \boldsymbol{\delta}_t \, \mathrm{d}t - 3, \tag{C.68a}$$

$$\tau_K(\boldsymbol{D}_\delta) = 4 \cdot \int_0^1 \boldsymbol{\delta}_t \, \mathrm{d}t - 3, \tag{C.68b}$$

which may provide a method of fitting the copulas of interest to the available data. The interesting point is that Eqs. (C.68) yield bounds for the values of Kendall's τ_K of a copula $\mathbf{C}$, once the values at the quartiles $\mathbf{C}(i/4, i/4)$, $i = 1, 2, 3$, are known [209].

Constructions of (non-symmetric) copulas with given diagonal section are recently considered in [78], where a method for constructing a family of copulas with given lower and upper tail dependence coefficients is also presented. Specifically, given λ_L and λ_U (the prescribed tail dependence coefficients), consider the following piecewise linear function

$$\delta(t) = \begin{cases} \lambda_L t, & t \in \left[0, \frac{\lambda_U - 1}{2 - \lambda_U - \lambda_L}\right], \\ (2 - \lambda_U)t - 1 + \lambda_U, & \text{otherwise.} \end{cases}$$

Set $l_\delta = \max\{\lambda_L, 2 - \lambda_U\}$. Then, it can be shown that, for every $\lambda \in \left(1 - \frac{1}{l_\delta}, \frac{1}{l_\delta}\right)$, all the copulas in the family $\{\mathbf{C}_{\delta,\lambda}\}$ defined by

$$\mathbf{C}_{\delta,\lambda}(u, v) = \min\{\lambda\, \delta(u) + (1 - \lambda)\, \delta(v), u, v\}$$

have a upper tail dependence coefficient λ_U and a lower tail dependence coefficient λ_L.

REFERENCES

1. Abdous B, Genest JC, Rémillard B (2005) Dependence properties of meta-elliptical distributions. In Duchesne P, Rémillard B, eds, *Statistical Modeling and Analysis for Complex Data Problems*, Springer-Verlag, New York, pp 1–15
2. Abdous B, Ghoudi K, Khoudraji A (1999) Non parametric estimation of the limit dependence function of multivariate extremes. *Extremes*, 2:3:245–268
3. Agnese C, D'Asaro F, Grossi G, Rosso R (1996) Scaling properties of topologically random channel networks. *J. Hydrology*, 187:183–193
4. Ancey C, Gervasoni C, Meunier M (2004) Computing extreme avalanches. *Cold Regions Science and Technology*, 39:161–180
5. Ancey C, Meunier M, Richard D (2003) Inverse problem in avalanche dynamics models. *Water Resour. Res.*, 39(4):1099
6. ASCE (1990) *Minimum Design Loads for Buildings and Other Structures*, vol 7–88. ASCE
7. Avérous J, Dortet-Bernadet J-L (2000) LTD and RTI dependence orderings. *Canad. J. Statist.*, 28:151–157
8. Bak P (1996) *How Nature Works: The Science of Self-organized Criticality*. Copernicus Press, New York
9. Barabasi AL (2002) *Linked. The New Science of Networks*. Perseus Publishing, Cambridge
10. Barbolini M, Cappabianca F, Savi F (2003) A new method for estimation of avalanche distance exceeded probability. *Surveys in Geophysics*, 24:587–601
11. Barbolini M, Gruber V, Keyloch CJ, Naaim M, Savi F (2000) Application and evaluation of statistical and hydraulic-continuum dense-snow avalanche models to five real European sites. *Cold Regions Science and Technology*, 31(2):133–149
12. Barbolini M, Natale L, Savi F (2002) Effect of release conditions uncertainty on avalanche hazard mapping. *Natural Hazards*, 25:225–244
13. Beirlant J, Teugels JL, Vynckier P (1996) *Practical Analysis of Extreme Values*, Leuven University Press, Leuven
14. Benson MA (1962) Evolution of methods for evaluating the occurrence of floods. *U.S. Geol. Surv. Water Supply Pap.*, 1580-A
15. Berg D, Bakken H (2005) A goodness-of-fit test for copulæ based on the probability integral transform. Technical Report SAMBA/41/05, Norsk Regnesentral, Oslo, Norway
16. Berman SM (1961–1962) Convergence to bivariate limiting extreme value distributions. *Ann. Inst. Statist. Math.*, 13:217–233
17. Bocchiola D, De Michele C, Rosso R (2003) Review of recent advances in index flood estimation. *Hydrology and Earth System Sciences*, 7(3):283–296

18. Bocchiola D, Megliani M, Rosso R (2006) Regional snow depth frequency curves for avalanche hazard mapping in Central Italian Alps. *Cold Regions Science and Technology* (in press)
19. Bocchiola D, Rosso R (2006) On the distribution of the daily Snow Water Equivalent in central Italian Alps. *Advan. in Water Resour.* (in press)
20. Bocchiola D, Rulli MC, Rosso R (2006) Transport of large woody debris in the presence of obstacles. *Geomorphology*, 76:166–178
21. Boccotti P (2000) *Wave Mechanics for Ocean Engineering*, Elsevier Oceanography Series. Elsevier Science, Amsterdam
22. Bolt BA (1999) *Earthquakes*, W.H. Freeman, New York
23. Bolt BA, Horn WL, Macdonald GA, Scott RF (1977) *Geological Hazards*, Springer-Verlag, New York
24. Breymann W, Dias A, Embrechts P (2003) Dependence structures for multivariate high-frequency data in finance. *Quant. Finance*, 3:1–14
25. Burkard A, Salm B (1992) Die bestimmung der mittleren anrissmachtigkeit d_0 zur berechnung von fliesslawinen (Estimate of the average release depth d_0 for the calculation of flowing avalanches). Technical Report 668, Swiss Federal Institute for Snow and Avalanche Research, Davos
26. Burlando P, Rosso R (1991) Comment on "Parameter estimation and sensitivity analysis for the modified Bartlett-Lewis rectangular pulses model of rainfall" by Islam S et al. *J. Geophys. Res. - Atmospheres*, 96(D5):9391–9395
27. Burlando P, Rosso R (1996) Scaling and multiscaling models of depth-duration-frequency curves of storm precipitation. *J. Hydrology*, 187:45–64
28. Burlando P, Rosso R (2002) Effects of transient climate scenarios on basin hydrology: 1. Precipitation scenarios for the Arno river, Central Italy. *Hydrolog. Proc.*, 16:1151–1175
29. Burlando P, Rosso R (2002) Effects of transient climate scenarios on basin hydrology: 2. Impact on runoff variability in the Arno river, Central Italy. *Hydrolog. Proc.*, 16:1177–1199
30. Burn DH (1990) Evaluation of regional flood frequency analysis with a region of influence approach. *Water Resour. Res.*, 26(10):2257–2265
31. Burn DH (1997) Catchments similarity for regional flood frequency analysis using seasonality measures. *J. Hydrology*, 202:212–230
32. Burroughs SM, Tebbens SF (2001) Upper-truncated power laws in natural systems. *Pure Appl. Geophys*, 158:741–757
33. Burroughs SM, Tebbens SF (2005) Power-law scaling and probabilistic forecasting of tsunami runup heights. *Pure Appl. Geophys*, 162:331–342
34. Cannon SH (2000) Debris-flow response of southern California watersheds burned by wildfire. In Wieczorek GF, Naeser ND, eds, *Debris-Flow Hazards Mitigation, Mechanics, Prediction and Assessment*, Balkema, Rotterdam, pp 45–52
35. Capéraà P, Fougères A-L, Genest C (1997) A nonparametric estimation procedure for bivariate extreme value copulas. *Biometrika*, 84(3):567–577
36. Castillo E (1988) *Extreme Value Theory in Engineering*, Academic Press, San Diego, CA
37. Castillo E, Hadi AS (1997) Estimating the parameters of the Generalized Pareto law. *J. Amer. Statist. Assoc.*, 92(440):159–173
38. Chakak A, Koehler KJ (1995) A strategy for constructing multivariate distributions. *Comm. Stat. – Simulation*, 24:537–550

39. Chan JC, Jan CD (2000) Debris-flow occurrence probability on hillslopes. In Wieczorek GF, Naeser ND, eds, *Debris-Flow Hazards Mitigation, Mechanics, Prediction and Assessment*, Balkema, Rotterdam, pp 411–416
40. Chapman CR (2004) The hazard of near-Earth asteroid impacts on Earth. *Earth and Planetary Science Letters*, 1:1–15
41. Chapman CR, Morrison D (1994) Impact on the Earth by asteroids and comets: assessing the hazard. *Nature*, 367:33–40
42. Charpentier A (2006) *Dependence Structures and Limiting Results, with Applications in Finance and Insurance*. PhD thesis, Katholieke Universiteit Leuven (Belgium)
43. Cheng E, Yeung C (2002) Generalized extreme gust wind speeds distributions. *J. Wind Eng. Ind. Aerod.*, 90:1657–1669
44. Chow VT, Maidment DR, Mays LW (1998) *Applied Hydrology*, McGraw-Hill, Singapore
45. Coles S (2001) *An Introduction to Statistical Modeling of Extreme Values*. Springer-Verlag
46. Coles SG, Heffernan J, Tawn JA (1990) Dependence measures for extreme value analyses. *Extremes*, 2(4):339–365
47. Coles SG, Tawn JA (1991) Modeling extreme multivariate events. *J. R. Stat. Soc. Ser. B - Stat. Methodol.*, 53(2):377–392
48. Coles SG, Tawn JA (1994) Statistical methods for multivariate extremes: An application to structural design. *Appl. Statistics*, 43:1–48
49. Committee on Techniques for Estimating Probabilities of Extreme Floods (1988) *Techniques for Estimating Probabilities of Extreme Floods. Methods and Recommended Research*. National Academy Press, Washington, D.C.
50. Cook NJ, Harris I (2001) Discussion on "Application of the generalized Pareto distribution to extreme value analysis in wind engineering" by Holmes JD, Moriarty WW. *J. Wind Eng. Ind. Aerod.*, 89:215–224
51. Cordova JR, Rodriguez-Iturbe I (1983) Geomorphoclimatic estimation of extreme flow probabilties. *J. Hydrology*, 65:159–173
52. Cornell CA (1968) Engineering Seismic Hazard and Risk Analysis. *Bull. Seis. Soc. Am.*, 58(5):1583–1606
53. Cornell CA, Winterstein SR (1988) Temporal and magnitude dependence in earthquake recurrence models. *Bull. Seismological Soc. Amer.*, 78:1522–1537
54. Cruden DM, Varnes DJ (1996) Landslide types and processes. In Turner AK, Schuster RL, eds, *Landslides. Investigation and Mitigation. Transportation Research Board Special Report 247*, National Academy Press, Washington D.C., pp 36–75
55. Cuadras CM, Augé J (1981) A continuous general multivariate distribution and its properties. *Comm. Stat. – Theory and Methods*, 10:339–353
56. Cunnane C (1993) Unbiased plotting positions – a review. *J. Hydrology*, 37(3–4): 205–222
57. Dalrymple T (1960) Flood frequency methods. *U.S. Geol. Surv. Water Supply Pap.*, 1543-A:11–51
58. David HA, Nagaraja HN (2003) *Order Statistics*. J. Wiley & Sons, New York
59. De Haan L (1976) Sample extremes: an elementary introduction. *Stat. Neerlandica*, 30:161–172
60. De Haan L (1985) Extremes in high dimensions: the model and some statistics. In *Proc. 45th Session of the Int. Statist. Inst.* Paper 26.3.

61. De Michele C, La Barbera P, Rosso R (2002) Power law distribution of catastrophic floods. In Finnsdòttir HP, Snorrason A, Moss M, eds, *The Extremes of the Extremes: Extraordinary Floods*, vol 271, IAHS, Wallingford, pp 277–282
62. De Michele C, Kottegoda NT, Rosso R (2001) The derivation of areal reduction factor of storm rainfall from its scaling properties. *Water Resour. Res.*, 37(12):3247–3252
63. De Michele C, Kottegoda NT, Rosso R (2002) IDAF curves of extreme storm rainfall: A scaling approach. *Water Science and Technology*, 45(2):83–90
64. De Michele C, Rosso R (2002) A multi-level approach to flood frequency regionalization. *Hydrology and Earth System Sciences*, 6(2):185–194
65. De Michele C, Salvadori G (2002) On the derived flood frequency distribution: analytical formulation and the influence of antecedent soil moisture condition. *J. Hydrology*, 262:245–258
66. De Michele C, Salvadori G (2003) A Generalized Pareto intensity-duration model of storm rainfall exploiting 2-copulas. *J. Geophys. Res. – Atmospheres*, 108(D2):4067
67. De Michele C, Salvadori G, Canossi M, Petaccia A, Rosso R (2005) Bivariate statistical approach to check adequacy of dam spillway. *ASCE – J. Hydrol. Eng.*, 10(1):50–57
68. De Michele C, Salvadori G, Passoni G, Vezzoli R (2006) A multivariate model of sea storms using copulas. (submitted)
69. Deheuvels P (1979) La fonction de dépendence empirique et ses propriétés. Un test non paramétrique d'indépendence. *Acad. Roy. Belg. Bull. Cl. Sci.*, 65(5):274–292
70. Deheuvels P (1984) Probabilistic aspects of multivariate extremes. In *Statistical Extremes and Applications*, pp 117–130. Reidel Publishing Company
71. Devroye L (1986) *Non-Uniform Random Variate Generation*. Springer-Verlag, New York
72. Diaz-Granados MA, Valdes JB, Bras LR (1984) A physically based flood frequency distribution. *Water Resour. Res.*, 20(7):995–1002
73. D'Odorico P, Fagherazzi S, Rigon R (2005) Potential for landsliding: Dependence on hyetograph characteristics. *J. Geophys. Res. – Earth Surface*, 110(F1):F01007
74. Draper L (1963) Derivation of a 'design wave' from instrumental records of sea waves. In *Proc. Inst. Civ. Eng.*, 26:291–304, London
75. Drouet-Mari D, Kotz S (2001) *Correlation and Dependence*. Imperial College Press, London
76. Durante F (2006) *New Results on Copulas and Related Concepts*. PhD thesis, Università di Lecce (Italy)
77. Durante F, Klement EP, Quesada Molina JJ (2007) Copulas: compatibility and Fréchet classes. (submitted)
78. Durante F, Kolesárová A, Mesiar R, Sempi C (2006) Copulas with given diagonal sections: novel constructions and applications. (submitted)
79. Durante F, Mesiar R, Sempi C (2006) On a family of copulas constructed from the diagonal section. *Soft Computing*, 10:490–494
80. Durante F, Sempi C (2005) Copula and semicopula transforms. *Int. J. Math. Math. Sci.*, 2005:645–655
81. Eagleson PS (1972) Dynamics of flood frequency. *Water Resour. Res.*, 8(4):878–898
82. Eagleson PS (1978) Climate, Soil and Vegetation 5. A derived distribution of storm surface runoff. *Water Resour. Res.*, 14(5):741–748
83. Embrechts P, Kluppelberg C, Mikosch T (1997) *Modelling Extremal Events*. Springer-Verlag, Berlin

84. Embrechts P, Lindskog F, McNeil AJ (2003) Modelling dependence with copulas and applications to risk management. In Rachev ST, ed, *Handbook of Heavy Tailed Distributions in Finance*, Elsevier, Amsterdam, pp 329–384
85. Embrechts P, McNeil AJ, Straumann D (2002) Correlation and dependence in risk management: properties and pitfalls. In *Risk Management: Value at Risk and Beyond*, Cambridge University Press, Cambridge, pp 176–223
86. Esary J, Proschan F, Walkup DW (1967) Association of random variables, with applications. *Ann. Math. Statist.*, 38:1466–1474
87. Falk M, Hüsler J, Reiss R (1994) *Laws of Small Numbers: Extremes and Rare Events*. Birkhäuser, Basel
88. Fang H-B, Fang K-T (2002) The meta-elliptical distributions with given marginals. *J. Multivar. Anal.*, 82:1–16
89. Fang K-T, Kotz S, Ng K-W (1990) *Symmetric Multivariate and Related Distributions*. Chapman and Hall, London
90. Favre A-C, El Adlouni S, Perreault L, Thiémonge N, Bobée B (2004) Multivariate hydrological frequency analysis using copulas. *Water Resour. Res.*, 40(W01101)
91. Federal Emergency Management Agency (1997) *Multi-Hazard Identification and Risk Assessment*. Federal Emergency Management Agency, Washington
92. Feller W (1971) *An Introduction to Probability and its Applications*, vol 2. J. Wiley & Sons, New York, second edition
93. FEMA (1990) *FAN: An Alluvial Fan Flooding Computer Program*. Federal Emergency Management Agency, Washington D.C.
94. Fermanian J-D (2005) Goodness-of-fit tests for copulas. *J. Multivar. Anal.*, 95:119–152
95. Fisher RA, Tippet HC (1928) Limiting forms of the frequency distribution of the largest or smallest member of a sample. *Proc. Cambridge Phil. Soc.*, 24:180–190
96. Frahm G (2006) On the extremal dependence coefficient of multivariate distributions. *Statist. Probabil. Lett.*, 76:1470–1481
97. Frahm G, Junker M, Schmidt R (2005) Estimating the tail-dependence coefficient: Properties and pitfalls. *Insurance: Mathematics and Economics*, 37:80–100
98. Francis P, Oppenheimer C (2004) *Volcanoes*, Oxford University Press, Oxford, second edition
99. Frank MJ (1979) On the simultaneous associativity of $f(x,y)$ and $x+y-f(x,y)$. *Aequationes Math.*, 19:194–226
100. Fréchet M (1927) Sur la loi de probabilité de l'écart maximum. *Annales de la Societé Polonaise de Mathématique*, 6:93–117
101. Fredricks GA, Nelsen RB (1997) Copulas constructed from diagonal sections. In Beneš V, Štepán J, eds, *Distributions with Given Marginals and Moment Problems*, Kluwer Academic Publishers, Dordrecht, pp 129–136
102. Fredricks GA, Nelsen RB (2002) The Bertino family of copulas. In Cuadras CM, Fortiana J, Rodríguez Lallena JA, eds, *Distributions with Given Marginals and Statistical Modelling*, Kluwer Academic Publishers, Dordrecht, pp 81–91
103. Fujita TT (1985) *The Downburst*, The University of Chicago, SMRP Research Paper n. 210
104. Galambos J (1975) Order statistics of samples from multivariate distributions. *J. Amer. Statist. Assoc.*, 70:674–680
105. Galambos J (1987) *The Asymptotic Theory of Extreme Order Statistics*, Kreiger Publishing Co., Melbourne (FL)

106. Garralda-Guillem AI (2000) Structure de dépendance des lois de valeurs extrêmes bivariées. *C.R. Acad. Sci. Paris*, 330:593–596
107. Garson RC, Morla-Catalan J, Cornell CA (1975) Tornado risk evaluation using wind speed profiles. *ASCE – J. Struct. Eng. Division*, 101:1167–1171
108. Genest C (1987) Frank's family of bivariate distributions. *Biometrika*, 74(3):549–555
109. Genest C, Favre A-C (2007) Everything you always wanted to know about copula modeling but were afraid to ask. *J. Hydrol. Eng.*, 12 (in press)
110. Genest C, Favre A-C, Béliveau J, Jacques C (2007) Meta-elliptical copulas and their use in frequency analysis of multivariate hydrological data. (submitted)
111. Genest C, Ghoudi K, Rivest L-P (1998) Discussion on the paper *Understanding relationships using copulas* by Frees EW, Valdez EA. *North Amer. Act. J.*, 2:143–149
112. Genest C, Quesada Molina JJ, Rodríguez Lallena JA (1995) De l'impossibilité de construire des lois à marges multidimensionnelles données à partir de copules. *C.R. Acad. Sci. Paris Sér. I Math.*, 320:723–726
113. Genest C, Quessy J-F, Rémillard B (2006) Goodness-of-fit procedures for copula models based on the probability integral transformation. *Scand. J. Statist.*, 33:337–366
114. Genest C, Rémillard B (2005) Validity of the parametric bootstrap for goodness-of-fit testing in semiparametric models. Technical Report GP2005P51, Les cahiers du GERAD, Montréal (Québec), Canada
115. Genest C, Rivest LP (1989) A characterization of Gumbel's family of extreme value distributions. *Statist. Probabil. Lett.*, 8:207–211
116. Ghilardi P, Rosso R (1990) Comment on "Chaos in rainfall" by Rodriguez-Iturbe I et al. *Water Resour. Res.*, 26(8):1837–1839
117. Ghoudi K, Khoudraji A, Rivest L-P (1998) Propriétés statistiques des copules de valeurs extrêmes bidimensionnelles. *Canad. J. Statist.*, 26:187–197
118. Gnedenko BV (1943) Sur la distribution limite du terme maximum d'une série aléatorie. *Ann. Math.*, 44:423–453
119. Gomes L, Vickery BJ (1977) Extreme wind speeds in mixed climates. *J. Wind Eng. and Ind. Aerodynamics*, 2:331–344
120. Gray DD, Katz PG, deMonsabert SM, Cogo NP (1982) Antecedent moisture condition probabilities. *ASCE – J. Irrig. Drain. Division*, 108(2):107–114
121. Grimaldi S, Serinaldi F (2006) Asymmetric copula in multivariate flood frequency analysis. *Advan. in Water Resour.*, 29(8):1155–1167
122. Grimmett GR, Stirzaker DR (1992) *Probability and Random Processes*, Oxford University Press, Oxford, second edition
123. Guillot P, Duband D (1967) La méthode du Gradex pour le calcul de la probabilité des crues á partir des pluies. *IAHS*, 84:560–569
124. Gumbel EJ (1958) *Statistics of Extremes*. Columbia University Press, New York
125. Gumbel EJ (1960) Distributions des valeurs extrémes en plusieurs dimensions. *Publ. Inst. Statist. Univ. Paris*, 9:171–173
126. Gupta VK, Dawdy DR (1994) Regional analysis of flood peaks: multi scaling theory and its physical basis. In Rosso R et al. ed, *Advances in Distributed Hydrology*, Water Resources Publications, Fort Collins pp 149–168
127. Gupta VK, Mesa OJ, Dawdy DR (1994) Multiscaling theory of flood peaks: Regional quantile analysis. *Water Resour. Res.*, 30(12):3405–3421
128. Gupta VK, Waymire EC (1990) Multiscaling Properties of Spatial Rainfall and River Flow Distributions. *J. Geophys. Res. – Atmospheres*, 95(D3):1999–2009

129. Gutenberg B, Richter CF (1944) Frequency of earthquakes in California. *Bull. Seismological Soc. Amer.*, 34:185–188
130. Gutenberg B, Richter CF (1954) *Seismicity of the Earth and Associated Phenomena*. Princeton University Press, Princeton, second edition
131. Harris I (2005) Generalised Pareto methods for wind extremes. Useful tool or mathematical mirage? *J. Wind Eng. Ind. Aerod.*, 93:341–360
132. Harris I (2006) Errors in GEV analysis of wind epoch maxima from Weibull parents. *Wind and Structures*, 9(3)
133. Hawkins RH, Hjelmfelt AT Jr, Zevenbergen AW (1985) Runoff probability, storm depth, and curve numbers. *ASCE – J. Irrig. Drain. Division*, 111(4):330–340
134. Hegan BD, Johnson JD, Severne CM (2003) Landslide risk from the Hipua geothermal area, Lake Taupo, New Zealand. In Picarelli L, ed, *Fast Slope Movements: Prediction and Prevention for Risk Mitigation*, vol 1. Patron Editore, Bologna
135. Hill BM (1975) A simple general approach to inference about the tail of a distribution. *Ann. Statist.*, 3(5):1163–1174
136. Holland PW, Wang YJ (1987) Regional dependence for continuous bivariate densities. *Comm. Stat. – Theory and Methods*, 16:193–206
137. Holmes JD, Moriarty WW (1999) Application of the generalized Pareto distribution to extreme value analysis in wind engineering. *J. Wind Eng. Ind. Aerod.*, 83:1–10
138. Horton RE (1945) Erosional Development of Streams and Their Drainage Basins: Hydrophysical Approach to Quantitative Morphology. *Bull. Geol. Soc. Am.*, 56:275–370
139. Hosking JRM (1990) L-moments: Analysis and estimation of distributions using linear combinations of order statistics. *J. R. Stat. Soc. Ser. B – Stat. Methodol.*, 52(1):105–124
140. Hosking JRM, Wallis JR (1988) The effect of intersite dependence on regional flood frequency analysis. *Water Resour. Res.*, 24(2):588–600
141. Hosking JRM, Wallis JR, Wood EF (1984) Estimation of the generalized extreme value distribution by the method of probability-weighted moments. MRC Technical Summary Report 2674, University of Wisconsin, Mathematics Research Center
142. Hürlimann W (2004) Multivariate Fréchet copulas and conditional value–at–risk. *Int. J. Math. Math. Sci.*, 2004:345–364
143. Hurst HE (1951) Long-term storage capacity of reservoir (with discussion). *Trans. Am. Soc. Civ. Eng.*, 116(2447):770–808
144. Hüsler J, Reiss R-D (1989) Maxima of normal random vectors: between independence and complete dependence. *Statist. Probabil. Lett.*, 7:283–286
145. Hutchinson TP, Lai CD (1990) *Continuous Bivariate Distributions, Emphasising Applications*. Rumsby Scientific Publishing, Adelaide
146. Iida T (1999) A stochastic hydro-geomorphological model for shallow landsliding due to rainstorm. *Catena*, 34(3–4):293–313
147. Institute of Hydrology (1999) *Flood Estimation Handbook*. Institute of Hydrology, Wallingford
148. ICC (2000) *International Building Code*. International Code Council, Falls Church
149. Iverson RM (2000) Landslide triggering by rain infiltration. *Water Resour. Res.*, 36(7):1897–1910
150. Janssen P (2004) *The Interaction of Ocean Waves and Wind*. Cambridge University Press, New York
151. Joe H (1990) Families of min-stable multivariate exponential and multivariate extreme value distributions. *Statist. Probabil. Lett.*, 9:75–81

152. Joe H (1990) Multivariate concordance. *J. Multivar. Anal.*, 35:12–30
153. Joe H (1993) Parametric families of multivariate distributions with given margins. *J. Multivar. Anal.*, 46:262–282
154. Joe H (1994) Multivariate extreme-value distributions with applications to environmental data. *Canad. J. Statist.*, 22:47–64
155. Joe H (1997) *Multivariate Models and Dependence Concepts*. Chapman & Hall, London
156. Johnson ME (1987) *Multivariate Statistical Simulation*. J. Wiley & Sons, New York
157. Juri A, Wüthrich M (2002) Copula convergence theorems for tail events. *Insurance Math. Econom.*, 24:139–148
158. Kaplan S, Garrick BJ (1981) On the quantitative definition of risk. *Risk Analysis*, 1(1):11–27
159. Kelman I (2003) Defining Risk. *FloodRiskNet Newsletter*, 2:6–8
160. Kendall MG (1937) A new measure of rank correlation. *Biometrika*, 6:83–93
161. Keylock CJ, Clung DM, Magnusson MM (1999) Avalanche risk mapping by simulation. *J. Glaciol.*, 45(150):303–315
162. Kimberling CH (1974) A probabilistic interpretation of complete monotonicity. *Aequationes Math.*, 10:152–164
163. Klement EP, Mesiar R, Pap E (2005) Archimax copulas and invariance under transformations. *C. R. Acad. Sci. Paris*, 240:755–758
164. Klement EP, Mesiar R, Pap E (2005) Transformations of copulas. *Kybernetika*, 41:425–434
165. Klemeš V (1978) Physically based stochastic hydrologic analysis. In Chow VT, ed, *Advances in Hydrosciences*, vol 11, Academic Press, New York, pp 285–352
166. Koehler KJ, Symanowski JT (1995) Constructing multivariate distributions with specific marginal distributions. *J. Multivar. Anal.*, 55:261–282
167. Kottegoda NT (1980) *Stochastic Water Resouces Technology*. Halsted Press (J. Wiley & Sons), New York
168. Kottegoda NT, Rosso R (1997) *Statistics, Probability and Reliability for Civil and Environmental Engineers*, McGraw-Hill, New York
169. Kotz S, Nadarajah S (2000) *Extreme Value Distributions, Theory and Applications*, Imperial College Press, London
170. Kruskal WH (1958) Ordinal measures of association. *J. Amer. Statist. Assoc.*, 53:814–861
171. Laaha G, Bloschl G (2005) Low flow estimates from short stream flow records-a comparison of methods. *J. Hydrology*, 150:409–432
172. La Barbera P, Rosso R (1989) On the fractal dimension of stream networks. *Water Resour. Res.* 25(4):735–741
173. Lagomarsino S, Piccardo G, Solari G (1982) Statistical analysis of high return period wind speeds. *J. Wind Eng. and Ind. Aerodynamics*, 41-44:485–496
174. Lai CD, Xie M (2000) A new family of positive quadrant dependent bivariate distributions. *Statist. Probabil. Lett.*, 46:359–364
175. Laternser M, Schneebeli M (2003) Long-term snow climate trends of the Swiss Alps. *Int. J. Climatol.*, 23:733–750
176. Lehmann EL (1966) Some concepts of dependence. *Ann. Math. Statist.*, 37:1137–1153
177. Lied K, Bakkehoi S (1980) Empirical calculations of snow-avalanche run-out distance based on topographic parameters. *J. Glaciol.*, 26(94):165–177

178. Liu H (1991) *Wind Engineering: A Handbook for Structural Engineers*. Prentice Hall, Englewood Cliffs
179. Lomnitz C, Rosenbleuth E (1976) eds. *Seismic Risk and Engineering Decisions*. Elsevier, Amsterdam
180. Lopes R (2005) *Volcano Adventure Guide*. Cambridge University Press, Cambridge
181. Luke YL (1969) *The Special Functions and their Approximations*, vol II. Academic Press, New York
182. Malamud BD, Morein G, Turcotte DL (1998) Forest fires: An example of self-organized critical behavior. *Science*, 281:1840–1842
183. Malamud BD, Turcotte DL (2006) The applicability of power-law frequency statistics to floods. *J. Hydrology*, 322:168–180
184. Mandelbrot B (1982) *The Fractal Geometry of Nature*. W. H. Freeman and Company, New York
185. Marshall A, Olkin I (1967) A generalized bivariate exponential distribution. *J. Appl. Prob.*, 4:291–302
186. Marshall A, Olkin I (1967) A multivariate exponential distribution. *J. Amer. Statist. Assoc.*, 62:30–44
187. Marshall A, Olkin I (1983) Domains of attraction of multivariate Extreme Value distributions. *Ann. Prob.*, 11(1):168–177
188. Marshall A, Olkin I (1988) Families of multivariate distributions. *J. Amer. Statist. Assoc.*, 83(403):834–841
189. Matalas NC, Benson MA (1961) Effects of interstation correlation on regression analysis. *J. Geophys. Res. - Atmospheres*, 66(10):3285–3293
190. Mathiesen MY, Goda Y, Hawkes PJ, Mansard E, Martin MJ, Pelthier E, Thomson EF, Van Vledder G (1994) Recommended practice for extreme wave analysis. *J. Hydr. Res.*, 32(6):803–814
191. McClung DM, Lied K (1987) Statistical and geometrical definition of snow avalanche runout. *Cold Regions Science and Technology*, 13(2):107–119
192. McClung DM, Mears A (1991) Extreme value prediction of snow avalanche runout. *Cold Regions Science and Technology*, 19:163–175
193. McGuire RK (2005) Sesmic Hazard and Risk Analysis. Earthquake Engineering Research Institute Publication n. MNO-10
194. McLelland L, Simkin T, Summers M, Nielson E, Stein TC (1989) *Global volcanism 1975–1985*. Prentice Hall, Englewood Cliffs
195. Menke W, Levin V (2005) A strategy to rapidly determine the magnitude of great earthquakes. *EOS Trans. AGU*, 86(19):185–189
196. Molenberghs G, Lesaffre E (1994) Marginal modeling of correlated ordinal data using a multivariate Plackett distribution. *J. Amer. Statist. Assoc.*, 89:633–644
197. Montgomery DR, Dietrich WE (1994) A physically based model for the topographic control on shallow landsliding. *Water Resour. Res.*, 30(4):1153–1171
198. Morillas PM (2005) A method to obtain new copulas from a given one. *Metrika*, 61:169–184
199. Muir LR, El-Shaarawi AH (1986) On the calculation of extreme wave heights. *Ocean Eng.*, 13(1):93–118
200. Murname RJ, Liu KB, eds. (2004) *Hurricanes and Typhoons*. Columbia University Press, New York
201. Nadarajah S (2003) Extreme Value Theory. Models and simulation. In Rao CR, ed, *Handbook of Statistics*, vol 21, Elsevier Science, pp 607–691

202. Natural Environmental Research Council (1975) *Flood Studies Report*, Natural Environmental Research Council, London
203. Nelsen RB (1986) Properties of a one-parameter family of bivariate distributions with specified marginals. *Comm. Stat. - Theory and Methods*, 15(11):3277–3285
204. Nelsen RB (1996) Nonparametric measures of multivariate association. In Rüschendorf L, Schweizer B, Taylor MD eds, *Distributions with Fixed Marginals and Related Topics*, pages 223–232. Institute of Mathematical Statistics, Hayward, CA
205. Nelsen RB (1997) Dependence and order in families of Archimedean copulas. *J. Multivar. Anal.*, 60:111–122
206. Nelsen RB (2005) Copulas and quasi-copulas: an introduction to their properties and applications. In Klement EP, Mesiar R eds, *Logical, Algebraic, Analytic, and Probabilistic Aspects of Triangular Norms*, Elsevier, Amsterdam, pp 391–413
207. Nelsen RB (2006) *An Introduction to Copulas*. Springer-Verlag, New York, second edition
208. Nelsen RB, Fredricks GA (1997) Diagonal copulas. In Beneš V, Štepán J eds, *Distributions with Given Marginals and Moment Problems*, Kluwer Academic Publishers, Dordrecht, pp 121–127
209. Nelsen RB, Quesada Molina JJ, Rodríguez Lallena JA, Úbeda Flores M (2004) Best possible bounds on sets of bivariate distribution functions. *J. Multivar. Anal.*, 90:348–358
210. Oakes D (1989) Bivariate survival models induced by frailties. *J. Amer. Statist. Assoc.*, 84:487–493
211. Oakes D (2005) On the preservation of copula structure under truncation. *Canad. J. Statist.*, 33:465–468
212. Panchenko V (2005) Goodness-of-fit test for copulas. *Physica A*, 355:176–182
213. Paz M ed. (1994) *International Handbook of Earthquake Engineering: Codes, Programs and Examples*. Chapman and Hall, London
214. Peters E, van Lanen HAJ, Torfs PJJF, Bier G (2005) Drought in groundwater-drought distribution and performance indicators. *J. Hydrology*, 306(1-4):302–317
215. Petruakas C, Aagaard PM (1971) Extrapolation of historical storm data for estimating design-wave heights. *J. Soc. Pet. Eng.*, 11:23–37
216. Pickands J (1975) Statistical inference using extreme order statistics. *Ann. Statist.*, 3(1):119–131
217. Pickands J (1981) Multivariate Extreme Value distributions. *Bull. Int. Statist. Inst.*, 49:859–878
218. Piock-Ellena U, Merz R, Blösch G, Gutknecht D (1999) On the regionalization of flood frequencies - Catchment similarity based on seasonalitry measures. In *Proc. XXVIII Congress IAHR*, page 434.htm
219. Plackett RL (1965) A class of bivariate distributions. *J. Amer. Statist. Assoc.*, 60:516–522
220. Ponce VM (1999) *Engineering Hydrology. Principles and Practices*. Prentice Hall, New Jersey
221. Pons FA (1992) *Regional Flood Frequency Analysis Based on Multivariate Lognormal Models*. PhD thesis, Colorado State University, Fort Collins (Colorado)
222. Poulin A, Huard D, Favre A-C, Pugin S (2006) On the importance of the tail dependence in bivariate frequency analysis. *ASCE - J. Hydrol. Eng.* (accepted in the Special Issue on Copulas in Hydrology)

223. Press WH, Flannery BP, Teukolsky SA, Vetterling WT (1992) *Numerical Recipes in C*. Cambridge University Press, Cambridge, second edition
224. Pugh D (2004) *Changing Sea Levels: Effects of Tides, Weather and Climate*. Cambridge University Press, Cambridge
225. Pyle DM (2003) Size of volcanic eruptions. In Sigurdosson H, Houghton B, Rymer H, Stix J, McNutt S, eds, *Encyclopedia of Volcanoes*, San Diego, Academic Press, pp 263–269
226. Quesada Molina JJ, Rodríguez Lallena JA (1994) Some advances in the study of the compatibility of three bivariate copulas. *J. Ital. Statist. Soc.*, 3:947–965
227. Raftery AE (1984) A continuous multivariate exponential distribution. *Comm. Stat. - Theory and Methods*, 13:947–965
228. Raines TH, Valdes JB (1993) Estimation of flood frequencies for ungaged catchments. *ASCE - J. Hydrol. Eng.*, 119(10):1138–1155
229. Ramachandra Rao A, Hamed KH (2000) *Flood Frequency Analysis*. CRC Press, Boca Raton
230. Reiter L (1990) *Earthquake Hazard Analysis: Issues and Insights*. Columbia University Press, New York
231. Rényi A (1959) On measures of dependence. *Acta Math. Acad. Sci. Hungar.*, 10:441–451
232. Resnick SI (1987) *Extreme Values, Regular Variation and Point Processes*. Springer-Verlag, New York
233. Robinson JS, Sivapalan M (1997) An investigation into the physical causes of scaling and heterogeneity of regional flood frequency. *Water Resour. Res.*, 33(5):1045–1059
234. Robinson JS, Sivapalan M (1997) Scaling of flood frequency: temporal scales and hydrologic regimes. *Water Resour. Res.*, 33(12):2981–2999
235. Roder H, Braun T, Schuhmann W, Boschi E, Buttner R, Zimanawski B (2005) Great Sumatra Earthquake Registers on Electrostatic Sensor. *EOS Trans. AGU*, 86(45):445–449
236. Rodriguez-Iturbe I (1997) Scale of fluctuation of rainfall models. *Water Resour. Res.*, 22(9):15S–37S
237. Rodriguez-Iturbe I, Rinaldo A (1997) *Fractal River Basins: Chance and Self-Organization*. Cambridge Univ. Press, Cambridge
238. Rodríguez Lallena JA, Úbeda Flores M (2004) A new class of bivariate copulas. *Statist. Probabil. Lett.*, 66:315–325
239. Rohatgi VK (1976) *An Introduction to Probability Theory and Mathematical Statistics*. J. Wiley & Sons, New York
240. Rosenblatt M (1952) Remarks on a multivariate transformation. *Ann. Math. Statist.*, 23:470–472
241. Ross S (1996) *Stochastic Processes*. J. Wiley & Sons, New York
242. Rossi F, Fiorentino M, Versace P (1984) Two component extreme value distribution for flood frequency analysis. *Water Resour. Res.*, 20(7):847–856
243. Rosso R, Bacchi B, La Barbera P (1991) Fractal relation of mainstream length to catchment area in river networks. *Water Resour. Res.*, 27(3):381–388
244. Rosso R, Rulli MC (2002) An integrated simulation approach for flash-flood risk assessment: 1. Frequency predictions in the Bisagno river by combining stochastic and deterministic methods. *Hydrology and Earth System Sciences*, 6(2):267–284
245. Rosso R, Rulli MC (2002) An integrated simulation approach for flash-flood risk assessment: 2. Effects of changes in land use a historical perspective. *Hydrology and Earth System Sciences*, 6(2):285–294

246. Rosso R, Rulli MC, Vannucchi G (2006) A physically based model for the hydrologic control on shallow landsliding. *Water Resour. Res.*, 42(6):W06410
247. Rulli MC, Bozzi S, Spada M, Rosso R (2006) Rainfall simulations on a fire disturbed Mediterranean area. *J. Hydrology*, 327:323–338
248. Rulli MC, Rosso R (2005) Modeling catchment erosion after wildfires in the Saint Gabriel Mountains of southern California. *Geophys. Res. Lett.*, 32(19):L19401
249. Salvadori G (2002a) Linear combinations of order statistics to estimate the position and scale parameters of the Generalized Pareto law. *Stoch. Environ. Resear. and Risk Assess.*, 16:1–17
250. Salvadori G (2002b) Linear combinations of order statistics to estimate the position and scale parameters of the Generalized Extreme Values law. *Stoch. Environ. Resear. and Risk Assess.*, 16:18–42
251. Salvadori G (2004) Bivariate return periods via 2-copulas. *Statist. Methodol.*, 1:129–144
252. Salvadori G, De Michele C (2001) From Generalized Pareto to Extreme Values laws: scaling properties and derived features. *J. Geophys. Res. - Atmospheres*, 106(D20):24063–24070
253. Salvadori G, De Michele C (2004) Analytical calculation of storm volume statistics with Pareto-like intensity-duration marginals. *Geophys. Res. Lett.*, 31(L04502)
254. Salvadori G, De Michele C (2004) Frequency analysis via copulas: theoretical aspects and applications to hydrological events. *Water Resour. Res.*, 40(W12511)
255. Salvadori G, De Michele C (2006) Statistical characterization of temporal structure of storms. *Advan. in Water Resour.*, 29(6):827–842
256. Samorodnitsky G, Taqqu MS (1994) *Stable Non-Gaussian Random Processes: Stochastic Models with Infinite Variance*, Chapman and Hall, New York
257. Scaillet O (2006) Kernel based goodness-of-fit tests for copulas with fixed smoothing parameters. *J. Multivar. Anal.*, 98(3):533–543
258. Scarsini M (1984) On measures of concordance. *Stochastica*, 8:201–218
259. Schertzer D, Lovejoy S (1987) Physical modelling and analysis of rain and clouds by anisotropic scaling multiplicative processes. *J. Geophys. Res. - Atmospheres*, 92(D8):9693–9714
260. Schmid F, Schmidt R (2007) Multivariate conditional versions of Spearman's rho and related measures of tail dependence. *J. Multivar. Anal.*, (in press)
261. Schmidt R (2002) Tail dependence for elliptically contoured distributions. *Math. Methods of Operations Research*, 55:301–327
262. Schmidt R (2003) *Dependencies of Extreme Events in Finance. Modelling, Statistics and Data Analysis*. PhD thesis, Universität Ulm, Fakultät für Mathematik und Wirtschaftswissenschaften
263. Schmitz V (2004) Revealing the dependence structure between $X_{(1)}$ and $X_{(n)}$. *J. Statist. Plann. Inference*, 123:41–47
264. Schweizer B, Sklar A (1983) *Probabilistic Metric Spaces*. North-Holland, New York
265. Schweizer B, Wolff EF (1981) On nonparametric measures of dependence for random variables. *Ann. Statist.*, 9:879–885
266. Scott DF (1993) The hydrological effects of fire in South African mountain catchments. *J. Hydrology*, 150(2–4):409–432
267. Sempi C (2003) Copulæ and their uses. In Lindqvist BH, Doksum KA, eds, *Quality, Reliability and Engineering Statistics*, vol 7, World Scientific, Singapore, pp 73–85

268. Shiau JT (2003) Return period of bivariate distributed extreme hydrological events. *Stoch. Environ. Resear. and Risk Assess.*, 17:42–57
269. Shiau JT (2006) Fitting drought duration and severity with two-dimensional copulas. *Water Resour. Managem.*, 20:795–815
270. Sibuya M (1960) Bivariate extreme statistics. *Ann. Inst. Statist. Math. Tokyo*, 11:195–210
271. Silveira L, Charbonnier F, Genta JL (2000) The antecedent soil moisture condition of the curve number procedure. *Hydrol. Sci. J.*, 45(1):3–12
272. Simiu E, Scanlan RH (1986) *Wind Effects on Structures: An Introduction to Wind Engineering*. J. Wiley & Sons, New York, second edition
273. Singh VP (1987) *Hydrologic Frequency Modeling*, Reidel Publishing Company, Boston
274. Sivapalan M, Wood EF, Beven KJ (1990) On hydrologic similarity 3. A dimensionless flood frequency model using a generalized geomorphic unit hydrograph and partial area ruonff generation. *Water Resour. Res.*, 26(1):43–58
275. Sklar A (1959) Fonctions de répartition à n dimensions et leurs marges. *Publ. Inst. Statist. Univ. Paris*, 8:229–231
276. Solari G (1986) Statistical analysis of extreme wind speed. In Lalas DP and Ratto CF, eds, *Modeling of Atmosphere Flow Fields*. World Scientific, Singapore
277. Stedinger JR (1983) Estimating a regional flood frequency distribution. *Water Resour. Res.*, 19(2):503–510
278. Stedinger JR, Vogel RM, Foufoula-Georgiou E (1993) Frequency analysis of extreme events. In Maidment DA, ed, *Handbook of Hydrology*, McGraw-Hill, New York, pp 18.1–18.66
279. Takahashi T (1981) Debris flow. *Annual Review of Fluid Mechanics*, 13:57–77
280. Tawn JA (1988) Bivariate extreme value theory: models and estimation. *Biometrika*, 75:397–415
281. Tawn JA (1990) Modelling multivariate extreme value distributions. *Biometrika*, 77(2):245–253
282. Tawn JA (1993) Extreme sea-levels. In Barnett V, Turkman KF, eds., *Statistics for the Environment*, J. Wiley & Sons, Chichester, pp 243–263
283. Tebbens SF, Burroughs SM, Barton CC, Naar DF (2001) Statistical self-similarity of hotspot seamount volumes modeled as self-similar criticality. *Geophys. Res. Lett.*, 28(14):2711–2714
284. Temez JR (1991) Extended and improved rational method. In *Proc.* XXIV Congress IAHR, vol A, Madrid, pp 33–40
285. Tiago de Oliveira J (1962–1963) Structure theory of bivariate extremes; extensions. *Estudos de Matemática , Estatistica e Econometria*, 7:165–195
286. Tiago de Oliveira J (1975a) Bivariate and multivariate extremal distributions. In *Statistical Distributions in Scientific Work*, vol 1, Riedel Publishing Company, pp 355–361
287. Tiago de Oliveira J (1975b) Bivariate extremes: Extensions. In *Proc. 40th Session of the Int. Statist. Inst.*, vol 2, pp 241–252
288. Tiago de Oliveira J (1984) Bivariate models for extremes. In *Statistical Extremes and Applications*, Riedel Publishing Company pp 131–153
289. Titterington DM, Smith AFM, Makov UE (1985) *Statistical Analysis of Finite Mixture Distributions*. J. Wiley & Sons, Hoboken, N.J.
290. Troen I, Petersen EL (1989) European wind atlas. Technical report, Commission of the European Community, Directorate-General for Science, Research and Development, Brussels

291. Turcotte DL (1997) *Fractals and Chaos in Geology and Geophysics.* Cambridge University Press, Cambridge
292. Turcotte DL (1999) Self-organized criticality. *Rep. Prog. Phys.*, 62:1377–1429
293. Twisdale LA, Dunn WL (1978) Tornado data characterization and wind speed risk. *ASCE - J. Struct. Eng. Division*, 104(10):1611–1630
294. U.S.D.A.-S.C.S (1986) *Nat. Eng. Handbook, Hydrology*. U.S. Dept. of Agriculture, Soil Conservation Service, Washington D.C.
295. Varnes DJ (1984) Commission on Landslides and Other Mass-Movements-IAEG *Landslide Hazard Zonation: A Review of Principles and Practice*. UNESCO, Paris
296. von Mises R (1936) La distribution de la plus grande de *n* valeurs. *Rev. Math. de L'Union Interbalkanique*, 1:141–160
297. Wang QJ (1997) LH-moments for statistical analysis of extreme events. *Water Resour. Res.*, 33(12):2841–2848
298. Wang Z, Ormsbee L (2005) Comparison between probabilistic seismic hazard analysis and flood frequency analysis. *EOS Trans. AGU*, 86(5):51–52
299. Waymire EC, Gupta VK (1981) The mathematical structure of rainfall representation 3. Some applications of the point process theory to rainfall process. *Water Resour. Res.*, 17(5):1287–1294
300. Widder DV (1941) *The Laplace Transform*, Princeton University Press, Princeton
301. Wolde-Tinsæ AM, Porter ML, McKeown DI (1985) Wind speed analysis of tornadoes based on structural damage. *J. Clim. Appl. Meteorol.*, 24:699–710
302. Wood EF (1976) An analysis of the effects of parameter uncertainty in deterministic hydrologic models. *Water Resour. Res.*, 12(5):925–932
303. Wood EF, Hebson C (1986) On hydrologic similarity 1. Derivation of the dimensionless flood frequency curve. *Water Resour. Res.*, 22(11):1549–1554
304. Woodworth PL (1987) Trends in UK mean sea level. *Marine Geodesy*, 11(1):57–87
305. Working Group on California Earthquakes Probabilities (1995) Seismic hazards in southern California: probable earthquakes, 1994 to 2024. *Bull. Seismological Soc. Amer.*, 85:379–439
306. Yang J, Cheng S, Zhang L (2006) Bivariate copula decomposition in terms of comonoticity, countermonoticity and independence. *Insurance Math. Econom.*, 39(2):267–284
307. Yue S (2001) Comments on "Bivariate extreme values distributions: an application of the Gibbs sampler to the analysis of floods" by Adamson PT, Metcalfe AV, Parmentier B *Water Resour. Res.*, 37(4):1107–1110
308. Yue S, Rasmussen P (2002) Bivariate frequency analysis: discussion of some useful concepts in hydrological applications. *Hydrol. Process*, 16:2881–2898
309. Yun S (1997) On domains of attraction of multivariate extreme value distributions under absolute continuity. *J. Multivar. Anal.*, 63:277–295

INDEX

GLOSSARY

- **Sets of numbers**

$\mathbb{N}$: naturals
$\mathbb{Z}$: integers
$\mathbb{R}$: reals
$\mathbb{I}$: unit interval $[0, 1]$

- **Copulas**

$\mathbf{C}$: copula
$\overline{\mathbf{C}}$: survival copula
$\mathbf{c}$: copula density
$\mathbf{W}_d$: d-dimensional Fréchet-Hoeffding lower bound
$\mathbf{M}_d$: d-dimensional Fréchet-Hoeffding upper bound
$\mathbf{\Pi}_d$: d-dimensional independence copula
$\boldsymbol{\delta}_{\mathbf{C}}$: diagonal section of $\mathbf{C}$
γ : Archimedean copula generator
$\gamma^{[-1]}$: pseudo-inverse of γ

- **Association measures**

τ_K : Kendall's τ
ρ_S : Spearman's ρ
ρ_P : Pearson's ρ

- **Probability**

X, Y, Z, W : random variables
U, V : random variables Uniform on $\mathbb{I}$
F_X : c.d.f. of X
f_X : p.d.f. of X
$\overline{F}_X$: survival distribution function of X
$F_{X_1 \cdots X_d}$: joint c.d.f. of $(X_1, \ldots, X_d)$
$f_{X_1 \cdots X_d}$: joint p.d.f. of $(X_1, \ldots, X_d)$
$\overline{F}_{X_1 \cdots X_d}$: joint survival distribution function of $(X_1, \ldots, X_d)$
$F^{-1}, F^{(-1)}$: ordinary inverse of F
$F^{[-1]}$: quasi-inverse of F
c.d.f.: cumulative distribution function

p.d.f. : probability density function
r.v. : random variable or vector
i.d. : identically distributed
i.i.d. : independent identically distributed

- **Statistical operators**

$\mathbb{P}\{\cdot\}$: probability
$\mathbb{E}(\cdot)$: expectation
$\mathbb{V}(\cdot)$: variance
$\mathbb{S}(\cdot)$: standard deviation
$\mathbb{C}(\cdot,\cdot)$: covariance

- **Order Statistics**

$X_{(i)}$: i-th order statistic
$F_{(i)}$: c.d.f. of i-th order statistic
$f_{(i)}$: p.d.f. of i-th order statistic
$F_{(i,j)}$: joint c.d.f. of $(X_{(i)}, X_{(j)})$
$f_{(i,j)}$: joint p.d.f. of $(X_{(i)}, X_{(j)})$

- **Miscellanea**

Dom : domain
Ran : range
Rank : rank
$\mathbf{R}(\cdot)$: risk function
$\zeta(\cdot)$: impact function
$\mathbf{1}_{\{\cdot\}}$: indicator function
$\#(\cdot)$: cardinality
$\Gamma(\cdot)$: Gamma function
$\Gamma_{\cdot}(\cdot)$: Incomplete Gamma function
$\mathrm{B}(\cdot,\cdot)$: Beta function
$\mathrm{B}_{\cdot}(\cdot,\cdot)$: Incomplete Beta function

- **Abbreviations**

EV : Extreme Value
EVC : Extreme Value Copula
MEV : Multivariate Extreme Value
GEV : Generalized Extreme Value
GP : Generalized Pareto
MDA : Maximum Domain of Attraction
CDA : Copula Domain of Attraction
POT : Peaks-Over-Threshold
AMC : Antecedent Moisture Condition

Water Science and Technology Library

1. A.S. Eikum and R.W. Seabloom (eds.): *Alternative Wastewater Treatment*. Low-Cost Small Systems, Research and Development. Proceedings of the Conference held in Oslo, Norway (7–10 September 1981). 1982 ISBN 90-277-1430-4
2. W. Brutsaert and G.H. Jirka (eds.): *Gas Transfer at Water Surfaces*. 1984 ISBN 90-277-1697-8
3. D.A. Kraijenhoff and J.R. Moll (eds.): *River Flow Modelling and Forecasting*. 1986 ISBN 90-277-2082-7
4. World Meteorological Organization (ed.): *Microprocessors in Operational Hydrology*. Proceedings of a Conference held in Geneva (4–5 September 1984). 1986 ISBN 90-277-2156-4
5. J. N.ĕmec: *Hydrological Forecasting*. Design and Operation of Hydrological Forecasting Systems. 1986 ISBN 90-277-2259-5
6. V.K. Gupta, I. Rodríguez-Iturbe and E.F. Wood (eds.): *Scale Problems in Hydrology*. Runoff Generation and Basin Response. 1986 ISBN 90-277-2258-7
7. D.C. Major and H.E. Schwarz: *Large-Scale Regional Water Resources Planning*. The North Atlantic Regional Study. 1990 ISBN 0-7923-0711-9
8. W.H. Hager: *Energy Dissipators and Hydraulic Jump*. 1992 ISBN 0-7923-1508-1
9. V.P. Singh and M. Fiorentino (eds.): *Entropy and Energy Dissipation in Water Resources*. 1992 ISBN 0-7923-1696-7
10. K.W. Hipel (ed.): *Stochastic and Statistical Methods in Hydrology and Environmental Engineering*. A Four Volume Work Resulting from the International Conference in Honour of Professor T.E. Unny (21–23 June 1993). 1994
    10/1: Extreme values: floods and droughts ISBN 0-7923-2756-X
    10/2: Stochastic and statistical modelling with groundwater and surface water applications ISBN 0-7923-2757-8
    10/3: Time series analysis in hydrology and environmental engineering ISBN 0-7923-2758-6
    10/4: Effective environmental management for sustainable development ISBN 0-7923-2759-4
    Set 10/1–10/4: ISBN 0-7923-2760-8
11. S.N. Rodionov: *Global and Regional Climate Interaction: The Caspian Sea Experience*. 1994 ISBN 0-7923-2784-5
12. A. Peters, G. Wittum, B. Herrling, U. Meissner, C.A. Brebbia, W.G. Gray and G.F. Pinder (eds.): *Computational Methods in Water Resources X*. 1994 Set 12/1–12/2: ISBN 0-7923-2937-6
13. C.B. Vreugdenhil: *Numerical Methods for Shallow-Water Flow*. 1994 ISBN 0-7923-3164-8
14. E. Cabrera and A.F. Vela (eds.): *Improving Efficiency and Reliability in Water Distribution Systems*. 1995 ISBN 0-7923-3536-8
15. V.P. Singh (ed.): *Environmental Hydrology*. 1995 ISBN 0-7923-3549-X
16. V.P. Singh and B. Kumar (eds.): *Proceedings of the International Conference on Hydrology and Water Resources* (New Delhi, 1993). 1996
    16/1: Surface-water hydrology ISBN 0-7923-3650-X
    16/2: Subsurface-water hydrology ISBN 0-7923-3651-8

16/3: Water-quality hydrology ISBN 0-7923-3652-6
16/4: Water resources planning and management ISBN 0-7923-3653-4
Set 16/1–16/4 ISBN 0-7923-3654-2

17. V.P. Singh: *Dam Breach Modeling Technology*. 1996 ISBN 0-7923-3925-8
18. Z. Kaczmarek, K.M. Strzepek, L. Somlyódy and V. Priazhinskaya (eds.): *Water Resources Management in the Face of Climatic/Hydrologic Uncertainties*. 1996 ISBN 0-7923-3927-4
19. V.P. Singh and W.H. Hager (eds.): *Environmental Hydraulics*. 1996 ISBN 0-7923-3983-5
20. G.B. Engelen and F.H. Kloosterman: *Hydrological Systems Analysis*. Methods and Applications. 1996 ISBN 0-7923-3986-X
21. A.S. Issar and S.D. Resnick (eds.): *Runoff, Infiltration and Subsurface Flow of Water in Arid and Semi-Arid Regions*. 1996 ISBN 0-7923-4034-5
22. M.B. Abbott and J.C. Refsgaard (eds.): *Distributed Hydrological Modelling*. 1996 ISBN 0-7923-4042-6
23. J. Gottlieb and P. DuChateau (eds.): *Parameter Identification and Inverse Problems in Hydrology, Geology and Ecology*. 1996 ISBN 0-7923-4089-2
24. V.P. Singh (ed.): *Hydrology of Disasters*. 1996 ISBN 0-7923-4092-2
25. A. Gianguzza, E. Pelizzetti and S. Sammartano (eds.): *Marine Chemistry*. An Environmental Analytical Chemistry Approach. 1997 ISBN 0-7923-4622-X
26. V.P. Singh and M. Fiorentino (eds.): *Geographical Information Systems in Hydrology*. 1996 ISBN 0-7923-4226-7
27. N.B. Harmancioglu, V.P. Singh and M.N. Alpaslan (eds.): *Environmental Data Management*. 1998 ISBN 0-7923-4857-5
28. G. Gambolati (ed.): *CENAS. Coastline Evolution of the Upper Adriatic Sea Due to Sea Level Rise and Natural and Anthropogenic Land Subsidence*. 1998 ISBN 0-7923-5119-3
29. D. Stephenson: *Water Supply Management*. 1998 ISBN 0-7923-5136-3
30. V.P. Singh: *Entropy-Based Parameter Estimation in Hydrology*. 1998 ISBN 0-7923-5224-6
31. A.S. Issar and N. Brown (eds.): *Water, Environment and Society in Times of Climatic Change*. 1998 ISBN 0-7923-5282-3
32. E. Cabrera and J. García-Serra (eds.): *Drought Management Planning in Water Supply Systems*. 1999 ISBN 0-7923-5294-7
33. N.B. Harmancioglu, O. Fistikoglu, S.D. Ozkul, V.P. Singh and M.N. Alpaslan: *Water Quality Monitoring Network Design*. 1999 ISBN 0-7923-5506-7
34. I. Stober and K. Bucher (eds): *Hydrogeology of Crystalline Rocks*. 2000 ISBN 0-7923-6082-6
35. J.S. Whitmore: *Drought Management on Farmland*. 2000 ISBN 0-7923-5998-4
36. R.S. Govindaraju and A. Ramachandra Rao (eds.): *Artificial Neural Networks in Hydrology*. 2000 ISBN 0-7923-6226-8
37. P. Singh and V.P. Singh: *Snow and Glacier Hydrology*. 2001 ISBN 0-7923-6767-7
38. B.E. Vieux: *Distributed Hydrologic Modeling Using GIS*. 2001 ISBN 0-7923-7002-3
39. I.V. Nagy, K. Asante-Duah and I. Zsuffa: *Hydrological Dimensioning and Operation of Reservoirs*. Practical Design Concepts and Principles. 2002 ISBN 1-4020-0438-9

Water Science and Technology Library

40. I. Stober and K. Bucher (eds.): *Water-Rock Interaction*. 2002 ISBN 1-4020-0497-4
41. M. Shahin: *Hydrology and Water Resources of Africa*. 2002 ISBN 1-4020-0866-X
42. S.K. Mishra and V.P. Singh: *Soil Conservation Service Curve Number (SCS-CN) Methodology*. 2003 ISBN 1-4020-1132-6
43. C. Ray, G. Melin and R.B. Linsky (eds.): *Riverbank Filtration*. Improving Source-Water Quality. 2003 ISBN 1-4020-1133-4
44. G. Rossi, A. Cancelliere, L.S. Pereira, T. Oweis, M. Shatanawi and A. Zairi (eds.): *Tools for Drought Mitigation in Mediterranean Regions*. 2003 ISBN 1-4020-1140-7
45. A. Ramachandra Rao, K.H. Hamed and H.-L. Chen: *Nonstationarities in Hydrologic and Environmental Time Series*. 2003 ISBN 1-4020-1297-7
46. D.E. Agthe, R.B. Billings and N. Buras (eds.): *Managing Urban Water Supply*. 2003 ISBN 1-4020-1720-0
47. V.P. Singh, N. Sharma and C.S.P. Ojha (eds.): *The Brahmaputra Basin Water Resources*. 2004 ISBN 1-4020-1737-5
48. B.E. Vieux: *Distributed Hydrologic Modeling Using GIS. Second Edition*. 2004 ISBN 1-4020-2459-2
49. M. Monirul Qader Mirza (ed.): *The Ganges Water Diversion: Environmental Effects and Implications*. 2004 ISBN 1-4020-2479-7
50. Y. Rubin and S.S. Hubbard (eds.): *Hydrogeophysics*. 2005 ISBN 1-4020-3101-7
51. K.H. Johannesson (ed.): *Rare Earth Elements in Groundwater Flow Systems*. 2005 ISBN 1-4020-3233-1
52. R.S. Harmon (ed.): *The Río Chagres, Panama*. A Multidisciplinary Profile of a Tropical Watershed. 2005 ISBN 1-4020-3298-6
53. To be published ISBN 1-4020-3965-4
54. V. Badescu, R.S. Cathcart and R.D. Schuiling (eds): Macro Engineering: A Challenge for the Future. 2006 ISBN 1-4020-3739-2
55. To be published ISBN 1-4020-5305-3
56. G. Salvadori, C. De Michele, N.T. Kottegoda and R. Rosso: *Extremes in Nature*. An Approach Using Copulas. 2007 ISBN 978-1-4020-4414-4
57. S.K. Jain, R.K. Agarwal and V.P. Singh: *Hydrology and Water Resources of India*. 2007 ISBN 978-1-4020-5179-1
58. To be published
59. M. Shahin: *Water Resources and Hydrometeorology of the Arab Region*. 2007 ISBN 978-1-4020-4577-6
60. To be published
61. R.S. Govindaraju and B.S. Das: *Moment Analysis for Subsurface Hydrologic Applications*. 2007 ISBN 978-1-4020-5751-9

Water Science and Technology Library